F. Borghese   P. Denti   R. Saija

Scattering from Model Nonspherical Particles

Theory and Applications
to Environmental Physics

With 95 Figures

 Springer

Professor Dr. Ferdinando Borghese
Università di Messina
Dipartimento di Fisica della Materia e Tecnologie Fisiche Avanzate
Salita Sperone 31
98166 Messina
Italy

Professor Dr. Paolo Denti
Università di Messina
Dipartimento di Fisica della Materia e Tecnologie Fisiche Avanzate
Salita Sperone 31
98166 Messina
Italy

Dr. Rosalba Saija
Università di Messina
Dipartimento di Fisica della Materia e Tecnologie Fisiche Avanzate
Salita Sperone 31
98166 Messina
Italy

Library of Congress Cataloging-in-Publication Data: Borghese, F. (Ferdinando), 1940– Scattering from model nonspherical particles : Theory and applications to environmental physics/ F. Borghese, P. Denti, R. Saija. p. cm. – (Physics of earth and space environments, ISSN 1610-1677) Includes bibliographical references. ISBN 3540440143 (alk. paper) 1. Electromagnetic waves–Scattering. 2 Particles–Optical properties. 3. Atmospheric physics–Mathematical models. I. Denti, P. (Paolo), 1942– II: Saija, R. (Rosalba), 1958– III. Title. IV. series. QC665.S3 B67 2002 539.2–dc21 2002030384

ISSN 1610-1677

ISBN 3-540-44014-3   Springer-Verlag Berlin Heidelberg New York

This work is subject to copyright. All rights are reserved, whether the whole or part of the material is concerned, specifically the rights of translation, reprinting, reuse of illustrations, recitation, broadcasting, reproduction on microfilm or in any other way, and storage in data banks. Duplication of this publication or parts thereof is permitted only under the provisions of the German Copyright Law of September 9, 1965, in its current version, and permission for use must always be obtained from Springer-Verlag. Violations are liable for prosecution under the German Copyright Law.

Springer-Verlag Berlin Heidelberg New York
a member of BertelsmannSpringer Science+Business Media GmbH

© Springer-Verlag Berlin Heidelberg 2003
Printed in Germany

The use of general descriptive names, registered names, trademarks, etc. in this publication does not imply, even in the absence of a specific statement, that such names are exempt from the relevant protective laws and regulations and therefore free for general use.

Typesetting by the authors
Print data prepared by Le-TeX, Leipzig
Cover design: Erich Kirchner, Heidelberg

Printed on acid-free paper   SPIN: 10847632   54/3141/ba - 5 4 3 2 1 0

To my wife ... for reasons that are quite evident
to whomever knew her
Nando

Preface

The Mie theory is known to be the first approach to the electromagnetic scattering from homogeneous spheres endowed with all the accuracy of the Maxwell electromagnetic theory. It applies to spheres of arbitrary radius and refractive index and marks, therefore, noticeable progress over the approximate approach of Rayleigh, which applies to particles much smaller than the wavelength. As a consequence, after the publication of the Mie theory in 1908, several scattering objects, even when their shape was known to be nonspherical, were described in terms of equivalent spherical scatterers. It soon became evident, however, that the morphological details of the actual particles were often too important to be neglected, especially in some wavelength ranges. On the other hand, setting aside some particular cases in which the predictions of the Mie theory were acceptable, no viable alternative for the description of scattering from particles of arbitrary shape was at hand. This situation lasted, with no substantial changes, until about 25 years ago, when the exact solution to the problem of dependent scattering from aggregates of spheres was devised. This solution is a real improvement over the Mie theory because several processes that occur, e.g., in the atmospheric aerosols and in the interstellar medium, can be interpreted in terms of clustering of otherwise spherical scatterers. Moreover, nonspherical particles may be so distributed (both in size and orientation) as to smooth out the individual scattering properties. In this case, aggregates of spherical scatterers may be designed to approximate the real objects through a wise choice of the geometry and of the refractive indices of their components.

In the last 20 years, we have been active in the study of the scattering properties of aggregates of spheres, to whose theory we and our coworkers made some significant contributions that are, however, badly scattered throughout the literature. For this reason, we resolved to collect in this book these contributions organized within a coherent formalism, namely, the formalism of the multipole expansions in the framework of the transition matrix approach. The theory is supplemented by several results of calculations performed by ourselves and by our coworkers, whose inclusion is meant to yield information on the possible practical applications of the theory to systems of interest. In order to achieve uniformity of representation, the calculations were renewed and the results replotted. Therefore, this book should not be understood as a treatise on scattering, although the general theory is described in some detail, but rather as a digest of the experience we have gained through our work in the field. The general topic of the comparison of our results with the experimental data or with the results yielded by other theoretical methods and approaches is pursued, but we do not pretend to be exhaustive because both the theoretical and the experimental study of scattering from actual nonspherical particles is a fast-moving field. Anyway, references to the relevant papers and books are given for all the cases of interest.

We start by summarizing in Chap. 1 the general theory of the spherical vector multipole fields, which are a basic tool for the development of the whole electromagnetic theory.

In Chap. 2 we recall the essentials of the theory of electromagnetic scattering and of the propagation of electromagnetic waves through dispersions of particles of arbitrary shape. The whole chapter is meant to establish a notation and to stress the need for introducing averages over the orientation of the particles in order to get predictions that are comparable to the observational data.

In Chap. 3 we expand the electromagnetic field in terms of spherical vector multipole fields and relate the expansion of the incident to that of the scattered field through the introduction of the transition matrix. The transformation properties of latter operator under rotation of the coordinate frame are exploited to show how orientational averages over an ensemble of particles can be performed. Explicit formulas for the averages of quantities of practical interest are given for a few instances of orientational distribution functions in which the averaging can be performed analytically.

In Chap. 4, after recalling the essentials of the Mie theory for homogeneous spheres and its extensions to layered and radially nonhomogeneous spheres, we introduce the equations that solve the problem for an aggregate of spherical scatterers of arbitrary geometry. The theory for homogeneous spheres containing one or more spherical inclusions is also expounded. In the last section of this chapter, we summarize the essentials of the discrete dipole approximation and of the finite difference time domain method. This summary is meant to help the reader to appreciate the advantages of the transition matrix approach and to better understand our considerations when comparing our results with those of other researchers in the field.

In Chap. 5 the topic of the scattering of particles in the presence both of a metallic and a dielectric plane interface is discussed. Even in this case the problem of the orientational average over the particles is dealt with in some detail.

The applications of the theory are described in Chap. 6, where the scattering properties of single and aggregated spheres and of spheres containing inclusions are considered, and in Chap. 7, where single and aggregated spheres and hemispheres in the presence of a plane interface are dealt with.

Finally, in Chaps. 8 and 9 we discuss how the theory can be applied in two specific contexts, the study of atmospheric ice crystals and of the interstellar dust grains, respectively.

We conclude this book with a short Appendix meant to review the mathematics implied in the calculations we presented. Actually, this appendix is rather cursory as we constantly resorted to mathematical tools that are quite standard in electromagnetics and in the theory of angular momentum. Our choice allowed us to use the mathematics that is reported in the existing literature instead of proving and reporting the needed significant relations. Anyway, in our opinion, the content of the mathematical appendix should be sufficient to give useful hints to whomever wants to face the problem of electromagnetic scattering through the transition matrix approach.

Before closing, we want to acknowledge the effort of our colleagues Dr. Orazio I. Sindoni, Prof. Giuditta Toscano, Dr. Enrico Fucile, Dr. Maria Antonia Iatì, Prof. Santi Aiello, and Dr. Cesare Cecchi-Pestellini who in time collaborated to the research on which this book is based. Finally, we want to express our sincere thanks to Prof. Michael I. Mishchenko for a critical reading of the manuscript and for several helpful suggestions and to Prof. Rodolfo Guzzi, whose advice and encouragement greatly helped us in writing this book.

Messina, Italy
July 2002

Ferdinando Borghese
Paolo Denti
Rosalba Saija

Contents

1. **Multipole Fields** 1
   1.1 Introduction 1
   1.2 Field Equations 2
   1.3 Vector Helmholtz Equation 4
   1.4 Transformation Properties of Vector Fields 5
       1.4.1 Cartesian Vectors 5
       1.4.2 Spin of a Vector Field 6
   1.5 Eigenvectors of the Angular Momentum 8
       1.5.1 Representations of the Rotation Operators 8
       1.5.2 Spherical Harmonics 11
       1.5.3 Spin Eigenvectors 12
   1.6 Spherical Tensors on the Unit Sphere 13
       1.6.1 Coupling of Angular Momenta 14
       1.6.2 Clebsch–Gordan Series 16
       1.6.3 Irreducible Spherical Tensors 18
       1.6.4 Vector Solutions to the Helmholtz Equation 19
       1.6.5 Divergence and Curl
             of the Vector Helmholtz Harmonics 20
   1.7 Multipole Fields 22
       1.7.1 Hansen's Vectors 22
       1.7.2 Vector Spherical Harmonics 23
       1.7.3 Spherical Multipole Fields 25
   1.8 Addition Theorem for Multipole Fields 26
       1.8.1 Nozawa's Theorem 26
       1.8.2 Addition Theorem
             for the Vector Helmholtz Harmonics 28
       1.8.3 Application to Hansen's Vectors 28

2. **Propagation Through an Assembly
   of Nonspherical Scatterers** 31
   2.1 Scattering Amplitude 31
   2.2 Cross Sections 33
   2.3 Plane of Scattering 34
       2.3.1 Polarization of the Incident Plane Wave 36
       2.3.2 Polarization of the Scattered Wave 38

XII   Contents

    2.4 Stokes Parameters 39
          2.4.1 Müller Transformation Matrix 40
    2.5 Optical Theorem 41
    2.6 Scattering by an Ensemble of Particles 43
    2.7 Refractive Index of a Dispersion of Particles 44
          2.7.1 Definition of the Refractive Index 44
          2.7.2 Features of the Refractive Index 47

3. Multipole Expansions and Transition Matrix 49
    3.1 Multipole Expansions 49
          3.1.1 Expansion of a Homogeneous Plane Wave 49
          3.1.2 Inhomogeneous Plane Wave 51
          3.1.3 Scattered Field 51
    3.2 Transition Matrix 53
    3.3 Refractive Index of a Polydispersion of Particles 55
    3.4 Orientational Averages 56
          3.4.1 Asymmetry Parameter 60
    3.5 Effect of an Electrostatic Field 62
          3.5.1 Polarizability of a Particle of Arbitrary Shape 63
    3.6 Effect of the Diffusive Motion 66

**4. Transition Matrix of Single
and Aggregated Spheres** 71
    4.1 Homogeneous Spheres 71
    4.2 Radially Nonhomogeneous Spheres 74
    4.3 Resonances 77
    4.4 Aggregates of Spheres 78
    4.5 Spheres Containing Spherical Inclusions 84
    4.6 Finite Elements Methods 88
          4.6.1 Discrete Dipole Approximation 89
          4.6.2 Finite Difference Time Domain Method 91

**5. Scattering from Particles
on a Plane Surface** 95
    5.1 Incident and Reflected Fields 96
    5.2 Perfectly Reflecting Surface 98
          5.2.1 Orientational Averages 100
    5.3 Dielectric Half-Space 103
          5.3.1 Reflection Rule for H-Multipole Fields 103
          5.3.2 Calculation of the Reflected Field 107
    5.4 Scattering from a Sphere on a Dielectric Substrate 109
          5.4.1 Reflection of the Incident and Scattered Wave 110
          5.4.2 Amplitudes of the Scattered Field 111
          5.4.3 Scattering Amplitude and Transition Matrix 112
    5.5 Aggregated Spheres on a Dielectric Substrate 114

		5.5.1 Multipole Expansion of the Fields 114
		5.5.2 Transition Matrix for an Aggregate in the Presence of a Surface 118
	5.6	Perfectly Reflecting vs. Dielectric Surface: Similarities and Differences 119

6. Applications: Aggregated Spheres, Layered Spheres, and Spheres Containing Inclusions 121

   6.1 General Features of Scattering from Aggregated Spheres 122
       6.1.1 Comparison with Experimental Data 125
       6.1.2 Effect of the Structural Changes.................... 128
   6.2 Clusters in an Electrostatic Field 135
   6.3 Extinction from Single and Aggregated Layered Spheres 141
       6.3.1 Metal Spheres with a Soft Surface 143
       6.3.2 Metal Spheres with a Dielectric Coating............. 144
       6.3.3 Dielectric Spheres with a Metal Coating............. 146
       6.3.4 Metal Spheres with a Metallic Coating 147
       6.3.5 Considerations on Convergence..................... 148
   6.4 Spheres Containing Inclusions 149
       6.4.1 Metallic Inclusion in a Dielectric Sphere 151
       6.4.2 Empty Cavity in a Dielectric Sphere 154
       6.4.3 Spheres Containing Two Metallic Inclusions 155
       6.4.4 Resonances of a Sphere Containing a Spherical Inclusion 161
   6.5 Correlation Spectroscopy 166

7. Applications: Single and Aggregated Spheres and Hemispheres on a Plane Interface 173

   7.1 Aggregated Spheres and Hemispheres on a Metallic Surface .. 173
   7.2 Inclusion-Containing Hemispheres on a Metallic Surface 179
   7.3 Resonance Suppression Mechanism 186
       7.3.1 Single Hemispheres 187
       7.3.2 Binary Clusters 189
   7.4 Particles on a Dielectric Substrate 192
       7.4.1 Single Spheres 192
       7.4.2 Aggregates of Spheres in Fixed Orientation 198
       7.4.3 Randomly Oriented Aggregates 202

8. Applications: Atmospheric Ice Crystals 207

   8.1 Properties of Atmospheric Ice Crystals in the Infrared....... 207
   8.2 Ice Crystals in the mm Wave Range 213

9. Applications: Cosmic Dust Grains ... 223
- 9.1 Introduction ... 223
- 9.2 Modeling Cosmic Dust Grains as Aggregates ... 224
- 9.3 Fluffy Particles ... 230
  - 9.3.1 Optical Properties of Porous Bare Grains ... 230
  - 9.3.2 Coated Grains and Clustering ... 234

A. Appendix ... 239
- A.1 Bessel and Hankel Functions ... 239
  - A.1.1 Mie Coefficients for Radially Nonhomogeneous Spheres ... 240
- A.2 Spherical Harmonics ... 240
- A.3 Clebsch–Gordan Coefficients ... 241
  - A.3.1 Translation of Origin ... 241
  - A.3.2 Orientational Averages ... 241
- A.4 Rotation Matrices $D^{(l)}$ ... 242
- A.5 Calculating M^{-1}, H^{-1}, and H_α^{-1} ... 242
- A.6 General Approach to the Computational Problem ... 242

References ... 245

Index ... 251

1. Multipole Fields

1.1 Introduction

The propagation of an electromagnetic wave through an assembly of particles embedded within a homogeneous matrix is known to depend, to a large extent, both on the distribution and on the specific features of the particles themselves. The simplest case is that of a plane wave that impinges on a particle. As a result of the field-particle interaction, the incident wave undergoes *extinction*, i.e., a detector along the path of the wave records a smaller intensity than would be recorded in the absence of the particle. Extinction itself, however, is the result of two concurrent processes: the particle, on one hand, *absorbs* and, on the other hand, *scatters* the impinging radiation. The absorption process implies the transformation of the electromagnetic energy into some other form, in general heat. This can be accounted for by attributing to the particle a complex index of refraction whose choice falls into the realm of the solid state physics and will not be further discussed here. In turn, the scattering process can be roughly described by stating that the incident field excites the particle to emit a wave that at a large distance looks like a more or less spherical outgoing wave superposed to the original incident field. The whole process is uniquely determined by the boundary conditions that the components of the field must satisfy across the surface of the particle and at infinity. The form of the equations that stem from the boundary conditions, however, may turn out to be unmanageably complicated depending on the shape of the particle, so it is not surprising that, until recently, the scattering has been exactly calculated only for particles of simple shape.

The case in which the plane wave propagates through a medium containing several particles is even more complex. The field that excites the scattering from each particle does not coincide with the primary incident wave but is rather the superposition of the latter with the field that has previously been scattered by all other particles in the dispersion. Therefore, describing the propagation through such a medium requires taking into account the *multiple scattering processes* that occur among all the particles present. This, in turn, implies knowing the position as well as the orientation of all the particles in the dispersion.

A great simplification is achieved by assuming the particles to be spherical and homogeneous; in this case the scattered field does not depend on the orientation of the particles and can be described by the so-called Mie theory. It is not surprising, however, that Mie theory calculations are often of little use for the interpretation of the experimental data, because the particles that form the most common inhomogeneous media, e.g., an aerosol, are far from spherical. Furthermore, the particles can undergo aggregation processes that yield complex nonspherical scatterers. Therefore, in order to get useful information on the propagation through an inhomogeneous medium, one needs to resort to mathematically solvable models both for the particles and for their dispersions.

In choosing a suitable model, we are guided by the remark that in several cases either nonspherical scatterers turn out to be aggregations of spherical monomers [1], or the distribution of the particles, both in size and in shape, may be such as to smooth out the specific scattering features that depend on the details of their shape. As a result, it is reasonable that the dispersion as a whole has a memory, at most, of the quantity of scattering material within the individual particles and of their overall shape. Therefore, we choose a *cluster of spherical scatterers* as a model to approximate the properties of the actual particles in the dispersion. The most striking advantage of this model is that no limitation of principle needs to be imposed on the radii and on the refractive indices of the component spheres. Furthermore, once the model for a given particle is set up, no further approximation is necessary.

The mathematical technique that is used to calculate the scattering of electromagnetic waves from a cluster of spheres is based on the expansion of the electromagnetic field in terms of a complete set of multipole fields and on the introduction of a *transition matrix* that transforms the expansion of the incident field into that of the scattered field. The transformation properties of the multipole fields under rotation of the coordinate axes can then be extensively used to assess the dependence of the scattered field and of other quantities of interest on the orientation of the scattering particle with respect to the incident field.

According to the considerations above, this chapter is devoted to summarizing the properties of the multipole fields with a particular emphasis on their transformation properties. A number of formulas will be established that will be useful both for setting up the theory and for establishing the relation among the scattering properties of the particles and the macroscopic observable quantities.

1.2 Field Equations

The behavior of the electromagnetic field is governed by the Maxwell equations that we write here in Gaussian units [2]

1.2 Field Equations

$$\nabla \times \bar{\boldsymbol{E}} = -\frac{1}{c}\frac{\partial \bar{\boldsymbol{B}}}{\partial t}, \tag{1.1a}$$

$$\nabla \times \bar{\boldsymbol{H}} = \frac{1}{c}\frac{\partial \bar{\boldsymbol{D}}}{\partial t} + \frac{4\pi}{c}\bar{\boldsymbol{j}}, \tag{1.1b}$$

$$\nabla \cdot \bar{\boldsymbol{D}} = 4\pi\bar{\rho}, \tag{1.1c}$$

$$\nabla \cdot \bar{\boldsymbol{B}} = 0. \tag{1.1d}$$

Equations (1.1a–d) must be supplemented by the constitutive equations that relate to each other $\bar{\boldsymbol{E}}$, $\bar{\boldsymbol{D}}$, $\bar{\boldsymbol{B}}$, and $\bar{\boldsymbol{H}}$ within a given medium. Here, we are concerned with *linear media* whose electromagnetic properties are described by the constitutive equations

$$\bar{\boldsymbol{D}} = \epsilon\bar{\boldsymbol{E}}, \qquad \bar{\boldsymbol{B}} = \mu\bar{\boldsymbol{H}}, \qquad \bar{\boldsymbol{j}} = \sigma\bar{\boldsymbol{E}},$$

where ϵ, μ, and σ are (Cartesian) tensors whose components may be space-dependent but, on account of the assumed linearity of the medium, are independent of the fields. Hereafter, we will deal with nonmagnetic, isotropic media so that $\mu \approx 1$, whereas ϵ and σ will reduce to scalars that are space-dependent as far as the medium of interest is nonhomogeneous. With a few exceptions, this book deals with homogeneous media, however. Therefore, unless the nonhomogeneity is explicitly stated, ϵ, μ and σ are to be considered as space-independent scalars.

The linearity of the field equations and of the constitutive equations makes them applicable to each of the monochromatic components of the time Fourier transform of the fields $\bar{\boldsymbol{E}}$ and $\bar{\boldsymbol{B}}$ as well as of the sources $\bar{\rho}$ e $\bar{\boldsymbol{j}}$. Therefore, no generality is lost by assuming that each of the quantities of interest depends on time through the factor $\exp(-i\omega t)$, so that, for instance, $\bar{\boldsymbol{E}}(\boldsymbol{r},t) = \boldsymbol{E}(\boldsymbol{r})\exp(-i\omega t)$. Accordingly, since μ has been assumed to be a constant scalar, (1.1a) and (1.1b) become

$$\nabla \times \boldsymbol{E} = ik_{\rm v}\boldsymbol{B}, \tag{1.2a}$$

$$\nabla \times \boldsymbol{B} = -ik_{\rm v}n^2\boldsymbol{E}, \tag{1.2b}$$

where we define the magnitude of the propagation vector in vacuo $k_{\rm v}$ and the complex refractive index n as

$$k_{\rm v} = \frac{\omega}{c}, \qquad n^2 = \mu\left(\epsilon + \frac{4\pi i\sigma}{\omega}\right).$$

Note that the positive sign in front of the imaginary part of n^2 stems from our choice of the time dependence of the fields through the factor $\exp(-i\omega t)$. This choice is opposite to that assumed, e.g., by van de Hulst [3] that leads to a negative sign in front of the imaginary part of n^2. We also remark that both ϵ and σ may be frequency dependent, so that the value of n in (1.2b) must be that appropriate to the frequency concerned.

From (1.2b) and the assumed homogeneity of the medium, it follows that

$$\nabla \cdot \boldsymbol{E} = 0 , \tag{1.3}$$

and, with the help of the identity

$$\operatorname{curl}\operatorname{curl} = \operatorname{grad}\operatorname{div} - \nabla^2 , \tag{1.4}$$

it is an easy matter to see that the electric and the magnetic field should satisfy the pair of (decoupled) vector Helmholtz equations

$$(\nabla^2 + n^2 k_{\mathrm{v}}^2)\boldsymbol{E} = 0 , \qquad (\nabla^2 + n^2 k_{\mathrm{v}}^2)\boldsymbol{B} = 0 . \tag{1.5}$$

Of course, across any surface that separates two (homogeneous) media of different dielectric properties, the solutions to (1.5) should satisfy the boundary conditions

$$\hat{\boldsymbol{n}} \times (\boldsymbol{E}_2 - \boldsymbol{E}_1) = 0 , \quad \hat{\boldsymbol{n}} \times (\boldsymbol{B}_2 - \boldsymbol{B}_1) = 0 ,$$

where $\hat{\boldsymbol{n}}$ is the unit normal to the separation surface. Note that the second equation above holds true for media of finite conductivity [4].

1.3 Vector Helmholtz Equation

Before proceeding further, a word of caution is necessary on the vector nature of (1.5). In fact, the operator ∇^2 is defined as

$$\nabla^2 \equiv \operatorname{div}\operatorname{grad}$$

and should therefore be applied to scalar functions only. Nevertheless, using (1.4), it is an easy matter to see that, provided rectangular coordinates are used, $\nabla^2 \boldsymbol{E}$ can be consistently defined as a vector whose rectangular components are $\nabla^2 E_x$, $\nabla^2 E_y$, and $\nabla^2 E_z$. This definition grants separation of the vector Helmholtz equation into three scalar equations, each of which involves only one rectangular component of the field. As an example, each of the rectangular components of the polarized plane wave

$$\boldsymbol{E} = E_0 \hat{\boldsymbol{e}} \exp(\mathrm{i}\boldsymbol{k} \cdot \boldsymbol{r}) ,$$

where $\hat{\boldsymbol{e}}$ is a constant vector and $\boldsymbol{k} = k\hat{\boldsymbol{k}}$, with $k = nk_{\mathrm{v}}$, satisfies the Helmholtz equation, provided that

$$\boldsymbol{k} \cdot \hat{\boldsymbol{e}} = 0 ,$$

as required by (1.3).

Using rectangular coordinates is undoubtedly the best choice when one is concerned with the solution of the Helmholtz equations that should satisfy the boundary conditions across a plane surface. Nevertheless, when the boundary conditions need to be imposed across a differently shaped surface, for instance across a spherical surface, the choice of a more appropriate system of coordinates is in order. In fact, the vector Helmholtz equation separates in several systems of orthogonal curvilinear coordinates, but each of the resulting scalar equations always involves more than one component of the field. Since this kind of separation can hardly be considered as a simplification, it is advisable to search for a complete set of elementary vector solutions to (1.5), in terms of which each field vector could be conveniently expanded.

1.4 Transformation Properties of Vector Fields

The distinctive features of the vector fields stem from the transformation properties of their components under rotation of the coordinate system. Since the Helmholtz operator $\Omega_k = \nabla^2 + k^2$ is invariant under rotations, a detailed investigation of the transformation properties of the vector fields will help us determining the most convenient choice for the vector solutions to the Helmholtz equation.

1.4.1 Cartesian Vectors

We consider a rectangular coordinate system $Ox_1x_2x_3$ that a rotation around the origin through a vector angle $\boldsymbol{\omega} = \omega\hat{\boldsymbol{n}}$ transforms into the system $Ox'_1x'_2x'_3$. These systems will be referred to as Σ and Σ', respectively. As usual, the amplitude ω is counted as positive in a counterclockwise sense with respect to the unit vector $\hat{\boldsymbol{n}}$ along the axis of rotation. The transformation of Σ into Σ' is uniquely defined by the real orthogonal matrix $\mathrm{R}(\boldsymbol{\omega})$ that transforms the coordinates in Σ of any arbitrarily chosen point P into those that the *same point* has in Σ'. The transformation rule reads

$$x'_i = \sum_j R_{ij} x_j , \quad \text{or, matrixwise,} \quad \boldsymbol{r}' = \mathrm{R}(\boldsymbol{\omega})\boldsymbol{r} ,$$

where $\boldsymbol{r}$ and $\boldsymbol{r}'$ must be understood as two different representations of the coordinates of the same point P. Then, a *vector* is defined as a set of three quantities v_1, v_2, and v_3, that are its components in Σ, whose transformation rule under rotation of the coordinate system is identical to that of the coordinates of a point. Thus

$$v'_i = \sum_j R_{ij} v_j , \quad \text{or} \quad \mathbf{v}' = \mathrm{R}(\boldsymbol{\omega})\mathbf{v} ,$$

where v'_1, v'_2, and v'_3 are to be understood as the components of the same vector $\mathbf{v}$ in Σ'. It may be worth stressing that the real orthogonal matrix R has an inverse R^{-1}, so that

$$\boldsymbol{r} = \mathrm{R}^{-1}(\boldsymbol{\omega})\boldsymbol{r}' . \tag{1.6}$$

1. Multipole Fields

When v is space dependent, both v and its argument transform alike under rotation of the coordinate system. Accordingly we write

$$v'(r') = R(\omega)v(r) ,$$

or, on account of (1.6),

$$v'(r') = R(\omega)v(R^{-1}(\omega)r') .$$

Now, since r' denotes the vector coordinate of any point, the prime can be dropped, and the preceding equation is accordingly rewritten as

$$v'(r) = R(\omega)v(R^{-1}(\omega)r) . \tag{1.7}$$

Then, we associate to the transformation effected by the matrix R the operator $O_R \equiv O_\omega$ defined by the equation

$$O_\omega v(r) = v'(r) ,$$

and, accordingly, rewrite (1.7) in the form

$$O_\omega v(r) = R(\omega)v(R^{-1}(\omega)r) , \tag{1.8}$$

which yields the general transformation rule of a vector field under rotation of the coordinate axes [5].

1.4.2 Spin of a Vector Field

We now apply (1.8) to investigate the transformation of the components of any vector field. To this end it suffices to consider only infinitesimal rotations $d\omega$ that are effected by the matrix $R(d\omega)$ with associated operator $O_{d\omega}$. It is an easy matter to see that, for general direction of $d\omega$, the transformation of r reads [6]

$$R(d\omega)r = r' = r - d\omega \times r . \tag{1.9}$$

In particular, for rotations around the z axis $d\omega = d\omega \, \hat{e}_z$, and (1.9) becomes

$$r' = r - d\omega \, \hat{e}_z \times r ,$$

which is equivalent to the scalar equations

$$x' = x + y \, d\omega ,$$
$$y' = y - x \, d\omega ,$$
$$z' = z ,$$

with inverses

1.4 Transformation Properties of Vector Fields

$$x = x' - y' \, d\omega ,$$
$$y = y' + x' \, d\omega ,$$
$$z = z' .$$

As a consequence, the components of any vector field $\boldsymbol{v}(\boldsymbol{r})$ transform according to the equations

$$v'_x(x,y,z) = v_x(x - d\omega\, y, y + d\omega\, x, z) + d\omega\, v_y(x - d\omega\, y, y + d\omega\, x, z) ,$$
$$v'_y(x,y,z) = v_y(x - d\omega\, y, y + d\omega\, x, z) - d\omega\, v_x(x - d\omega\, y, y + d\omega\, x, z) ,$$
$$v'_z(x,y,z) = v_z(x - d\omega\, y, y + d\omega\, x, z) .$$

Series expansion to first order in $d\omega$ then yields

$$v'_x(x,y,z) = v_x(x,y,z) - d\omega \left(y\frac{\partial}{\partial x} - x\frac{\partial}{\partial y} \right) v_x(x,y,z) + d\omega\, v_y(x,y,z) ,$$
$$v'_y(x,y,z) = v_y(x,y,z) - d\omega \left(y\frac{\partial}{\partial x} - x\frac{\partial}{\partial y} \right) v_y(x,y,z) - d\omega\, v_x(x,y,z) ,$$
$$v'_z(x,y,z) = v_z(x,y,z) - d\omega \left(y\frac{\partial}{\partial x} - x\frac{\partial}{\partial y} \right) v_z(x,y,z) ,$$

i.e., in operator form

$$O_{d\omega}\boldsymbol{v}(x,y,z) = (\boldsymbol{1} - \mathrm{i} d\omega\, \mathrm{J}_z)\boldsymbol{v}(x,y,z) , \qquad (1.10)$$

where

$$\mathrm{J}_z = \mathrm{L}_z + \mathrm{s}_z ,$$

with

$$\mathrm{L}_z = -\mathrm{i}\left(y\frac{\partial}{\partial x} - x\frac{\partial}{\partial y} \right)\boldsymbol{1} \quad \text{and} \quad \mathrm{s}_z = \begin{pmatrix} 0 & -\mathrm{i} & 0 \\ \mathrm{i} & 0 & 0 \\ 0 & 0 & 0 \end{pmatrix} .$$

Equation (1.10) shows that the vector fields transform according to a rule that is different from that appropriate for scalar fields. In fact, for rotations around the z axis, the generator of the transformations of a scalar field is known to be L_z, the z component of the orbital angular momentum operator. According to (1.10), the generator of the transformation of a vector field turns out to be J_z, which is the sum of L_z and of a new angular momentum operator s_z that is represented by a 3×3 matrix and has therefore the quantum number $s = 1$. In other words any vector field, by its very nature, has an intrinsic *spin* of 1. As a check, we can effect two further infinitesimal rotations around the x and the y axis and find that

$$\mathrm{s}_x = \begin{pmatrix} 0 & 0 & 0 \\ 0 & 0 & -\mathrm{i} \\ 0 & \mathrm{i} & 0 \end{pmatrix} , \quad \mathrm{s}_y = \begin{pmatrix} 0 & 0 & \mathrm{i} \\ 0 & 0 & 0 \\ -\mathrm{i} & 0 & 0 \end{pmatrix} .$$

It is then an easy matter to see that the matrices s_x, s_y, and s_z satisfy the commutation rule

$$[s_x, s_y] = i s_z ,$$

and cyclical permutations, which are the defining relations of the angular momentum operators.

1.5 Eigenvectors of the Angular Momentum

The results in Sect. 1.4 suggest that, from the point of view of the transformation properties, the most suitable set of vector basis functions for the expansion of any vector field, and in particular of the electromagnetic field, is composed of the simultaneous vector eigenfunctions of J^2 and J_z. In order to build these eigenfunctions, it is convenient to start from their transformation properties that will now be investigated in some detail. In the course of our study we will also establish a number of formulas that will prove useful for later applications.

1.5.1 Representations of the Rotation Operators

In the preceding section we associated to the transformation of coordinates defined by the matrix $R(\omega)$ the operator O_ω that transforms the vector fields according to (1.8). For finite rotations, the form of O_ω is [6]

$$O_\omega = \exp[-i\boldsymbol{\omega} \cdot \mathbf{J}] ,$$

which can be easily checked by expanding O_ω to first order in ω and letting $\omega \to d\omega$. Note that the hermiticity of $\mathbf{J}$ grants the unitarity of O_ω that characterizes any rotation operator. In principle, O_ω can be represented on any orthonormal basis, but the most convenient choice proves to be the set of the eigenvectors $|j, m\rangle$, i.e., the solutions of the simultaneous eigenvalue equations

$$J^2 |j\, m\rangle = j(j+1)|j\, m\rangle ,$$
$$J_z |j\, m\rangle = m|j\, m\rangle ,$$

with $j \geq 0$ and $-j \leq m \leq j$. Operating on the eigenvectors $|j\, m\rangle$, O_ω yields a linear combination of them with the same j but different m, i.e.,

$$O_R |j\, m\rangle = \sum_{m'} |j\, m'\rangle \mathcal{D}^{(j)}_{m'm}(\boldsymbol{\omega}) ,$$

where we define

$$\mathcal{D}^{(j)}_{m'm}(\boldsymbol{\omega}) = \langle j\, m'| O_\omega |j\, m\rangle .$$

1.5 Eigenvectors of the Angular Momentum

The coefficients $\mathcal{D}^{(j)}_{m'm}$ are the elements of the matrices $D^{(j)}$ that form a representation of order $2j+1$ of the three-dimensional rotation operators on the basis of the eigenvectors $|j\,m\rangle$. Their explicit form is customarily given in terms of the Euler rotations that are described below (see also Fig. 1.1) and, according to commonly accepted conventions, are considered as positive when effected counterclockwise.

First, the rotation through an angle α around the z axis of Σ transforms the axes $Oxyz$ into the axes $Ox'y'z'$. The operator associated to this rotation is

$$O_\alpha = \exp[-i\alpha J_z]\,.$$

Second, the rotation through an angle β around the y' axis transforms the axes $Ox'y'z'$ into $Ox''y''z''$. The associated operator is

$$O_\beta = \exp[-i\beta J_{y'}]\,.$$

Third, the rotation through an angle γ around the z'' axis transforms the axes $Ox''y''z''$ into the axes $Ox'''y'''z'''$. The associated operator is

$$O_\gamma = \exp[-i\gamma J_{z''}]\,.$$

It may be worth noticing that the preceding definition of the Euler angles is such that, in Σ, the polar angles ϑ and φ of the z'' axis around which the rotation γ is effected coincide with β and α, respectively.

The rotation defined by the vector angle $\boldsymbol{\omega}$ that corresponds to Eulerian angles α, β and γ is thus effected by the operator

$$O_{\boldsymbol{\omega}} = \exp[-i\gamma J_{z''}]\,\exp[-i\beta J_{y'}]\,\exp[-i\alpha J_z]\,, \tag{1.11}$$

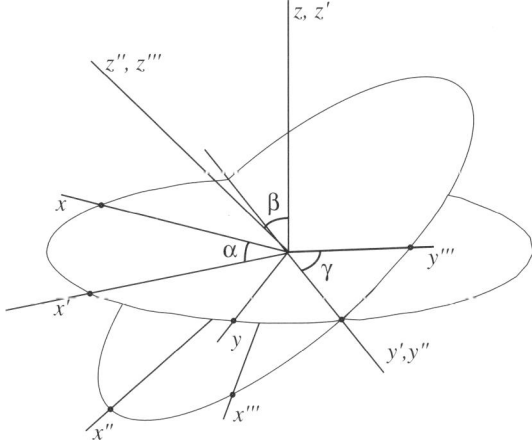

Fig. 1.1. The Eulerian angles

whose representation on the basis of the $|j\,m\rangle$ vectors yield the required $\mathrm{D}^{(j)}$ matrices. We remark, however, that the form of O_ω in (1.11) is not suitable for finding its representation because the three Eulerian rotations are effected around axes belonging to different reference frames. Nevertheless, it can be shown that an alternative form of O_ω is [5]

$$\mathrm{O}_\omega = \exp[-i\alpha \mathrm{J}_z]\,\exp[-i\beta \mathrm{J}_y]\,\exp[-i\gamma \mathrm{J}_z]\,, \tag{1.12}$$

i.e., the Eulerian rotations can be effected around the axes of Σ, provided that their order is reversed. Using (1.12) it is an easy matter to show that the matrix elements of O_ω are

$$\mathcal{D}^{(j)}_{m'm}(\alpha,\beta,\gamma) = \langle j\,m'|\exp[-i\alpha \mathrm{J}_z]\,\exp[-i\beta \mathrm{J}_y]\,\exp[-i\gamma \mathrm{J}_z]|j\,m\rangle$$
$$= \exp(-im'\alpha)\,d^{(j)}_{m'm}(\beta)\,\exp(-im\gamma)\,,$$

where we use the notation

$$d^{(j)}_{m'm}(\beta) = \langle j\,m'|\exp[-i\beta \mathrm{J}_y]|j\,m\rangle$$

for the matrix elements of O_β. The explicit expression of the matrix elements $d^{(j)}_{m'm}(\beta)$ has been found by Wigner and turns out to be [5]

$$d^{(j)}_{m'm}(\beta) = (-)^{m'-m}\bigl[(j+m)!(j-m)!(j+m')!(j-m')!\bigr]^{1/2}$$
$$\times \sum_\kappa (-)^\kappa \frac{(\cos\beta/2)^{2j+m-m'-2\kappa}(\sin\beta/2)^{m'-m+2\kappa}}{(j-m'-\kappa)!(j+m-\kappa)!(\kappa+m'-m)!\kappa!}\,,$$

which holds true for $m' \geq m$ only, however. For $m' < m$ we resort to the properties

$$d^{(j)}_{m'm}(\beta) = d^{(j)}_{mm'}(-\beta) = (-)^{m-m'}d^{(j)}_{mm'}(\beta)\,,$$

which stem from the unitarity of $\mathrm{d}^{(j)}$. Close examination of the expression of $d^{(j)}_{m'm}$ yields the relations

$$d^{(j)}_{m'm}(\beta) = d^{(j)}_{-m,-m'}(\beta) = (-)^{m'-m}d^{(j)}_{-m',-m}(\beta)\,,$$

which, when inserted into the definition of $\mathrm{D}^{(j)}$ imply

$$\mathcal{D}^{(j)}_{m'm}(-\gamma,-\beta,-\alpha) = \mathcal{D}^{(j)*}_{mm'}(\alpha,\beta,\gamma) = (-)^{m-m'}\mathcal{D}^{(j)}_{-m,-m'}(\alpha,\beta,\gamma)\,.$$

Since the matrices $\mathrm{D}^{(j)}$ transform a set of $2j+1$ orthonormal vectors into a new set of $2j+1$ rotated orthonormal vectors, their elements $\mathcal{D}^{(j)}_{m'm}$ must satisfy the customary orthonormality conditions

$$\sum_m \mathcal{D}^{(j)*}_{m'm}(\alpha,\beta,\gamma)\mathcal{D}^{(j)}_{m''m}(\alpha,\beta,\gamma) = \delta_{m'm''}\,,$$

$$\sum_m \mathcal{D}^{(j)*}_{mm'}(\alpha,\beta,\gamma)\mathcal{D}^{(j)}_{mm''}(\alpha,\beta,\gamma) = \delta_{m'm''}\,.$$

1.5 Eigenvectors of the Angular Momentum

It can also be proved that the elements $\mathcal{D}^{(j)}_{m'm}$ satisfy the integral relation

$$\int \mathcal{D}^{(j_1)*}_{\mu_1 m_1}(\alpha,\beta,\gamma)\, \mathcal{D}^{(j_2)}_{\mu_2 m_2}(\alpha,\beta,\gamma)\, \mathrm{d}\Omega = \frac{8\pi^2}{2j_1+1} \delta_{\mu_1\mu_2}\delta_{m_1 m_2}\delta_{j_1 j_2}\,, \quad (1.13)$$

where $\mathrm{d}\Omega$ is a shorthand for

$$\mathrm{d}\Omega = \mathrm{d}\alpha\, \sin\beta\, \mathrm{d}\beta\, \mathrm{d}\gamma\,,$$

with

$$0 \le \alpha \le 2\pi\,,\quad 0 \le \beta \le \pi\,,\quad 0 \le \gamma \le 2\pi\,.$$

Equation (1.13) thus states the orthogonality of the rotation matrix elements on the surface of the unit sphere.

The transformation law of the eigenvectors of the angular momentum allows us to define uniquely the so called *irreducible spherical tensors* as follows:

> An irreducible spherical tensor of rank j is a set of $2j+1$ quantities that under rotation of the coordinate axes transform according to $\mathrm{D}^{(j)}$.

The $2j+1$ quantities that form an irreducible spherical tensor of rank j are its *spherical components*, and each one of them may be a Cartesian tensors of any rank. For instance, a set of three *vectors* that under rotation of the coordinate system transform among themselves according to $\mathrm{D}^{(1)}$ form an irreducible spherical tensor of rank 1. We shall see later that each of the three eigenvectors of the spin of a vector field is a Cartesian vector with complex components, but the whole set transforms under rotation according to $\mathrm{D}^{(1)}$, so that they are also the components of an irreducible spherical tensor of rank 1. In any case, since the transformation of the vector fields is governed by the angular momentum $\mathrm{J} = \mathrm{L} + \mathrm{s}$, it is convenient to consider in some detail the eigenstates of L and those of s.

1.5.2 Spherical Harmonics

The spherical harmonics, i.e., the simultaneous eigenfunctions of L^2 and L_z, are defined by the simultaneous eigenvalue equations

$$\mathrm{L}^2 Y_{lm}(\hat{\boldsymbol{r}}) = l(l+1) Y_{lm}(\hat{\boldsymbol{r}})\,,$$
$$\mathrm{L}_z Y_{lm}(\hat{\boldsymbol{r}}) = m Y_{lm}(\hat{\boldsymbol{r}})\,,$$

with $l \ge 0$ and $-l \le m \le l$, and their explicit form is [2]

$$Y_{lm}(\hat{\boldsymbol{r}}) \equiv Y_{lm}(\vartheta,\varphi) = \sqrt{\frac{2l+1}{4\pi}\frac{(l-m)!}{(l+m)!}}\, P_{lm}(\cos\vartheta)\, \mathrm{e}^{\mathrm{i}m\varphi}\,. \quad (1.14)$$

In (1.14) the quantities P_{lm} are the associated Legendre functions of the first kind

$$P_{lm}(z) = \frac{(-)^m}{2^l l!} (1-z^2)^{m/2} \frac{\mathrm{d}^{l+m}}{\mathrm{d}z^{l+m}} (z^2-1)^l , \qquad (1.15)$$

with

$$P_{l,-m}(z) = (-)^m \frac{(l-m)!}{(l+m)!} P_{lm}(z) ,$$

whose argument z may be either real or complex. The spherical harmonics form a complete orthonormal set on the surface of the unit sphere

$$\int Y_{lm}^* Y_{l'm'} \, \mathrm{d}\Omega = \delta_{ll'} \delta_{mm'} ,$$

and, under rotation of the coordinate system, transform according to $\mathrm{D}^{(l)}$

$$O_{\alpha\beta\gamma} Y_{lm}(\vartheta, \varphi) = \sum_{m'} Y_{lm'}(\vartheta, \varphi) \mathcal{D}_{m'm}^{(l)}(\alpha, \beta, \gamma) . \qquad (1.16)$$

The spherical harmonics with a given l are thus the components of an irreducible spherical tensor of rank l. For real argument they satisfy the complex conjugation equation

$$Y_{lm}^* = (-)^m Y_{l,-m} ,$$

and their parity is defined by the equation

$$Y_{lm}(\hat{\boldsymbol{r}}) = (-)^l Y_{lm}(-\hat{\boldsymbol{r}}) .$$

Hereafter, we will find useful the relation

$$\frac{4\pi}{2l+1} \sum_m Y_{lm}^*(\vartheta_1, \varphi_1) Y_{lm}(\vartheta_2, \varphi_2) = P_l(\cos\omega) , \qquad (1.17)$$

where P_l is a Legendre polynomial of degree l and ω is the angle between the directions defined by the polar angles (ϑ_1, φ_1) and (ϑ_2, φ_2). Equation (1.17) is customarily referred to as the *addition theorem for the spherical harmonics*.

1.5.3 Spin Eigenvectors

The simultaneous eigenvectors of s^2 and s_z belonging to $s=1$ are less known but no less important than the spherical harmonics. They are, in fact, essential for a correct description of the transformation properties of the vector fields. According to the results in Sect. 1.4.2, they should be three-dimensional vectors and should satisfy the eigenvalue equations

$$s^2 \bm{v} = s(s+1)\bm{v} \,, \qquad \text{with } s = 1 \,,$$
$$s_z \bm{v} = \mu \bm{v} \,, \qquad \text{with } \mu = 0, \pm 1 \,.$$

With the help of the explicit form of s_x, s_y and s_s given in Sect. 1.4.2, it is an easy matter to see that the spin eigenvectors should be proportional to linear combinations of the unit vectors that characterize the rectangular axes. It can be checked that the required eigenvectors are

$$\bm{\xi}_{\pm 1} = \mp \tfrac{1}{\sqrt{2}} (\hat{\bm{e}}_x \pm i \hat{\bm{e}}_y) \,, \qquad \bm{\xi}_0 = \hat{\bm{e}}_z \,.$$

For our purposes it may be useful to stress that these eigenvectors, under rotation of the coordinate system, transform according to $D^{(1)}$ and are therefore the components of an irreducible spherical tensor of rank 1. This is easily understood by observing that the vectors $\bm{\xi}_\mu$ are proportional to the spherical harmonics with $l = 1$.

The vectors $\bm{\xi}_\mu$ satisfy the conjugation condition

$$\bm{\xi}_\mu^* = (-)^\mu \bm{\xi}_{-\mu} \,,$$

and the orthonormality relation

$$\bm{\xi}_\mu \cdot \bm{\xi}_{\mu'}^* = (-)^{\mu'} \bm{\xi}_\mu \cdot \bm{\xi}_{-\mu'} = \delta_{\mu\mu'} \,.$$

Any three-dimensional vector can be represented on the basis of the vectors $\bm{\xi}_\mu$ rather than on the basis of the unit vectors of the rectangular coordinate axes as

$$\bm{B} = \sum_\mu (-)^\mu B_\mu \bm{\xi}_{-\mu} \,,$$

where

$$B_\mu = \bm{\xi}_\mu \cdot \bm{B}$$

defines the *spherical components* of $\bm{B}$. Then, the dot product of any pair of vectors in terms of their spherical components can be written as

$$\bm{A} \cdot \bm{B} = \sum_\mu (-)^\mu A_\mu B_{-\mu} \,.$$

The usefulness of this representation stems from the fact that the spherical components of any Cartesian vector are the components of an irreducible spherical tensor of rank 1.

1.6 Spherical Tensors on the Unit Sphere

The angular momentum eigenvectors that were determined in the preceding section are a good basis for the description of any vector field. In fact, since the transformation properties of the latter are governed by the total angular momentum $J = L + s$, we can resort to the general theory of the coupling of angular momenta to combine the eigenvectors of L and those of s to yield the eigenvectors of J. The resulting eigenvectors will transform according to $D^{(j)}$ and thus according to a rule that decrees their identification with the components of an irreducible spherical tensor of rank j.

1.6.1 Coupling of Angular Momenta

Let us consider two angular momentum operators, J_1 and J_2, whose respective eigenvectors we denote as $|j_1, m_1\rangle$ and $|j_2, m_2\rangle$. Since the vector spaces spanned by these two sets are linearly independent, the vector sum of the operators J_1 and J_2 is the operator

$$J = J_1 + J_2,$$

which is easily verified to be itself an angular momentum. Our present task is therefore to build the eigenvectors of J out of the sets of eigenvectors $|j_1, m_1\rangle$ and $|j_2, m_2\rangle$.

We remark that the set of the direct product vectors $|j_1, m_1\rangle|j_2, m_2\rangle$ are not eigenvectors of the square of the total angular momentum J^2 but are eigenvectors of J_z. It suffices to state that the set of the product eigenvectors form a complete set that spans the space of the eigenvectors of J. As a consequence, the simultaneous eigenvectors of J^2 and J_z can be built as linear combinations of the product vectors. In fact, we can write

$$|j, m\rangle = \sum_{m_1 m_2} C(j_1, j_2, j; m_1, m_2, m) |j_1, m_1\rangle|j_2, m_2\rangle, \quad (1.18)$$

where the C's are the so-called Clebsch–Gordan coefficients. The latter can be understood as the elements of the unitary matrix that transforms the set of the $(2j_1 + 1)(2j_2 + 1)$ product vectors into the set of the simultaneous eigenvectors of J^2 and J_z. The explicit expression of the Clebsch–Gordan coefficients can be determined in the framework of group theory and turns out to be [5]

$$\begin{aligned}
C(j_1, j_2, j; m_1, m_2, m) &= \delta_{m, m_1 + m_2} \\
&\times \left[(2j+1) \frac{(j_1 + j_2 - j)!(j + j_1 - j_2)!(j + j_2 - j_1)!}{(j_1 + j_2 + j + 1)!} \right. \\
&\quad \left. \times (j_1 + m_1)!(j_1 - m_1)!(j_2 + m_2)!(j_2 - m_2)!(j + m)!(j - m)! \right]^{1/2} \\
&\times \sum_\nu \frac{(-)^\nu}{\nu!} \left[(j_1 + j_2 - j - \nu)!(j_1 - m_1 - \nu)!(j_2 + m_2 - \nu)! \right. \\
&\quad \left. \times (j - j_2 + m_1 + \nu)!(j - j_1 - m_2 + \nu)! \right]^{-1}. \quad (1.19)
\end{aligned}$$

The Clebsch–Gordan coefficients possess symmetry properties that relate to each other the coefficients that differ for the order of the quantum numbers j_1, j_2, and j or for the sign of the projection quantum numbers. These relations, which can be inferred from the explicit expression (1.19), are

$$C(j_1, j_2, j; m_1, m_2, m)$$
$$= (-)^{j_1+j_2-j} C(j_1, j_2, j; -m_1, -m_2, -m)$$
$$= (-)^{j_1+j_2-j} C(j_2, j_1, j; m_2, m_1, m)$$
$$= (-)^{j_1-m_1} \left(\frac{2j+1}{2j_2+1} \right)^{1/2} C(j_1, j, j_2; m_1, -m, -m_2) \,.$$

The preceding relations are the fundamental ones and can help in finding the further relations

$$C(j_1, j_2, j; m_1, m_2, m)$$
$$= (-)^{j_2+m_2} \left(\frac{2j+1}{2j_1+1} \right)^{1/2} C(j, j_2, j_1; -m, m_2, -m_1)$$
$$= (-)^{j_1-m_1} \left(\frac{2j+1}{2j_2+1} \right)^{1/2} C(j, j_1, j_2; m, -m_1, m_2)$$
$$= (-)^{j_2+m_2} \left(\frac{2j+1}{2j_1+1} \right)^{1/2} C(j_2, j, j_1; -m_2, m, m_1) \,,$$

which complete the set of the symmetry properties.

We remark that, because of the factor δ_{m, m_1+m_2} in (1.19), the double sum in (1.18) is actually a single sum. Accordingly, we also denote the Clebsch–Gordan coefficients as $C(j_1, j_2, j; m_1, m_2)$ and, since $m_2 = m - m_1$, we rewrite the coupling equation as

$$|j, m\rangle = \sum_{m_1} C(j_1, j_2, j; m_1, m - m_1) |j_1, m_1\rangle |j_2, m - m_1\rangle \,. \tag{1.20}$$

The Clebsch–Gordan coefficients vanish unless their arguments satisfy the so called triangular relation

$$|j_1 - j_2| \leq j \leq j_1 + j_2 \,,$$

which is denoted for short as $\Delta(j_1, j_2, j)$. Furthermore, the unitarity of the matrix of the C-coefficients imply that they should satisfy the orthogonality relations

$$\sum_{m_1} C(j_1, j_2, j; m_1, m - m_1) C(j_1, j_2, j'; m_1, m - m_1) = \delta_{jj'} \,, \tag{1.21a}$$

$$\sum_{j} C(j_1, j_2, j; m_1, m - m_1) C(j_1, j_2, j; m'_1, m' - m'_1) = \delta_{m_1 m'_1} \delta_{mm'} \,.$$

$$\tag{1.21b}$$

Equation (1.21b) can be used to show that the inverse to (1.20) reads

$$|j_1,m_1\rangle|j_2,m-m_1\rangle = \sum_j C(j_1,j_2,j;m_1,m-m_1)|j,m\rangle. \qquad (1.22)$$

1.6.2 Clebsch–Gordan Series

The coupling of the angular momentum applies not only to the eigenvectors $|j,m\rangle$ but also to the elements of the $D^{(j)}$-matrices. In fact, under rotation of the coordinate system, equation (1.22) becomes

$$\sum_{\mu_1}\sum_{\mu_2} \mathcal{D}^{(j_1)}_{\mu_1 m_1}\mathcal{D}^{(j_2)}_{\mu_2 m_2}|j_1,\mu_1\rangle|j_2,\mu_2\rangle$$
$$= \sum_j\sum_\mu C(j_1,j_2,j;m_1,m_2)\mathcal{D}^{(j)}_{\mu,m_1+m_2}|j,\mu\rangle,$$

where all the $\mathcal{D}$-matrix elements should have the same argument. By using in place of $|j,\mu\rangle$ its expression given by (1.20) we get

$$\sum_{\mu_1}\sum_{\mu_2} \mathcal{D}^{(j_1)}_{\mu_1 m_1}\mathcal{D}^{(j_2)}_{\mu_2 m_2}|j_1,\mu_1\rangle|j_2,\mu_2\rangle$$
$$= \sum_j\sum_{\mu\mu'_1} C(j_1,j_2,j;m_1,m_2)C(j_1,j_2,j;\mu'_1,\mu-\mu'_1)$$
$$\times \mathcal{D}^{(j)}_{\mu,m_1+m_2}|j_1,\mu'_1\rangle|j_2,\mu-\mu'_1\rangle.$$

The sum over μ can be replaced by a sum over $\mu'_2 = \mu - \mu'_1$, and the prime on the resulting sum index can be dropped. Doing this and comparing the coefficients of $|j_1,\mu_1\rangle|j_2,\mu_2\rangle$ in both sides leads to the result

$$\mathcal{D}^{(j_1)}_{\mu_1 m_1}\mathcal{D}^{(j_2)}_{\mu_2 m_2} = \sum_j C(j_1,j_2,j;m_1,m_2)C(j_1,j_2,j;\mu_1,\mu_2)\mathcal{D}^{(j)}_{\mu_1+\mu_2,m_1+m_2}, \qquad (1.23)$$

where the sum index j must satisfy the triangular relation $\Delta(j_1 j_2 j)$. A similar procedure can be used to prove the inverse coupling law

$$\mathcal{D}^{(j)}_{\mu m} = \sum_{\mu_1}\sum_{m_1} C(j_1,j_2,j;\mu_1,\mu-\mu_1)C(j_1,j_2,j;m_1,m-m_1)$$
$$\times \mathcal{D}^{(j_1)}_{\mu_1 m_1}\mathcal{D}^{(j_2)}_{\mu-\mu_1,m-m_1},$$

which can be used for calculating any $D^{(j)}$ matrix by coupling D-matrices of lower rank.

As a last item, we prove the so-called Gaunt formula for the integral of the product of three spherical harmonics. We start from the addition theorem for the spherical harmonics (1.17) that, with the help of the relation

1.6 Spherical Tensors on the Unit Sphere

$$Y_{l0}(\omega, 0) = \left(\frac{2l+1}{4\pi}\right)^{1/2} P_l(\cos\omega),$$

can be rewritten as

$$Y_{l0}(\omega, 0) = \left(\frac{4\pi}{2l+1}\right)^{1/2} \sum_m Y^*_{lm}(\vartheta_1, \varphi_1) Y_{lm}(\vartheta_2, \varphi_2). \quad (1.24)$$

Transforming $Y_{lm}(\vartheta_2, \varphi_2)$ by means of (1.16) in which the rotation is chosen so that the polar angles of the z'' axis are ϑ_1 and φ_1, one gets

$$Y_{l0}(\omega, 0) = \sum_m Y_{lm}(\vartheta_2, \varphi_2) \mathcal{D}^{(l)}_{m0}(\varphi_1, \vartheta_1, 0).$$

Comparison of the latter equation with (1.24) leads to the conclusion that

$$\mathcal{D}^{(l)}_{m0}(\alpha, \beta, 0) = \left(\frac{4\pi}{2l+1}\right)^{1/2} Y^*_{lm}(\beta, \alpha). \quad (1.25)$$

We now write (1.23) with $j_1 = L$, $j_2 = l$ and $m_1 = m_2 = 0$ and make use of (1.25) with the result

$$Y_{LM}(\vartheta, \varphi) Y_{lm}(\vartheta, \varphi) = \sum_\lambda \left[\frac{(2L+1)(2l+1)}{4\pi(2\lambda+1)}\right]^{1/2}$$
$$\times C(L, l, \lambda; M, m) C(L, l, \lambda; 0, 0) Y_{\lambda, M+m}(\vartheta, \varphi). \quad (1.26)$$

Multiplication of (1.26) by $Y^*_{l'm'}$ and integration over the full solid angle, on account of the orthogonality of the spherical harmonics, yields the result

$$\int Y_{LM} Y_{lm} Y^*_{l'm'} \, d\Omega = I(LM, l, l'm')$$
$$= \left[\frac{(2L+1)(2l+1)}{4\pi(2l'+1)}\right]^{1/2} C(L, l, l'; M, m' - M) C(L, l, l'; 0, 0),$$
$$(1.27)$$

which is the required Gaunt integral. It is worth noticing that the properties of the Clebsch–Gordan coefficients decree the vanishing of the Gaunt integrals unless

$$|L - l| \leq l' \leq L + l, \quad \text{and} \quad L + l + l' = \text{even integer}.$$

The Gaunt integrals will be used to demonstrate an addition theorem for the multipole fields.

1.6.3 Irreducible Spherical Tensors

We are now able to build the simultaneous eigenvectors of the total angular momentum of a vector field by coupling the eigenfunctions of L^2 and L_z with those of s^2 and s_z. The required eigenvectors turn out to be

$$\boldsymbol{T}_{jlm}(\hat{\boldsymbol{r}}) = \sum_\mu C(1,l,j;-\mu,m+\mu)\, Y_{l,m+\mu}(\hat{\boldsymbol{r}})\, \boldsymbol{\xi}_{-\mu}\,, \tag{1.28}$$

and each of them is clearly a vector even though they are the components of an irreducible spherical tensor of rank j. We will find useful the inverse to (1.28) that reads

$$Y_{lm}\boldsymbol{\xi}_{-\mu} = \sum_j C(1,l,j;-\mu,m)\boldsymbol{T}_{jl,m-\mu}\,. \tag{1.29}$$

Each of the vectors $\boldsymbol{T}_{jlm}$ satisfies the simultaneous eigenvalue equations

$$J^2\,\boldsymbol{T}_{jlm} = j(j+1)\boldsymbol{T}_{jlm}\,,$$
$$J_z\,\boldsymbol{T}_{jlm} = m\boldsymbol{T}_{jlm}\,,$$
$$L^2\,\boldsymbol{T}_{jlm} = l(l+1)\boldsymbol{T}_{jlm}\,,$$
$$s^2\,\boldsymbol{T}_{jlm} = 2\boldsymbol{T}_{jlm}\,.$$

Their parity is given by the equation

$$\boldsymbol{T}_{jlm}(-\hat{\boldsymbol{r}}) = (-)^l \boldsymbol{T}_{jlm}(\hat{\boldsymbol{r}})\,,$$

and the conjugation relation reads

$$\boldsymbol{T}^*_{jlm} = (-)^{1+l-j+m} \boldsymbol{T}_{jl,-m}\,.$$

The triangular relation $\Delta(1lj)$ implies that for any value of j there exist three mutually independent vectors $\boldsymbol{T}$ with $l = j, j \pm 1$ and that j cannot take the value 0. Furthermore, from their very definition it follows that, under rotation of the coordinate system, the $\boldsymbol{T}_{jlm}$ transform according to the equation

$$O_{\boldsymbol{\omega}}\boldsymbol{T}_{jlm}(\hat{\boldsymbol{r}}) = \sum_{m'} \boldsymbol{T}_{jlm'}(\hat{\boldsymbol{r}})\mathcal{D}^{(j)}_{m'm}(\boldsymbol{\omega})\,,$$

were we used $\boldsymbol{\omega}$ as a shorthand for the eulerian angles that define the coordinate rotation that transforms $\boldsymbol{r}$ into $\boldsymbol{r}'$. The latter equation confirms our previous statement that the $2j+1$ vectors $\boldsymbol{T}_{jlm}$ are the components of an irreducible spherical tensor of rank j and parity $(-)^l$. The $\boldsymbol{T}_{jlm}$ have the property

$$\int \boldsymbol{T}^*_{j'l'm'} \cdot \boldsymbol{T}_{jlm}\, d\Omega = \delta_{jj'}\delta_{ll'}\delta_{mm'}\,,$$

so that they form a complete orthonormal set on the surface of the unit sphere. Note that

$$\boldsymbol{r} \cdot \boldsymbol{T}_{llm} = 0 \, ,$$

so that this vector is tangent to the surface of the unit sphere, whereas the vectors $\boldsymbol{T}_{l,l\pm1,m}$ are, in general, neither tangent nor orthogonal to such a surface. Finally, by recalling the identity

$$\nabla = \hat{\boldsymbol{r}} \frac{\partial}{\partial r} - \frac{i}{r} \hat{\boldsymbol{r}} \times \mathbf{L} \, , \tag{1.30}$$

it is an easy matter to show that

$$\nabla \cdot \boldsymbol{T}_{llm} = 0 \, . \tag{1.31}$$

1.6.4 Vector Solutions to the Helmholtz Equation

The considerations in the preceding sections led us to write the vector solutions to the Helmholtz equation for a homogeneous medium as the product of an irreducible spherical tensor of rank j by a function that depends on r only. Accordingly, we write

$$\boldsymbol{A}_{jlm}(\boldsymbol{r}) = R(r) \boldsymbol{T}_{jlm}(\hat{\boldsymbol{r}}) \, ,$$

where the functions $R(r)$ should be appropriately chosen. On account of the identity

$$\nabla^2 \equiv \frac{d^2}{dr^2} + \frac{2}{r} \frac{d}{dr} - \frac{\mathbf{L}^2}{r^2} \, ,$$

and of the fact that the vectors $\boldsymbol{T}_{jlm}$ are eigenfunctions of $\mathbf{L}^2$ with eigenvalue $l(l+1)$, it is an easy matter to see that the vector functions $\boldsymbol{A}_{jlm}$ satisfy the Helmholtz equation provided that the radial function satisfy the equation

$$\frac{1}{r^2} \frac{d}{dr} \left(r^2 \frac{dR}{dr} \right) + \left[k^2 - \frac{l(l+1)}{r^2} \right] R = 0 \, ,$$

which is immediately identified with the spherical Bessel equation of order l. Hereafter, the radial function will be denoted as $f_l(kr)$ with the understanding that it should be a spherical Bessel function or a spherical Hankel function of the first kind according to whether the required solution must be regular at the origin or must satisfy the radiation condition at infinity.

For the sake of convenience we report here the fundamental properties of the spherical Bessel and Hankel functions [7]

$$f_{l+1}(kr) + f_{l-1}(kr) = \frac{2l+1}{kr} f_l(kr) , \qquad (1.32a)$$

$$f'_l(kr) = \frac{1}{2l+1}[l f_{l-1}(kr) - (l+1) f_{l+1}(kr)] , \qquad (1.32b)$$

$$\left(\frac{\partial}{\partial r} - \frac{l}{r}\right) f_l(kr) = k\left[f'_l(kr) - \frac{l}{kr} f_l(kr)\right] = -k f_{l+1}(kr) , \qquad (1.32c)$$

$$\left(\frac{\partial}{\partial r} + \frac{l+1}{r}\right) f_l(kr) = k\left[f'_l(kr) + \frac{l+1}{kr} f_l(kr)\right] = k f_{l-1}(kr) , \qquad (1.32d)$$

which are needed for further developments. We will also need the limiting expressions of the Bessel and Hankel functions. For $kr \to 0$ we have

$$j_l(kr) \approx \frac{(kr)^l}{(2l+1)!!} ,$$

$$h_l(kr) \approx \frac{(2l-1)!!}{(kr)^{l+1}} ,$$

whereas for $kr \to \infty$

$$j_l(kr) \sim \frac{1}{kr} \sin(kr - l\pi/2) , \qquad (1.33a)$$

$$h_l(kr) \sim (-\mathrm{i})^{l+1} \frac{e^{\mathrm{i}kr}}{kr} . \qquad (1.33b)$$

The Wronskian

$$W[j_l(kr), h_l(kr)] = \mathrm{i}/(kr)^2$$

will also be useful for algebraic manipulations.

1.6.5 Divergence and Curl of the Vector Helmholtz Harmonics

The vector functions $\boldsymbol{A}_{jlm}$ that were defined in the preceding section, will be referred to as vector Helmholtz harmonics (VHH). These vector functions form a complete orthonormal set that is suitable as a basis for the expansion of the electromagnetic field. To perform these expansions, we will often need the explicit expression for the divergence and curl of these functions that stem from the so-called gradient formula whose demonstration has been reported, for instance, by Rose [5]:

$$\nabla f_l(kr) Y_{lm}(\hat{\boldsymbol{r}}) = -\sqrt{\frac{l+1}{2l+1}} \left(\frac{\partial}{\partial r} - \frac{l}{r}\right) f_l(kr) \boldsymbol{T}_{l,l+1,m}(\hat{\boldsymbol{r}})$$

$$+ \sqrt{\frac{l}{2l+1}} \left(\frac{\partial}{\partial r} + \frac{l+1}{r}\right) f_l(kr) \boldsymbol{T}_{l,l-1,m}(\hat{\boldsymbol{r}})$$

$$= \sqrt{\frac{l+1}{2l+1}} k f_{l+1}(kr) \boldsymbol{T}_{l,l+1,m}(\hat{\boldsymbol{r}})$$

$$+ \sqrt{\frac{l}{2l+1}} k f_{l-1}(kr) \boldsymbol{T}_{l,l-1,m}(\hat{\boldsymbol{r}}) .$$

The expressions for the curl are [8]

$$\nabla \times f_{l+1}(kr)\boldsymbol{T}_{l,l+1,m}(\hat{\boldsymbol{r}}) = -\mathrm{i}\left(\frac{\partial}{\partial r} + \frac{l+2}{r}\right)f_{l+1}(kr)\sqrt{\frac{l}{2l+1}}\boldsymbol{T}_{llm}(\hat{\boldsymbol{r}})$$

$$= -\mathrm{i}kf_l(kr)\sqrt{\frac{l}{2l+1}}\boldsymbol{T}_{llm}(\hat{\boldsymbol{r}}) \,, \quad (1.34\mathrm{a})$$

$$\nabla \times f_l(kr)\boldsymbol{T}_{llm}(\hat{\boldsymbol{r}}) = -\mathrm{i}\left(\frac{\partial}{\partial r} - \frac{l}{r}\right)f_l(kr)\sqrt{\frac{l}{2l+1}}\boldsymbol{T}_{l,l+1,m}(\hat{\boldsymbol{r}})$$

$$-\mathrm{i}\left(\frac{\partial}{\partial r} + \frac{l+1}{r}\right)f_l(kr)\sqrt{\frac{l+1}{2l+1}}\boldsymbol{T}_{l,l-1,m}(\hat{\boldsymbol{r}})$$

$$= \mathrm{i}k\left[\sqrt{\frac{l}{2l+1}}f_{l+1}(kr)\boldsymbol{T}_{l,l+1,m}(\hat{\boldsymbol{r}}) \right.$$

$$\left. -\sqrt{\frac{l+1}{2l+1}}f_{l-1}(kr)\boldsymbol{T}_{l,l-1,m}(\hat{\boldsymbol{r}})\right] \,, \quad (1.34\mathrm{b})$$

$$\nabla \times f_{l-1}(kr)\boldsymbol{T}_{l,l-1,m}(\hat{\boldsymbol{r}}) = -\mathrm{i}\left(\frac{\partial}{\partial r} - \frac{l-1}{r}\right)f_{l-1}(kr)\sqrt{\frac{l+1}{2l+1}}\boldsymbol{T}_{llm}(\hat{\boldsymbol{r}})$$

$$= \mathrm{i}kf_l(kr)\sqrt{\frac{l+1}{2l+1}}\boldsymbol{T}_{llm}(\hat{\boldsymbol{r}}) \,, \quad (1.34\mathrm{c})$$

and the formulas for the divergence turn out to be

$$\nabla \cdot f_{l+1}(kr)\boldsymbol{T}_{l,l+1,m}(\hat{\boldsymbol{r}}) = -\sqrt{\frac{l+1}{2l+1}}\left(\frac{\partial}{\partial r} + \frac{l+2}{r}\right)f_{l+1}(kr)Y_{lm}(\hat{\boldsymbol{r}})$$

$$= -\sqrt{\frac{l+1}{2l+1}}kf_l(kr)Y_{lm}(\hat{\boldsymbol{r}}) \,, \quad (1.35\mathrm{a})$$

$$\nabla \cdot f_l(kr)\boldsymbol{T}_{llm}(\hat{\boldsymbol{r}}) = 0 \,, \quad (1.35\mathrm{b})$$

$$\nabla \cdot f_{l-1}(kr)\boldsymbol{T}_{l,l-1,m}(\hat{\boldsymbol{r}}) = \sqrt{\frac{l}{2l+1}}\left(\frac{\partial}{\partial r} - \frac{l-1}{r}\right)f_{l-1}(kr)Y_{lm}(\hat{\boldsymbol{r}})$$

$$= -\sqrt{\frac{l}{2l+1}}kf_l(kr)Y_{lm}(\hat{\boldsymbol{r}}) \,. \quad (1.35\mathrm{c})$$

Note that, because of (1.31), (1.35b) hold true for any choice of the radial function.

1.7 Multipole Fields

Although the VHH solve the problem of expanding the electromagnetic field, there are cases in which their direct use is not advisable. We refer to the fact that the VHH do not form a set of vector functions that are orthogonal or tangential to the surface of the unit sphere. Moreover, the VHH are neither solenoidal nor irrotational, a property that would be desirable because the most general vector field is a superposition of an irrotational field and of a solenoidal field [9]. In particular, in a homogeneous medium both $\boldsymbol{E}$ and $\boldsymbol{B}$ are solenoidal and so they can be expanded in terms of solenoidal vector functions only. These considerations call for the determination of alternative forms for the solutions of the vector Helmholtz equation that are either solenoidal or irrotational. Of course, since the VHH form a complete set, the alternative solutions, once determined, can be related to the VHH.

1.7.1 Hansen's Vectors

We start by remarking that there should exist three mutually independent vector solutions to the Helmholtz equation because, in three-dimensional space, any vector can always be expressed as a linear combination of three mutually independent vectors. Now, let ψ be a scalar solution to the Helmholtz equation and $\boldsymbol{e}$ a constant vector, not necessarily of unit magnitude. Then it is an easy matter to verify that the vector functions $\boldsymbol{e}\psi$ and $\nabla\psi$ actually satisfy the Helmholtz equation. We further remark that if $\boldsymbol{A}$ is a vector solution to the Helmholtz equation then $\nabla \times \boldsymbol{A}$ is also a solution. Accordingly, we can write three mutually independent vector solutions to (1.5) as [4]

$$\boldsymbol{L} = \nabla\psi , \quad \boldsymbol{M} = \nabla \times \boldsymbol{e}\psi , \quad \boldsymbol{N} = \frac{1}{k}\nabla \times \boldsymbol{M} , \qquad (1.36)$$

where the factor $1/k$ in the definition of $\boldsymbol{N}$ has been introduced so that

$$\frac{1}{k}\nabla \times \boldsymbol{N} = \boldsymbol{M} . \qquad (1.37)$$

Equations (1.36) define the so-called Hansen's vectors [10] that are easily verified to be actually linearly independent even though they do not form an orthonormal set because, for general choice of $\boldsymbol{e}$ we have

$$\boldsymbol{L} \cdot \boldsymbol{M} = \boldsymbol{M} \cdot \boldsymbol{N} = 0 , \text{ but } \boldsymbol{L} \cdot \boldsymbol{N} \neq 0 .$$

From the definition in (1.36) it also turns out that

$$\nabla \cdot \boldsymbol{M} = \nabla \cdot \boldsymbol{N} = 0 , \qquad \nabla \times \boldsymbol{L} = 0 , \qquad (1.38)$$

so that Hansen's vectors are either solenoidal or irrotational as required. As a consequence, any solenoidal solution to the Helmholtz equation can be projected into the subspace spanned by $\boldsymbol{M}$ and $\boldsymbol{N}$, whereas any irrotational solution can be expressed in terms of $\boldsymbol{L}$ only.

The definition (1.36) of the Hansen's vectors is appropriate when the boundary conditions must be imposed across a plane surface; in this case, in fact, choosing $\boldsymbol{e}$ parallel to the surface greatly simplifies the algebraic manipulations. Nevertheless, in many optical problems one has to impose the boundary conditions across spherical surfaces, and this task is hardly manageable in terms of the vectors $\boldsymbol{M}$ and $\boldsymbol{N}$ defined in (1.36). The definition (1.36) is not unique, however, so that we can resort to the following alternative definition of Hansen's vectors.

Let ψ be a scalar solution to the Helmholtz equation and let O be a vector operator that, when applied to any space-dependent scalar function, yields a vector function $\boldsymbol{A}$. Then, by operating with O on the scalar Helmholtz equation we get

$$\mathrm{O}(\nabla^2 + k^2)\psi = (\nabla^2 + k^2)\mathrm{O}\psi + [\mathrm{O}, \nabla^2]\psi = 0 \ .$$

Therefore, if O commutes with ∇^2, any vector function of the form $\boldsymbol{A} = \mathrm{O}\psi$ is a solution to the Helmholtz equation. Now the operator $\Omega_k = \nabla^2 + k^2$ is invariant under translations as well as under rotations of the coordinate system. This implies that Ω_k commutes with the operators of the linear momentum $\mathrm{p} = -\mathrm{i}\nabla$ and of the angular momentum $\mathrm{L} = -\mathrm{i}\boldsymbol{r} \times \nabla$. Furthermore, it is easily verified that Ω_k also commutes with the operator $\nabla \times \mathrm{L}$. As a consequence, from any scalar solution ψ we get the three linearly independent vector solutions

$$-\mathrm{i}\nabla\psi \ , \quad \mathrm{L}\psi \ , \quad \nabla \times \mathrm{L}\psi \ ,$$

which will be particularly useful when

$$\psi = \psi_{lm}(\boldsymbol{r}) = [l(l+1)]^{-1/2} f_l(kr) Y_{lm}(\hat{\boldsymbol{r}}) \ ,$$

i.e., when ψ is an eigenfunction of the angular momentum. Note that the factor $[l(l+1)]^{-1/2}$ in the definition of ψ has been inserted for convenience in further manipulations.

1.7.2 Vector Spherical Harmonics

We are now able to write the following useful form of Hansen's vectors. With the help of (1.30) we find

$$\boldsymbol{L}_{lm} = \frac{1}{k}\nabla\psi_{lm} = [l(l+1)]^{-1/2} f_l'(kr) Y_{lm}\hat{\boldsymbol{r}} - \frac{\mathrm{i}}{kr} f_l(kr) \hat{\boldsymbol{r}} \times \boldsymbol{X}_{lm} \ , \quad (1.39\mathrm{a})$$

$$\boldsymbol{M}_{lm} = \mathrm{L}\psi_{lm} = f_l(kr) \boldsymbol{X}_{lm} \ , \quad (1.39\mathrm{b})$$

$$\boldsymbol{N}_{lm} = \frac{1}{k}\nabla \times \mathrm{L}\psi_{lm} = \frac{\mathrm{i}}{kr}[l(l+1)]^{1/2} f_l(kr) Y_{lm}\hat{\boldsymbol{r}}$$
$$+ \frac{1}{kr}[kr f_l(kr)]' \hat{\boldsymbol{r}} \times \boldsymbol{X}_{lm} \ , \quad (1.39\mathrm{c})$$

where the prime denotes differentiation with respect to the argument. In (1.39) we define the vector functions

$$\boldsymbol{X}_{lm} = [l(l+1)]^{-1/2} \mathbf{L} Y_{lm} \qquad (1.40)$$

that are customarily called *vector spherical harmonics* [2], and the factor $1/k$ has been included in the definition of $\boldsymbol{N}$ to ensure the validity of (1.37) whereas in the definition of $\boldsymbol{L}$ it has been included for normalization purposes. It may be useful to note that, in terms of spherical components, (1.40) reads

$$\begin{aligned}\boldsymbol{X}_{lm} &= [l(l+1)]^{-1/2} \sum_\mu (-)^\mu \mathrm{L}_\mu Y_{lm} \boldsymbol{\xi}_{-\mu} \\ &= -\sum_\mu C(1,l,l; -\mu, m+\mu) Y_{l, m+\mu} \boldsymbol{\xi}_{-\mu} \,. \end{aligned} \qquad (1.41)$$

The advantage of the definition (1.39) stems from the fact that the vector spherical harmonics $\hat{\boldsymbol{r}} Y_{lm}$, $\boldsymbol{X}_{lm}$, and $\hat{\boldsymbol{r}} \times \boldsymbol{X}_{lm}$ constitute an orthonormal set of eigenfunctions of the angular momentum with the following orthogonality properties

$$\int \boldsymbol{X}^*_{l'm'} \cdot \boldsymbol{X}_{lm} \, d\Omega = \int \hat{\boldsymbol{r}} \times \boldsymbol{X}^*_{l'm'} \cdot \hat{\boldsymbol{r}} \times \boldsymbol{X}_{lm} \, d\Omega = \delta_{ll'} \delta_{mm'} \,, \qquad (1.42)$$

$$\int \boldsymbol{X}^*_{l'm'} \cdot \hat{\boldsymbol{r}} \times \boldsymbol{X}_{lm} \, d\Omega = 0 \,. \qquad (1.43)$$

Moreover, the harmonic $\hat{\boldsymbol{r}} Y_{lm}$ is clearly orthogonal to the surface of the unit sphere, whereas the harmonics $\boldsymbol{X}_{lm}$ and $\hat{\boldsymbol{r}} \times \boldsymbol{X}_{lm}$ are tangent so that

$$\hat{\boldsymbol{r}} \cdot \boldsymbol{X}_{lm} = \hat{\boldsymbol{r}} \cdot \hat{\boldsymbol{r}} \times \boldsymbol{X}_{lm} = 0 \,.$$

We also remark that $\boldsymbol{L}_{lm}$, $\boldsymbol{M}_{lm}$ and $\boldsymbol{N}_{lm}$ are easily verified to satisfy (1.38).

For later developments, it is convenient to define the *transverse vector harmonics* [11]

$$\boldsymbol{Z}^{(1)}_{lm}(\hat{\boldsymbol{r}}) = \boldsymbol{X}_{lm}(\hat{\boldsymbol{r}}) \,, \quad \boldsymbol{Z}^{(2)}_{lm}(\hat{\boldsymbol{r}}) = \boldsymbol{X}_{lm}(\hat{\boldsymbol{r}}) \times \hat{\boldsymbol{r}} \,, \qquad (1.44)$$

which will be extensively used to expand any transverse field. In fact, they are mutually orthogonal in the usual vector sense

$$\boldsymbol{Z}^{(1)}_{lm}(\hat{\boldsymbol{r}}) \cdot \boldsymbol{Z}^{(2)}_{lm}(\hat{\boldsymbol{r}}) = 0 \,,$$

and orthogonal to their argument

$$\boldsymbol{Z}^{(1)}_{lm}(\hat{\boldsymbol{r}}) \cdot \hat{\boldsymbol{r}} = \boldsymbol{Z}^{(2)}_{lm}(\hat{\boldsymbol{r}}) \cdot \hat{\boldsymbol{r}} = 0 \,.$$

It is worth stressing that the definitions (1.44) imply that the transverse harmonics are irreducible spherical tensors of rank l so that, under rotation of the coordinate system, they transform according to $\mathrm{D}^{(l)}$.

1.7.3 Spherical Multipole Fields

We are now able to give the general expression of the vector functions that we are going to use. To this end we give the following definition:

> The *multipole fields* are the solutions to the Maxwell equations that are eigenvectors of L^2 and L_z as well as of the parity.

According to this definition both the Hansen's vectors and the VHHs are a complete set of multipole fields. Using the former or the latter is a matter of convenience for the algebraic manipulations. Anyway, we will show presently that Hansen's vectors can be written as linear combinations of the appropriate VHHs. In fact, we note that, according to (1.39b) the vector function $\boldsymbol{M}_{lm}$ is solenoidal and that according to (1.35b) $f_l \boldsymbol{T}_{llm}$ too is solenoidal. Actually, from (1.28) and (1.41)

$$\boldsymbol{M}_{lm} = f_l \boldsymbol{X}_{lm} = -f_l \boldsymbol{T}_{llm} \ . \tag{1.45}$$

Then, according to (1.36) and (1.34b) we find

$$\boldsymbol{N}_{lm} = i \left[\sqrt{\frac{l+1}{2l+1}} f_{l-1}(kr) \boldsymbol{T}_{l,l-1,m} - \sqrt{\frac{l}{2l+1}} f_{l+1}(kr) \boldsymbol{T}_{l,l+1,m} \right] \ , \tag{1.46}$$

which is easily verified through (1.35a) and (1.35c) to have vanishing divergence. Note that (1.46) is a combination of $\boldsymbol{T}_{l,l+1,m}$ and $\boldsymbol{T}_{l,l-1,m}$ as is required so that $\boldsymbol{N}_{lm}$ have definite parity. Since $\boldsymbol{L}_{lm}$ too must have a definite parity, with the help of (1.34a) and (1.34c), we find that the correct combination is

$$\boldsymbol{L}_{lm} = \left[\sqrt{\frac{l}{2l+1}} f_{l-1}(kr) \boldsymbol{T}_{l,l-1,m} + \sqrt{\frac{l+1}{2l+1}} f_{l+1}(kr) \boldsymbol{T}_{l,l+1,m} \right] \ ,$$

which is easily verified to have vanishing curl.

Hereafter, we need to distinguish the multipole fields that are regular at the origin from those that satisfy the radiation condition at infinity. Accordingly, we will use in place of $\boldsymbol{M}_{lm}$ and $\boldsymbol{N}_{lm}$, the notation

$$\boldsymbol{J}_{lm}^{(1)} = j_l(kr) \boldsymbol{X}_{lm}(\hat{\boldsymbol{r}}) \ , \quad \boldsymbol{J}_{lm}^{(2)} = \frac{1}{k} \nabla \times \boldsymbol{J}_{lm}^{(1)} \ ,$$

respectively, when their radial function is a spherical Bessel function $j_l(kr)$. We will use the notation

$$\boldsymbol{H}_{lm}^{(1)} = h_l(kr) \boldsymbol{X}_{lm}(\hat{\boldsymbol{r}}) \ , \quad \boldsymbol{H}_{lm}^{(2)} = \frac{1}{k} \nabla \times \boldsymbol{H}_{lm}^{(1)} \ ,$$

for the case in which the radial function of the multipole fields is a spherical Hankel function of the first kind $h_l(kr)$. The superscript $p = 1, 2$ that has been attached to J and H is a parity index that distinguishes the magnetic multipole fields M, $(p = 1)$ from the electric ones N, $(p = 2)$.

The equations [12]

$$\hat{r} Y_{lm}(\hat{r}) = -\sqrt{\frac{l+1}{2l+1}} T_{l,l+1,m}(\hat{r}) + \sqrt{\frac{l}{2l+1}} T_{l,l-1,m}(\hat{r}) , \quad (1.47a)$$

$$-i\hat{r} \times X_{lm}(\hat{r}) = +\sqrt{\frac{l}{2l+1}} T_{l,l+1,m}(\hat{r}) + \sqrt{\frac{l+1}{2l+1}} T_{l,l-1,m}(\hat{r}) , \quad (1.47b)$$

relate the vector spherical harmonics to the irreducible spherical tensors. Moreover, one may find useful the equations

$$H^{(1)}_{lm} = h_l Z^{(1)}_{lm} , \quad H^{(2)}_{lm} = \frac{i}{kr}[l(l+1)]^{1/2} h_l Y_{lm} \hat{r} - \frac{1}{kr}[krh_l]' Z^{(2)}_{lm} , \quad (1.48)$$

which give the explicit expression of the H-multipole fields in terms of transverse vector harmonics. Of course, similar equations hold also for the J-multipole fields.

1.8 Addition Theorem for Multipole Fields

We will now use the translational invariance of Helmholtz equation to establish an addition theorem that relates the VHHs referred to two systems of spherical coordinates that differ for a translation of their origins. To achieve our task we will at first consider the case of the scalar harmonics [13] and will then extend our findings to the case of the vector harmonics.

1.8.1 Nozawa's Theorem

Let us consider two frames of reference, say Σ and Σ', in which we define two systems of rectangular coordinates whose axes are mutually parallel and whose origins differ by a translation. A point P has vector position r in Σ and r' in Σ', so that we can write $r = r' + R$, where R is the vector position of the origin of Σ' with respect to Σ. Then, for a scalar plane wave the identity holds

$$\exp(i k \cdot r) = \exp(i k \cdot R) \exp(i k \cdot r') .$$

We use the Bauer formula to expand each of the exponentials that appear in both sides of this equation with the result

$$4\pi \sum_{l''m''} i^{l''} j_{l''}(kr) Y_{l''m''}(\hat{r}) Y^*_{l''m''}(\hat{k})$$
$$= 4\pi \sum_{l'm'} i^{l'} j_{l'}(kr') Y_{l'm'}(\hat{r}') Y^*_{l'm'}(\hat{k}) 4\pi \sum_{LM} i^L j_L(kR) Y^*_{LM}(\hat{R}) Y_{LM}(\hat{k}) .$$

1.8 Addition Theorem for Multipole Fields

Multiplication of both sides by $Y_{lm}(\hat{\boldsymbol{k}})$ and integration over the solid angle leads to the equation

$$j_l(kr)Y_{lm}(\hat{\boldsymbol{r}}) = \sum_{l'm'} j_{l'}(kr')Y_{l'm'}(\hat{\boldsymbol{r}}')$$
$$\times 4\pi \sum_{LM} i^{l'-l+L} j_L(kR) Y^*_{LM}(\hat{\boldsymbol{R}}) \int Y_{lm} Y_{LM} Y^*_{l'm'} \, d\Omega_{\boldsymbol{k}} ,$$

which can be rewritten as

$$j_l(kr)Y_{lm}(\hat{\boldsymbol{r}}) = \sum_{l'm'} j_{l'}(kr')Y_{l'm'}(\hat{\boldsymbol{r}}') G_{l'm'lm}(\boldsymbol{R},k) ,$$

where we define

$$G_{l'm'lm}(\boldsymbol{R},k) = 4\pi \sum_L i^{l'-l+L} I(lm, L, l'm') j_L(kR) Y^*_{L,m'-m}(\hat{\boldsymbol{R}}) ,$$

and the I are Gaunt integrals whose expression has been given in Sect. 1.6.2. Using the Neuman expansion of the function [7]

$$\frac{e^{ik|\boldsymbol{r}'+\boldsymbol{R}|}}{ik|\boldsymbol{r}'+\boldsymbol{R}|},$$

it is possible to show that an analogous formula holds for the function $h_l^{(1)}(kr)Y_{lm}(\hat{\boldsymbol{r}})$, i.e., for the case in which the radial function is a Hankel function of the first kind. Ultimately we can write

$$f_l(kr)Y_{lm}(\hat{\boldsymbol{r}}) = \sum_{l'm'} g_{l'}(kr')Y_{l'm'}(\hat{\boldsymbol{r}}') G_{l'm'lm}(\boldsymbol{R},k) , \qquad (1.49)$$

with

$$G_{l'm'lm}(\boldsymbol{R},k) = 4\pi \sum_L i^{l'-l+L} I(lm, L, l'm') \psi_L(kR) Y^*_{L,m'-m}(\hat{\boldsymbol{R}}) , \qquad (1.50)$$

where, when $f_l = j_l$, $\psi_l = j_l$ and $g_l = j_l$; when $f_l = h_l$ then

$$\psi_l = h_l , \quad g_l = j_l , \quad \text{for } r' < R ,$$
$$\psi_l = j_l , \quad g_l = h_l , \quad \text{for } r' > R .$$

Equations (1.49) and (1.50) express Nozawa's addition theorem for scalar Helmholtz harmonics.

28   1. Multipole Fields

1.8.2 Addition Theorem for the Vector Helmholtz Harmonics

Let us consider the VHH

$$\boldsymbol{A}_{jlm}(\boldsymbol{r}) = f_l(kr)\boldsymbol{T}_{jlm}(\hat{\boldsymbol{r}}) ,$$

which is centered at the origin, and let us express this function in terms of vector Helmholtz harmonics centered at a point of vector coordinate $\boldsymbol{R}$ [14]. To this end we recall that the definition of $\boldsymbol{A}_{jlm}$ is

$$\boldsymbol{A}_{jlm}(\boldsymbol{r}) = \sum_\mu C(1,l,j;-\mu,m+\mu) f_l(kr) Y_{l,m+\mu}(\hat{\boldsymbol{r}}) \boldsymbol{\xi}_{-\mu} ,$$

so that Nozawa's theorem applies to each of the terms $f_l(kr)Y_{l\,m+\mu}(\hat{\boldsymbol{r}})$, whereas the spherical basis vectors $\boldsymbol{\xi}_{-\mu}$ are left unchanged by the translation. Therefore we get

$$\begin{aligned}\boldsymbol{A}_{jlm}(\boldsymbol{r}) = &\sum_\mu C(1,l,j;-\mu,m+\mu) \sum_{l'm''} G_{l'm''l,m+\mu}(\boldsymbol{R},k) \\ &\times \sum_{j'} C(1,l',j';-\mu,m'') g_{l'}(kr')\boldsymbol{T}_{j'l',m''-\mu}(\hat{\boldsymbol{r}}') ,\end{aligned} \quad (1.51)$$

in which we used the inverse coupling relation (1.29). Then, the replacement $m' = m'' - \mu$ and the definition

$$\begin{aligned}\mathcal{G}_{j'l'm'jlm} = &\sum_\mu C(1,l',j';-\mu,m'+\mu) \\ &\times G_{l',m'+\mu,l,m+\mu}(\boldsymbol{R},k) C(1,l,j;-\mu,m+\mu) ,\end{aligned}$$

yield the equation

$$\boldsymbol{A}_{jlm}(\boldsymbol{r}) = \sum_{j'l'm'} g_{l'}(kr')\boldsymbol{T}_{j'l'm'}(\hat{\boldsymbol{r}}') \mathcal{G}_{j'l'm'jlm} ,$$

which is the required addition theorem [14].

1.8.3 Application to Hansen's Vectors

In view of (1.45) and (1.46), theorem (1.51) can be applied to each of the VHHs that appear in the definitions of $\boldsymbol{M}_{lm}$ and $\boldsymbol{N}_{lm}$. Accordingly we get, for instance,

$$\begin{aligned}\boldsymbol{M}_{lm}(\boldsymbol{r}) = -\sum_{l'm'} \big[&g_{l'}(kr')\boldsymbol{T}_{l'l'm'}(\hat{\boldsymbol{r}}')\mathcal{G}_{l'l'm'llm} \\ &+ g_{l'-1}(kr')\boldsymbol{T}_{l',l'-1,m'}(\hat{\boldsymbol{r}}')\mathcal{G}_{l',l'-1,m'llm} \\ &+ g_{l'+1}(kr')\boldsymbol{T}_{l',l'+1,m'}(\hat{\boldsymbol{r}}')\mathcal{G}_{l',l'+1,m'llm} \big] .\end{aligned} \quad (1.52)$$

1.8 Addition Theorem for Multipole Fields

Since $\boldsymbol{M}_{lm}$ and $\boldsymbol{N}_{lm}$ are solenoidal, the preceding equation yields the relation

$$\sqrt{l'+1}\mathcal{G}_{l',l'+1,m'llm} + \sqrt{l'}\mathcal{G}_{l',l'-1,m'llm} = 0 \,, \tag{1.53}$$

that can also by proved by direct use of the properties of the Clebsch–Gordan coefficients. On account of (1.53), we can rewrite (1.52) as

$$\boldsymbol{M}_{lm}(\boldsymbol{r}) = \sum_{l'm'} \left[\bar{\boldsymbol{M}}_{l'm'}(\boldsymbol{r}')A(l'm',lm) + \bar{\boldsymbol{N}}_{l'm'}(\boldsymbol{r}')B(l'm',lm) \right] \,, \tag{1.54}$$

where we define

$$A(l'm',lm) = \mathcal{G}_{l'l'm'llm} \,, \quad B(l'm',lm) = \mathrm{i}\sqrt{\frac{2l'+1}{l'+1}} \mathcal{G}_{l',l'-1,m'llm} \,,$$

and $\bar{\boldsymbol{M}}_{l'm'}$ and $\bar{\boldsymbol{N}}_{l'm'}$ are identical to $\boldsymbol{M}_{l'm'}$ and $\boldsymbol{N}_{l'm'}$, respectively, except for the substitution of g_l in place of f_l. On account of (1.36) and (1.37), we take the curl of (1.54) and get

$$\boldsymbol{N}_{lm}(\boldsymbol{r}) = \sum_{l'm'} \left[\bar{\boldsymbol{N}}_{l'm'}(\boldsymbol{r}')A(l'm',lm) + \bar{\boldsymbol{M}}_{l'm'}(\boldsymbol{r}')B(l'm',lm) \right] \,. \tag{1.55}$$

Hereafter, when the addition theorem is applied to the multipole fields $\boldsymbol{H}_{lm}^{(p)}$ and $\boldsymbol{J}_{lm}^{(p)}$ we will use the notation

$$\boldsymbol{H}_{lm}^{(p)}(\boldsymbol{r},k) = \sum_{p'l'm'} \boldsymbol{J}_{l'm'}^{(p')}(\boldsymbol{r}',k)\mathcal{H}_{l'm'lm}^{(p'p)}(\boldsymbol{R},k), \quad \text{for } r' < R \,, \tag{1.56a}$$

$$\boldsymbol{H}_{lm}^{(p)}(\boldsymbol{r},k) = \sum_{p'l'm'} \boldsymbol{H}_{l'm'}^{(p')}(\boldsymbol{r}',k)\mathcal{J}_{l'm'lm}^{(p'p)}(\boldsymbol{R},k), \quad \text{for } r' > R \,, \tag{1.56b}$$

and

$$\boldsymbol{J}_{lm}^{(p)}(\boldsymbol{r},k) = \sum_{p'l'm'} \boldsymbol{J}_{l'm'}^{(p')}(\boldsymbol{r}',k)\mathcal{J}_{l'm'lm}^{(p'p)}(\boldsymbol{R},k) \,, \tag{1.57}$$

respectively. Comparison with equations (1.54) and (1.55) leads to the conclusion that the quantities $\mathcal{H}_{lml'm'}^{(1\,1)}$ and $\mathcal{H}_{lml'm'}^{(1\,2)}$ are identical to $A(l'm',lm)$ and $B(l'm',lm)$ with $\psi_l = h_l$, whereas in the quantities $\mathcal{J}_{lml'm'}^{(1\,1)}$ and $\mathcal{J}_{lml'm'}^{(1\,2)}$ we have $\psi_l = j_l$. Moreover,

$$\mathcal{H}_{l'm'lm}^{(1\,1)} = \mathcal{H}_{l'm'lm}^{(2\,2)} \,, \quad \mathcal{H}_{l'm'lm}^{(pp')} = \mathcal{H}_{l'm'lm}^{(p'p)} \,,$$

$$\mathcal{J}_{l'm'lm}^{(1\,1)} = \mathcal{J}_{l'm'lm}^{(2\,2)} \,, \quad \mathcal{J}_{l'm'lm}^{(pp')} = \mathcal{J}_{l'm'lm}^{(p'p)} \,.$$

The explicit expression of the elements of matrices H and J is

$$\mathcal{H}^{(p'p)}_{l'm'lm}(\boldsymbol{R},k) = (1-\delta_{R,0})\left[\delta_{pp'} + i\sqrt{\frac{2l'+1}{l'+1}}(1-\delta_{pp'})\right]$$

$$\times \sum_{\mu} C(1,l'-1+\delta_{pp'},l';-\mu,m'+\mu)$$

$$\times G_{l'-1+\delta_{pp'},m'+\mu,l,m+\mu}(\boldsymbol{R},k) C(1,l,l;-\mu,m+\mu) \,,$$

(1.58a)

$$\mathcal{J}^{(p'p)}_{l'm'lm}(\boldsymbol{R},k) = \left[\delta_{pp'} + i\sqrt{\frac{2l'+1}{l'+1}}(1-\delta_{pp'})\right]$$

$$\times \sum_{\mu} C(1,l'-1+\delta_{pp'},l';-\mu,m'+\mu)$$

$$\times G_{l'-1+\delta_{pp'},m'+\mu,l,m+\mu}(\boldsymbol{R},k) C(1,l,l;-\mu,m+\mu) \,, \quad (1.58b)$$

where G is still given by (1.50). In particular, when $\boldsymbol{R} = \pm a\,\hat{\boldsymbol{e}}_z$, i.e., when the translation takes place along the z axis, the matrices H and J become diagonal in m. Indeed, the spherical harmonics that occur in G have the property

$$Y_{lm}(0,\varphi) = \delta_{m0}\left(\frac{2l+1}{4\pi}\right)^{1/2}, \quad Y_{lm}(\pi,\varphi) = (-)^l Y_{lm}(0,\varphi) \,, \quad (1.59)$$

so that

$$G_{lml'm'} = G_{ll';m}\delta_{mm'} \,,$$

and, as a consequence,

$$\mathcal{H}^{(pp')}_{lml'm'} = \mathcal{H}^{(pp')}_{ll';m}\delta_{mm'} \,, \quad \mathcal{J}^{(pp')}_{lml'm'} = \mathcal{J}^{(pp')}_{ll';m}\delta_{mm'} \,. \quad (1.60)$$

In case we have to translate multipole fields from an origin O_α at $\boldsymbol{R}_\alpha$ to another origin $O_{\alpha'}$ at $\boldsymbol{R}_{\alpha'}$ the transfer vector is $\boldsymbol{R}_{\alpha\alpha'} = \boldsymbol{R}_{\alpha'} - \boldsymbol{R}_\alpha$, and we can alternatively use the notation

$$\mathcal{H}^{(p'p)}_{\alpha'l'm'\alpha lm} = \mathcal{H}^{(p'p)}_{l'm'lm}(\boldsymbol{R}_{\alpha\alpha'},k) \,, \quad \mathcal{J}^{(p'p)}_{\alpha'l'm'\alpha lm} = \mathcal{J}^{(p'p)}_{l'm'lm}(\boldsymbol{R}_{\alpha\alpha'},k) \,. \quad (1.61)$$

The addition theorem we discussed so far helps to clarify the meaning of the customary classification of the multipole fields into magnetic and electric. In fact, this classification is deceiving because (1.56) and (1.57) show that a multipole field that is purely magnetic or electric with respect to a given origin becomes a superposition of both magnetic and electric multipole fields with respect to a different origin. In other words, the classification of the multipole fields is meaningful only when the frame of reference is specified. The preceding consideration should be borne in mind when discussing, e.g., the optical resonances of nonspherical particles.

2. Propagation Through an Assembly of Nonspherical Scatterers

In this chapter, we study the propagation of electromagnetic waves through a dispersion of nonspherical particles embedded within an otherwise homogeneous isotropic matrix. Our purpose is to relate the scattering features of the individual particles to the macroscopic optical properties of the medium as a whole. This relation can be established in general terms under conditions that are met by several media of interest, such as atmospheric aerosols and the interstellar medium.

2.1 Scattering Amplitude

In Sect. 1.1 we stated that, when a wave impinges on a particle, the latter emits a field that, at a large distance, looks like an outgoing spherical wave. In order to give a more detailed description of this process, let us assume that the particle is embedded into a homogeneous, isotropic indefinite medium of (real) refractive index n and that the incident field is the plane wave

$$\boldsymbol{E}_\mathrm{I} = E_0 \hat{\boldsymbol{e}}_\mathrm{I} \exp(\mathrm{i}\boldsymbol{k}_\mathrm{I} \cdot \boldsymbol{r}) ,$$

of (unit) polarization vector $\hat{\boldsymbol{e}}_\mathrm{I}$ and propagation vector $\boldsymbol{k}_\mathrm{I} = k\hat{\boldsymbol{k}}_\mathrm{I}$, with $k = nk_\mathrm{v}$. The scattering process depends both on the geometry and on the physical features of the particle, but the form of the scattered wave at a large distance from the particle can be established on general grounds. Let us recall that, since the scattered field $\boldsymbol{E}_\mathrm{S}$ satisfies the vector Helmholtz equation (1.7), each of its rectangular components should be a solution of the scalar Helmholtz equation. Therefore, denoting with ψ any of the rectangular components of $\boldsymbol{E}_\mathrm{S}$, we can write

$$(\nabla^2 + k^2)\psi(\boldsymbol{r}, k) = 0 . \tag{2.1}$$

The solutions to (2.1) that describe an actual scattering process must satisfy appropriate boundary conditions at the surface of the particle and should depend both on the direction of the incident wavevector $\hat{\boldsymbol{k}}_\mathrm{I}$ and on the direction of observation $\hat{\boldsymbol{k}}_\mathrm{S} = \hat{\boldsymbol{r}}$. By choosing the frame of reference so that the particle is at the origin, the solution $\psi(\boldsymbol{r}, k)$ can be expanded in a series of spherical harmonics in the form

$$\psi(\boldsymbol{r},k) = \sum_{lm} C_{lm}(\hat{\boldsymbol{k}}_\mathrm{I}) Y_{lm}(\hat{\boldsymbol{r}}) R_l(r,k)/r \ . \tag{2.2}$$

Substitution of expansion (2.2) into equation (2.1) leads to the equation for the radial functions

$$\frac{\mathrm{d}^2 R_l}{\mathrm{d}r^2} + \left[k^2 - \frac{l(l+1)}{r^2}\right] R_l = 0 \ ,$$

whereas the amplitudes C_{lm}, which depend on $\hat{\boldsymbol{k}}_\mathrm{I}$, are determined by the boundary conditions at the surface of the particle. Therefore, for small values of r, $\psi(\boldsymbol{r},k)$ has, in general, a complicated form that depends both on the geometry and on the physical features of the particle. Nevertheless, when we go to consider the radial equation in the far zone, i.e., for large values of r, we get

$$\frac{\mathrm{d}^2 R_l}{\mathrm{d}r^2} + k^2 R_l = 0 \ , \tag{2.3}$$

whose solutions are clearly independent of l. Since the scattered field should satisfy the *radiation condition*, i.e., it should include only outgoing waves at infinity [4, 15], the only acceptable solution to (2.3) is

$$R_l = E_0 \exp(\mathrm{i}kr) \ ,$$

for all values of l, so that the asymptotic form of $\psi(\boldsymbol{r},k)$ results

$$\psi(\boldsymbol{r},k) = E_0 f(\hat{\boldsymbol{k}}_\mathrm{S}, \hat{\boldsymbol{k}}_\mathrm{I}) \frac{\mathrm{e}^{\mathrm{i}kr}}{r} \ ,$$

where we define

$$f(\hat{\boldsymbol{k}}_\mathrm{S}, \hat{\boldsymbol{k}}_\mathrm{I}) = \sum_{lm} C_{lm}(\hat{\boldsymbol{k}}_\mathrm{I}) Y_{lm}(\hat{\boldsymbol{r}}) \ ,$$

with $\hat{\boldsymbol{k}}_\mathrm{S} = \hat{\boldsymbol{r}}$. Since $\psi(\boldsymbol{r},k)$ is any of the rectangular components of $\boldsymbol{E}_\mathrm{S}$, the whole vector should have the asymptotic form

$$\boldsymbol{E}_\mathrm{S} = E_0 \frac{\mathrm{e}^{\mathrm{i}kr}}{r} \boldsymbol{f}(\hat{\boldsymbol{k}}_\mathrm{S}, \hat{\boldsymbol{k}}_\mathrm{I}) \tag{2.4}$$

that defines the so-called *normalized scattering amplitude* $\boldsymbol{f}(\hat{\boldsymbol{k}}_\mathrm{S}, \hat{\boldsymbol{k}}_\mathrm{I})$, whose vector nature gives a correct description of the state of polarization of the scattered wave. For particles of general shape the scattering amplitude depends both on the direction of propagation of the incident wave, $\hat{\boldsymbol{k}}_\mathrm{I}$, and on the direction of observation, $\hat{\boldsymbol{k}}_\mathrm{S}$, whereas, for spherical particles, it depends on the angle of scattering, i. e. on the angle between $\hat{\boldsymbol{k}}_\mathrm{I}$ and $\hat{\boldsymbol{k}}_\mathrm{S}$, only. It may be worth remarking that the equation $\nabla \cdot \boldsymbol{E}_\mathrm{S} = 0$ implies

$$\hat{\boldsymbol{k}}_\mathrm{S} \cdot \boldsymbol{f} = 0 \ .$$

In other words, in the far zone the scattered field is transverse to the direction of observation, so that, when necessary, the scattered wave can be approximated by a locally plane wave.

2.2 Cross Sections

The scattering amplitude encompasses all the observable features of the scattering process, as it is related to the flux of the electromagnetic energy that the particle scatters within the unit solid angle around the direction of observation $\hat{\boldsymbol{k}}_\mathrm{S}$.

The flux of the electromagnetic energy that impinges on the particle is given by the (complex) Poynting vector of the incident field

$$\boldsymbol{S}_\mathrm{I} = \frac{c}{8\pi}(\boldsymbol{E}_\mathrm{I} \times \boldsymbol{B}_\mathrm{I}^*) = \frac{nc}{8\pi}|E_0|^2 \hat{\boldsymbol{k}}_\mathrm{I} \ .$$

In turn, the flux of the scattered energy in the direction $\hat{\boldsymbol{k}}_\mathrm{S}$ is

$$\boldsymbol{S}_\mathrm{S} = \frac{c}{8\pi}(\boldsymbol{E}_\mathrm{S} \times \boldsymbol{B}_\mathrm{S}^*) = \frac{nc}{8\pi}|E_\mathrm{S}|^2 \hat{\boldsymbol{k}}_\mathrm{S} \ .$$

We can then define the differential scattering cross section as

$$\frac{\mathrm{d}\sigma_\mathrm{S}}{\mathrm{d}\Omega} = \lim_{r\to\infty} \frac{r^2|\boldsymbol{S}_\mathrm{S}|}{|\boldsymbol{S}_\mathrm{I}|} = |\boldsymbol{f}(\hat{\boldsymbol{k}}_\mathrm{S}, \hat{\boldsymbol{k}}_\mathrm{I})|^2 \ ,$$

i.e., as the ratio of the energy scattered by the particle into the unit solid angle to the incident energy. On the whole, the energy that is scattered by the particle per unit incident energy is

$$\sigma_\mathrm{S} = \int \frac{\mathrm{d}\sigma_\mathrm{S}}{\mathrm{d}\Omega}\,\mathrm{d}\Omega = \int |\boldsymbol{f}(\hat{\boldsymbol{k}}_\mathrm{S} = \hat{\boldsymbol{r}}, \hat{\boldsymbol{k}}_\mathrm{I})|^2 \,\mathrm{d}\Omega \ , \tag{2.5}$$

where the quantity σ_S has the dimension of an area and is called the *scattering cross section*. This result could have been obtained by integrating the flux of the real part of the Poynting vector on any surface that includes the particle.

We can also define the *absorption* cross section σ_A by considering the total flux of electromagnetic energy that *enter* the particle according to the equation

$$\sigma_\mathrm{A} = -\frac{1}{|\boldsymbol{S}_\mathrm{I}|}\frac{c}{8\pi}\int \mathrm{Re}(\boldsymbol{E} \times \boldsymbol{B}^*) \cdot \hat{\boldsymbol{n}}\,\mathrm{d}S \ , \tag{2.6}$$

where $\hat{\boldsymbol{n}}$ is the unit outward normal to the surface of the particle. The same result is obtained by calculating the flux that enter any surface that includes the particle. The definition in (2.6) implies the vanishing of σ_A when the refractive index of the particle is real. Let us, indeed, use the Gauss theorem to transform (2.6) into

$$\sigma_\mathrm{A} = -\frac{1}{|\boldsymbol{S}_\mathrm{I}|}\frac{c}{8\pi}\int \mathrm{Re}\big[\nabla \cdot (\boldsymbol{E} \times \boldsymbol{B}^*)\big]\,\mathrm{d}V \ .$$

This equation can be further transformed with the help of the vector identity

$$\nabla \cdot (\boldsymbol{E} \times \boldsymbol{B}^*) = \boldsymbol{B}^* \cdot \nabla \times \boldsymbol{E} - \boldsymbol{E} \cdot \nabla \times \boldsymbol{B}^* \, .$$

Substituting in place of $\nabla \times \boldsymbol{B}^*$ and $\nabla \times \boldsymbol{E}$ their respective expressions given by (1.2) we get

$$\sigma_\mathrm{A} = \frac{1}{|\boldsymbol{S}_I|} \frac{c}{8\pi} \int \mathrm{Re}\left[\mathrm{i}k_\mathrm{v} n_0^{*2}|\boldsymbol{E}|^2 - \mathrm{i}k_\mathrm{v}|\boldsymbol{B}|^2\right] \mathrm{d}V \, ,$$

where n_0 denotes the refractive index of the particle. Clearly, the second term is purely imaginary and gives no contribution to the integral. The first term, in turn, has a nonvanishing real part only when n_0 has a nonvanishing imaginary part. This proves our earlier statement that complex refractive indexes take into account the absorption properties of the particles.

The sum of the scattering and of the absorption cross sections gives the total cross sections that in optical studies is customarily called *extinction cross section*,

$$\sigma_\mathrm{T} = \sigma_\mathrm{S} + \sigma_\mathrm{A} \, .$$

Finally, it is worth stressing that none of the cross sections defined above coincides with the geometrical cross section of the particle. Actually, σ_S and σ_A are the areas that would block from the incident beam as much energy as the particle removes by scattering and absorption, respectively. σ_T is thus the area that would block the energy that would be removed by the combined effect of scattering and absorption. Therefore, it is often convenient to characterize a scattering particle through *efficiencies* that are defined by the ratios

$$Q_\mathrm{T} = \frac{\sigma_\mathrm{T}}{G}, \quad Q_\mathrm{S} = \frac{\sigma_\mathrm{S}}{G}, \quad Q_\mathrm{A} = \frac{\sigma_\mathrm{A}}{G} \, , \tag{2.7}$$

where G is the geometrical cross section of the particle in a plane orthogonal to the direction of incidence.

2.3 Plane of Scattering

The kinematics of the scattering process is uniquely determined by the unit vectors $\hat{\boldsymbol{k}}_\mathrm{I}$ and $\hat{\boldsymbol{k}}_\mathrm{S}$ that define the directions of propagation of the incident and of the scattered wave, respectively. By introducing a system of rectangular coordinates, say Σ, with the particle at the origin and axes characterized by the unit vectors $\hat{\boldsymbol{e}}_x$, $\hat{\boldsymbol{e}}_y$, and $\hat{\boldsymbol{e}}_z$, the unit vectors $\hat{\boldsymbol{k}}_\mathrm{I}$ and $\hat{\boldsymbol{k}}_\mathrm{S}$ can be defined by the polar angles $\vartheta_\mathrm{I}, \varphi_\mathrm{I}$, and $\vartheta_\mathrm{S}, \varphi_\mathrm{S}$, respectively, i.e., by the direction cosines

$$\hat{\boldsymbol{k}}_\mathrm{I} \equiv (\alpha_\mathrm{I} = \sin \vartheta_\mathrm{I} \cos \varphi_\mathrm{I}, \, \beta_\mathrm{I} = \sin \vartheta_\mathrm{I} \sin \varphi_\mathrm{I}, \, \gamma_\mathrm{I} = \cos \vartheta_\mathrm{I}) \, ,$$
$$\hat{\boldsymbol{k}}_\mathrm{S} \equiv (\alpha_\mathrm{S} = \sin \vartheta_\mathrm{S} \cos \varphi_\mathrm{S}, \, \beta_\mathrm{S} = \sin \vartheta_\mathrm{S} \sin \varphi_\mathrm{S}, \, \gamma_\mathrm{S} = \cos \vartheta_\mathrm{S}) \, ,$$

and the *scattering angle*, i.e., the angle Φ between $\hat{\boldsymbol{k}}_\mathrm{I}$ and $\hat{\boldsymbol{k}}_\mathrm{S}$ is given by the equation

$$\cos\Phi = \hat{\boldsymbol{k}}_\mathrm{I} \cdot \hat{\boldsymbol{k}}_\mathrm{S} = \alpha_\mathrm{I}\alpha_\mathrm{S} + \beta_\mathrm{I}\beta_\mathrm{S} + \gamma_\mathrm{I}\gamma_\mathrm{S} \ .$$

The plane defined by $\hat{\boldsymbol{k}}_\mathrm{I}$ and $\hat{\boldsymbol{k}}_\mathrm{S}$ (see Fig. 2.1) is the scattering plane, whose equation is

$$\hat{\boldsymbol{n}} \cdot \boldsymbol{r} = 0 \ , \qquad (2.8)$$

where the unit normal $\hat{\boldsymbol{n}}$ is defined by the equation

$$\hat{\boldsymbol{k}}_\mathrm{I} \times \hat{\boldsymbol{k}}_\mathrm{S} = \hat{\boldsymbol{n}} \sin\Phi \ .$$

Accordingly, we can rewrite (2.8) in the form

$$\hat{\boldsymbol{k}}_\mathrm{I} \times \hat{\boldsymbol{k}}_\mathrm{S} \cdot \boldsymbol{r}/\sin\Phi = 0 \ ,$$

or, in terms of the direction cosines,

$$\begin{aligned}\hat{\boldsymbol{k}}_\mathrm{I} \times \hat{\boldsymbol{k}}_\mathrm{S} \cdot \boldsymbol{r}/\sin\Phi &= x(\beta_\mathrm{I}\gamma_\mathrm{S} - \beta_\mathrm{S}\gamma_\mathrm{I})/\sin\Phi \\ &+ y(\gamma_\mathrm{I}\alpha_\mathrm{S} - \gamma_\mathrm{S}\alpha_\mathrm{I})/\sin\Phi \\ &+ z(\alpha_\mathrm{I}\beta_\mathrm{S} - \alpha_\mathrm{S}\beta_\mathrm{I})/\sin\Phi \ . \end{aligned} \qquad (2.9)$$

In (2.9) the ratios between the terms within parenthesis and $\sin\Phi$ are respectively equal to the direction cosines of the normal to the plane of scattering. Therefore we have

$$\begin{aligned}\alpha_\mathrm{n} &= (\beta_\mathrm{I}\gamma_\mathrm{S} - \beta_\mathrm{S}\gamma_\mathrm{I})/\sin\Phi \ , \\ \beta_\mathrm{n} &= (\gamma_\mathrm{I}\alpha_\mathrm{S} - \gamma_\mathrm{S}\alpha_\mathrm{I})/\sin\Phi \ , \\ \gamma_\mathrm{n} &= (\alpha_\mathrm{I}\beta_\mathrm{S} - \alpha_\mathrm{S}\beta_\mathrm{I})/\sin\Phi \ , \end{aligned}$$

and the equation for the plane of scattering reads

$$x\alpha_\mathrm{n} + y\beta_\mathrm{n} + z\gamma_\mathrm{n} = 0 \ .$$

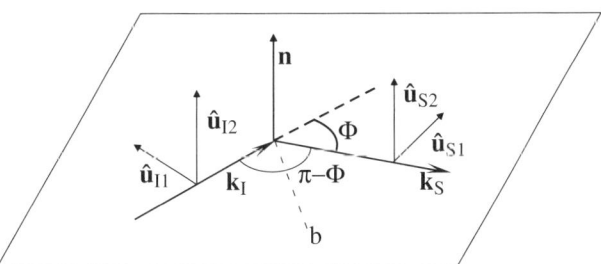

Fig. 2.1. The scattering plane. The pairs of basis vectors $\hat{\boldsymbol{u}}_{\mathrm{I}\eta}$ and $\hat{\boldsymbol{u}}_{\mathrm{S}\eta}$ for the description of the state of polarization are shown. b denotes the so-called bisectrix

In view of its definition the plane of scattering yields a good frame of reference for the description of the polarization both of the incident and of the scattered field in the far zone. Both fields are, indeed, transverse to the respective propagation vectors, and can therefore be decomposed into their components parallel and perpendicular to the plane of scattering. To this end we define two pairs of real unit vectors $\hat{\boldsymbol{u}}_{I\eta}$ and $\hat{\boldsymbol{u}}_{S\eta}$ ($\eta = 1, 2$) so oriented that

$$\hat{\boldsymbol{u}}_{I1} \times \hat{\boldsymbol{u}}_{I2} = \hat{\boldsymbol{k}}_I , \quad \hat{\boldsymbol{u}}_{S1} \times \hat{\boldsymbol{u}}_{S2} = \hat{\boldsymbol{k}}_S , \quad (2.10)$$

and with $\hat{\boldsymbol{u}}_{I1}$ and $\hat{\boldsymbol{u}}_{S1}$ lying in the plane of scattering. These pairs of vectors will be hereafter referred to as the *linear basis*; we also take $\hat{\boldsymbol{u}}_{I2} \equiv \hat{\boldsymbol{u}}_{S2} \equiv \hat{\boldsymbol{n}}$. The direction cosines of $\hat{\boldsymbol{u}}_{I1}$ and $\hat{\boldsymbol{u}}_{S1}$ can be easily inferred from the definitions (2.10).

The above definition of the plane of scattering holds true for any value of Φ except for forward scattering, $\Phi = 0°$, and backscattering, $\Phi = 180°$. In these cases, any plane through the common direction of $\hat{\boldsymbol{k}}_I$ and $\hat{\boldsymbol{k}}_S$ can be chosen as the plane of scattering. However, once one such plane has been chosen, a single pair of unit vectors, e.g. $\hat{\boldsymbol{u}}_{I\eta}$, can be used to analyze the polarization both of the incident and of the scattered field.

2.3.1 Polarization of the Incident Plane Wave

Let us consider the field

$$\boldsymbol{E} = E_0 \hat{\boldsymbol{e}} \exp(\mathrm{i}\boldsymbol{k} \cdot \boldsymbol{r} - \omega t) , \quad (2.11)$$

with

$$\hat{\boldsymbol{e}} \cdot \boldsymbol{k} = 0 .$$

For real $\hat{\boldsymbol{e}}$ (2.11) represents a linearly polarized plane wave. In particular, when $\hat{\boldsymbol{e}} = \hat{\boldsymbol{u}}_1$ the field is parallel to the plane of scattering, and we say that the wave is l-polarized. When $\hat{\boldsymbol{e}} = \hat{\boldsymbol{u}}_2$ the field is orthogonal to the plane of scattering, and we say that the wave is r-polarized (or, from the German senkrecht, s-polarized). The nomenclature above follows the one introduced by van de Hulst [3] that uses the final letters of the words parallel and perpendicular, respectively. Actually, an appropriate linear combination of two waves, such as those in (2.11), is able to describe a plane wave for whatever state of polarization. Let us thus consider three mutually orthogonal real unit vectors, $\hat{\boldsymbol{l}}_1$, $\hat{\boldsymbol{l}}_2$, and $\hat{\boldsymbol{k}}$, so oriented that

$$\hat{\boldsymbol{l}}_1 \times \hat{\boldsymbol{l}}_2 = \hat{\boldsymbol{k}} ,$$

and the plane wave

$$\boldsymbol{E} = \sum_\eta E_{0\eta} \hat{\boldsymbol{l}}_\eta \exp[\mathrm{i}(\boldsymbol{k} \cdot \boldsymbol{r} - \omega t)] . \quad (2.12)$$

Putting $\phi = \boldsymbol{k}\cdot\boldsymbol{r}$, $E_{0\eta} = a_\eta e^{i\alpha_\eta}$ with real a_η and α_η and $a_\eta \geq 0$, (2.12) can be rewritten as

$$\boldsymbol{E} = \sum_\eta \hat{\boldsymbol{l}}_\eta a_\eta \exp[i(\alpha_\eta + \phi - \omega t)]. \tag{2.13}$$

A look to (2.12) will convince the reader that, according to whether $a_1 = 0$ or $a_2 = 0$, the field is linearly polarized along $\boldsymbol{l}_2$ or along $\boldsymbol{l}_1$, respectively. When $a_1 = a_2$ and $|\alpha_2 - \alpha_1| = \pi/2$ the field is circularly polarized. More precisely, it is left-circularly polarized, with positive helicity, when $\alpha_2 - \alpha_1 = +\pi/2$, whereas it is right-circularly polarized, with negative helicity, when $\alpha_2 - \alpha_1 = -\pi/2$. Any state of circular polarization can thus be described by means of a single complex polarization vector. It may be useful to define the three vectors $\hat{\boldsymbol{c}}_1$, $\hat{\boldsymbol{c}}_2$, and $\hat{\boldsymbol{k}}$, where

$$\hat{\boldsymbol{c}}_1 = \tfrac{1}{\sqrt{2}}(\hat{\boldsymbol{l}}_1 + i\hat{\boldsymbol{l}}_2) \quad\text{and}\quad \hat{\boldsymbol{c}}_2 = \tfrac{1}{\sqrt{2}}(\hat{\boldsymbol{l}}_1 - i\hat{\boldsymbol{l}}_2)$$

describe left-polarized and right-polarized waves, respectively. As a consequence, we have

$$\hat{\boldsymbol{l}}_1 = \tfrac{1}{\sqrt{2}}(\hat{\boldsymbol{c}}_1 + \hat{\boldsymbol{c}}_2), \quad \hat{\boldsymbol{l}}_2 = -\tfrac{i}{\sqrt{2}}(\hat{\boldsymbol{c}}_1 - \hat{\boldsymbol{c}}_2),$$

which recall that any linearly-polarized state can be obtained by superposing a left- and a right-circularly polarized state. With the exception of the simple cases discussed above, (2.12) represents an elliptically polarized wave. Therefore, it may be useful to use, in place of the basis vectors $\hat{\boldsymbol{l}}_1$, $\hat{\boldsymbol{l}}_2$, and $\hat{\boldsymbol{k}}$, the vectors $\hat{\boldsymbol{e}}_1$, $\hat{\boldsymbol{e}}_2$, and $\hat{\boldsymbol{k}}$ with real $\hat{\boldsymbol{e}}_1$ and $\hat{\boldsymbol{e}}_2$, and $\hat{\boldsymbol{e}}_1$ lying along the major axis of the polarization ellipse. Using this basis we can rewrite (2.12) as

$$\boldsymbol{E} = \sum_\eta \bar{a}_\eta \hat{\boldsymbol{e}}_\eta \exp[i(\bar{\alpha}_\eta + \phi - \omega t)], \tag{2.14}$$

with

$$\bar{a}_1 = a_0 \cos\beta, \quad \bar{a}_2 = \pm a_0 \sin\beta,$$

$$\bar{\alpha}_1 = \alpha, \quad \bar{\alpha}_2 = \alpha \pm \pi/2,$$

where the upper or the lower sign apply according to whether $\beta > 0$ or $\beta < 0$, respectively. By defining $\hat{\boldsymbol{e}}_1 \cdot \hat{\boldsymbol{l}}_1 = \cos\chi$, (2.14) yields

$$\boldsymbol{E} = \big[(\bar{a}_1 \cos\chi\, e^{i\bar{\alpha}_1} - \bar{a}_2 \sin\chi\, e^{i\bar{\alpha}_2})\hat{\boldsymbol{l}}_1 \\ + (\bar{a}_1 \sin\chi\, e^{i\bar{\alpha}_1} + \bar{a}_2 \cos\chi\, e^{i\bar{\alpha}_2})\hat{\boldsymbol{l}}_2\big] \exp[i(\phi - \omega t)].$$

At this stage, we recall that what is actually observable is the real part of the field, so that we write

$$\mathrm{Re}(\boldsymbol{E}) = a_0 \big[\hat{\boldsymbol{e}}_1 \cos\beta \cos(\phi - \omega t) - \hat{\boldsymbol{e}}_2 \sin\beta \sin(\phi - \omega t) \big]$$

and the ellipticity, i.e., the ratio of the minor to the major axis of the ellipse described by the field, is given by $\tan\beta$, whose sign is positive or negative according to whether the wave is left- or right-polarized.

Ultimately, the preceding considerations make it clear that (2.13) can be rewritten with respect to any basis, in particular with respect the basis $\hat{\boldsymbol{c}}_1$, $\hat{\boldsymbol{c}}_2$, $\hat{\boldsymbol{k}}$ when convenient. Hereafter, we will refer with the notation $\hat{\boldsymbol{u}}_1$, $\hat{\boldsymbol{u}}_2$, $\hat{\boldsymbol{k}}$ to any of the bases described above with $\hat{\boldsymbol{u}}_\eta$ either real or complex according to the need.

2.3.2 Polarization of the Scattered Wave

In Sect. 2.1 it has been stressed that the scattered field in the far zone is transverse to the direction of observation. As a consequence, the state of polarization of the scattered wave can be analyzed by projecting $\boldsymbol{E}_S$ onto the unit vectors $\hat{\boldsymbol{u}}_{S\eta}$. Since the scattered field may depend on the polarization of the incident wave we use the notation

$$\boldsymbol{E}_{S\eta} = E_{0\eta} \frac{e^{ikr}}{r} \boldsymbol{f}_\eta(\hat{\boldsymbol{k}}_S, \hat{\boldsymbol{k}}_I)$$

to recall that the incident field is polarized along $\hat{\boldsymbol{u}}_{I\eta}$. In other words, we associate both to the scattered field and to the scattering amplitude the index η that recalls the polarization of the incident field. In terms of the components along $\hat{\boldsymbol{u}}_{S\eta'}$ we will write

$$\boldsymbol{E}_{S\eta} = \sum_{\eta'} E_{S\eta'\eta} \hat{\boldsymbol{u}}_{S\eta'} = E_{0\eta} \frac{e^{ikr}}{r} \sum_{\eta'} f_{\eta'\eta} \hat{\boldsymbol{u}}_{S\eta'} ,$$

where we define the components of the scattering amplitude as

$$f_{\eta'\eta} = \boldsymbol{f}_\eta \cdot \hat{\boldsymbol{u}}^*_{S\eta'} .$$

We stress that we dot multiplied by $\hat{\boldsymbol{u}}^*_{S\eta'}$ to allow for the case of elliptical polarization. Of course, nothing prevents us from using the circular basis to describe the scattered field. Finally, it is important to realize that, in general, the components of the scattering amplitude are complex and nonvanishing for both values of η'. Therefore, in analogy with the incident plane wave, we can write

$$f_{\eta'\eta} = a_{\eta'\eta} e^{i\alpha_{\eta'\eta}}$$

to stress that the components of the scattered wave may have a different phase. In fact, even when the incident wave is linearly polarized, α_1 may differ from α_2 and, as a consequence, the scattered wave may be elliptically polarized.

2.4 Stokes Parameters

In Sect. 2.3.1 we showed that the most general monochromatic plane wave is fully characterized by the amplitudes a_1 and a_2 and by the phase difference $\delta = \alpha_2 - \alpha_1$. The discussion of Sect. 2.3.2 stressed that similar quantities also characterize the scattered wave. The measurement of these quantities cannot, however, be performed directly, as optical measurements allow for the determination of *intensities*. Nevertheless, there exist physical procedures to measure a set of parameters, the so-called *Stokes parameters*, that have the dimension of intensities and yield the maximum information on the state of any optical wave. Since there is no unanimity either in the notation or in the definition of these parameters we stick to the choice of Mishchenko, Hovenier, and Travis [16].

If a linear basis is used, the Stokes parameters are defined by the equations

$$I = |E_1|^2 + |E_2|^2 = a_1^2 + a_2^2 , \tag{2.15a}$$

$$Q = |E_1|^2 - |E_2|^2 = a_1^2 - a_2^2 , \tag{2.15b}$$

$$U = -2\mathrm{Re}[E_1 E_2^*] = -2a_1 a_2 \cos\delta , \tag{2.15c}$$

$$V = 2\mathrm{Im}[E_1 E_2^*] = -2a_1 a_2 \sin\delta , \tag{2.15d}$$

from which it is an easy matter to get the relevant quantities

$$a_1^2 = \tfrac{1}{2}(I+Q) , \tag{2.16a}$$

$$a_2^2 = \tfrac{1}{2}(I-Q) , \tag{2.16b}$$

$$\tan\delta = V/U . \tag{2.16c}$$

Sometimes it may be convenient to use, instead of I and Q, the modified parameters

$$I_1 = |\hat{\boldsymbol{u}}_1 \cdot \boldsymbol{E}|^2 = a_1^2 ,$$

$$I_2 = |\hat{\boldsymbol{u}}_2 \cdot \boldsymbol{E}|^2 = a_2^2 ,$$

that give the intensity of the component parallel and perpendicular to the plane of scattering and are related to I and Q by (2.16a) and (2.16b). Note that we can also write

$$I = a_0^2 , \quad Q = a_0^2 \cos 2\beta \cos 2\chi ,$$

$$I_1 = \tfrac{1}{2}a_0^2(1 + \cos 2\beta \cos 2\chi) ,$$

$$I_2 = \tfrac{1}{2}a_0^2(1 - \cos 2\beta \cos 2\chi) ,$$

$$U = -a_0^2 \cos 2\beta \sin 2\chi, \quad V = -a_0^2 \sin 2\beta ,$$

so that the geometry of any monochromatic optical wave is fully defined.

When using the circular basis is preferable, the Stokes parameters can still be calculated through (2.15), provided a_η and δ are those appropriate for the $\hat{c}_\eta$ basis. A more usual definition to which we do not stick, however, scrambles the newly defined parameters in such a way that

$$I_c = I\,, V_c = Q\,, Q_c = U\,, U_c = V\,.$$

In any case, whatever the basis, the Stokes parameters of a monochromatic wave satisfy the relation

$$I^2 = Q^2 + U^2 + V^2\,,$$

so that for a single wave they are never independent. It should be borne in mind, however, that actual optical beams, even when monochromatic to a high degree, still contain a large number of frequencies, i.e., they are quasi-monochromatic. As a result, the Stokes parameters of an actual optical beam are, in fact, an average over wavepackets within which the polarization may change fairly fast in an unpredictable manner. Actually, an optical beam can be interpreted as the propagation of an assembly of photons, whose polarization and time of arrival on the detecting instrument are predictable only within the laws of quantum statistics. Anyway, measured Stokes parameters of actual optical beams do not satisfy the relation reported above, but in general

$$I^2 \geq Q^2 + U^2 + V^2\,.$$

A beam for which the equality sign holds is said to be fully polarized because the equality implies that all the component waves have the same value of χ and β. In the opposite case in which $Q = U = V = 0$, the beam is completely incoherent: it is natural light. According to these considerations it is useful to define the degree of polarization of an optical beam as

$$P = (Q^2 + U^2 + V^2)^{1/2}/I\,.$$

2.4.1 Müller Transformation Matrix

As the components of the scattered field are related to those of the incident field by the scattering amplitude matrix, it is possible to define a matrix that transforms the Stokes parameters of the incident field into those of the scattered field. To this end we define the Stokes vector $\underline{\mathcal{I}}$ as the one-column vector whose components are the Stokes parameters either of the incident or of the scattered wave. Thus,

$$\underline{\mathcal{I}}_\mathrm{S} = \frac{1}{r^2}\underline{\mathcal{M}}\,\underline{\mathcal{I}}_\mathrm{I}\,,$$

where the subscripts I and S refer to the incident and to the scattered wave, respectively, and $\underline{\mathcal{M}}$ is the so-called *Müller transformation matrix* or *phase matrix*. The elements of $\underline{\mathcal{M}}$ have a simpler expression in terms of the modified Stokes parameters [15]

$$\begin{vmatrix} |f_{11}|^2 & |f_{12}|^2 & -\text{Re}\,[f_{11}f_{12}^*] & -\text{Im}\,[f_{11}f_{12}^*] \\ |f_{21}|^2 & |f_{22}|^2 & -\text{Re}\,[f_{21}f_{22}^*] & -\text{Im}\,[f_{21}f_{22}^*] \\ -2\,\text{Re}\,[f_{11}f_{21}^*] & -2\,\text{Re}\,[f_{12}f_{22}^*] & \text{Re}\,[f_{11}f_{22}^* + f_{12}f_{21}^*] & \text{Im}\,[f_{11}f_{22}^* - f_{12}f_{21}^*] \\ 2\,\text{Im}\,[f_{11}f_{21}^*] & 2\,\text{Im}\,[f_{12}f_{22}^*] & \text{Im}\,[-f_{11}f_{22}^* - f_{12}f_{21}^*] & \text{Re}\,[f_{11}f_{22}^* - f_{12}f_{21}^*] \end{vmatrix}.$$

It has been shown by Hovenier and van der Mee [17] that the elements of the Müller matrix must satisfy a number of relations that may help the observer to check the consistency of the recorded data. Bottiger [18] has discussed how, once the plane of scattering has been chosen, the Müller matrix gives information on the shape of the particles.

2.5 Optical Theorem

We will now establish a fundamental theorem, the so-called *optical theorem*, that relates the extinction cross section to the scattering amplitude. In our derivation, that as far as possible follows the arguments of van de Hulst [3], we assume that the scattering particle is embedded in a medium with a real refractive index. Nevertheless, when the medium is absorptive, i.e., when the refractive index has a nonvanishing imaginary part, the theorem can be reformulated along the lines indicated by Bohren and Gilra [19].

We assume, with no loss of generality, that the scattering particle lies at the origin of the frame of reference and that the incident wave is linearly polarized and propagating along the z axis. With these assumptions the incident field reads

$$\boldsymbol{E}_\text{I} = E_0 \hat{\boldsymbol{e}}_\text{I} \text{e}^{\text{i}(kz-\omega t)}. \tag{2.17}$$

If $\hat{\boldsymbol{k}}_\text{S} = \hat{\boldsymbol{k}}_\text{I}$, i.e., if we consider the so-called forward scattering, the scattering amplitude will be denoted as $\boldsymbol{f}(0)$ and the scattered field, according to (2.4), will be written

$$\boldsymbol{E}_\text{S} = E_0 \frac{\text{e}^{\text{i}(kr-\omega t)}}{r} \boldsymbol{f}(0). \tag{2.18}$$

We already remarked that in the forward direction any plane through the common direction of $\hat{\boldsymbol{k}}_\text{I}$ and $\hat{\boldsymbol{k}}_\text{S}$ can be chosen as the plane of scattering and the unit vectors $\hat{\boldsymbol{u}}_{\text{I}\eta}$ can be taken to coincide with the vectors $\hat{\boldsymbol{u}}_{\text{S}\eta}$. Then, by dot multiplying (2.17) and (2.18) by $\hat{\boldsymbol{e}}_\text{I} = \hat{\boldsymbol{e}}_\text{S}$ we get the equations

$$\boldsymbol{E}_\text{I} \cdot \hat{\boldsymbol{e}}_\text{I} = E_0 \text{e}^{\text{i}(kz-\omega t)},$$

$$\boldsymbol{E}_\text{S} \cdot \hat{\boldsymbol{e}}_\text{I} = E_0 \frac{\text{e}^{\text{i}(kr-\omega t)}}{r} \boldsymbol{f}(0) \cdot \hat{\boldsymbol{e}}_\text{I},$$

that can be combined into

$$\boldsymbol{E}_\mathrm{S} \cdot \hat{\boldsymbol{e}}_\mathrm{I} = \frac{e^{ik(r-z)}}{r} \boldsymbol{f}(0) \cdot \hat{\boldsymbol{e}}_\mathrm{I} (\boldsymbol{E}_\mathrm{I} \cdot \hat{\boldsymbol{e}}_\mathrm{I}) .$$

We now remark that, in the forward direction, at a distance z from the scattering particle, an optical instrument records the combined intensity of the incident and the scattered wave. The plane of the receiving lens, whose diameter is d, has the equation $z = \mathrm{cost}$ so that, for $z \gg d$, we can write

$$r = (z^2 + x^2 + y^2)^{1/2} \approx z + \frac{x^2 + y^2}{2z} ,$$

and the total field on the lens is

$$\boldsymbol{E}_\mathrm{S} \cdot \hat{\boldsymbol{e}}_\mathrm{I} + \boldsymbol{E}_\mathrm{I} \cdot \hat{\boldsymbol{e}}_\mathrm{I} = \left[1 + \frac{e^{ik(x^2+y^2)/2z}}{z} \boldsymbol{f}(0) \cdot \hat{\boldsymbol{e}}_\mathrm{I} \right] (\boldsymbol{E}_\mathrm{I} \cdot \hat{\boldsymbol{e}}_\mathrm{I}) .$$

The corresponding intensity is

$$|\boldsymbol{E}_\mathrm{S} \cdot \hat{\boldsymbol{e}}_\mathrm{I} + \boldsymbol{E}_\mathrm{I} \cdot \hat{\boldsymbol{e}}_\mathrm{I}|^2 = \left\{ 1 + \frac{2}{z} \mathrm{Re} \left[e^{ik(x^2+y^2)/2z} \boldsymbol{f}(0) \cdot \hat{\boldsymbol{e}}_\mathrm{I} \right] \right\} |\boldsymbol{E}_\mathrm{I} \cdot \hat{\boldsymbol{e}}_\mathrm{I}|^2 ,$$

which has been obtained by neglecting the term $|\boldsymbol{f}(0) \cdot \hat{\boldsymbol{e}}_\mathrm{I}|^2/z^2$ that is actually negligible for large z. The total energy recorded by the instrument is given by integrating the preceding expression on the surface of the front lens, i.e., letting $x, y \leq d/2$. Close examination of the integrand shows, however, that when $\sqrt{z\lambda} \ll d \ll z$, as in the present case, the integration limits can be extended from $-\infty$ to ∞ without appreciably affecting the result. Now,

$$\int_{-\infty}^{\infty} dx \int_{-\infty}^{\infty} e^{ik(x^2+y^2)/2z} \, dy = \frac{2\pi i z}{k} ,$$

so that the energy recorded by the instrument is

$$W = \left[A - \frac{4\pi}{k} \mathrm{Im}(\boldsymbol{f}(0) \cdot \hat{\boldsymbol{e}}_\mathrm{I}) \right] I_0 ,$$

where $I_0 = |\boldsymbol{E}_\mathrm{I} \cdot \hat{\boldsymbol{e}}_\mathrm{I}|^2$. This equation shows that, due to the interference of the incident and the scattered wave, the front lens of the receiving instrument has seemingly an area smaller than A and is therefore able to collect a smaller quantity of energy than in the absence of the particle. Actually, since I_0 is the intensity of the incident wave, the term AI_0 is the energy that the instrument would collect in the absence of the particle. The second term

$$I_\mathrm{T} = \frac{4\pi}{k} \mathrm{Im}(\boldsymbol{f}(0) \cdot \hat{\boldsymbol{e}}_\mathrm{I}) I_0$$

measures the energy that is lacking because of the presence of the particle. The lacking energy should be equal to the energy that the particle removes from the incident beam by absorption and scattering, so that, according to the argument at the end of Sect. 2.2, the coefficient of I_0 must coincide with extinction cross section. We are thus led to the relation

$$\sigma_T = \frac{4\pi}{k} \operatorname{Im}[\boldsymbol{f}(0) \cdot \hat{\boldsymbol{e}}_I] \tag{2.19}$$

that is just the expression of the *optical theorem*.

We recall that for nonspherical particles the scattering amplitude depends on the orientation with respect to the incident field. Therefore, it is not surprising that the cross section depends on the polarization of the incident wave.

2.6 Scattering by an Ensemble of Particles

We are now able to relate the scattering properties of the single particles with the macroscopic optical properties of a dispersion of them. To this end, we need to assume that the *multiple scattering processes* can be neglected. More precisely, we will deal with dispersions in which each particle scatters independently of the other particles present and in which it is quite unlikely that a beam would be scattered more than once. Actually, both conditions amount to requiring that the field that illuminates each particle be the original incident field. In practice, we can speak of independent scattering when the particles are well separated, e.g. when their mutual distance is more or less twice their size for the particles bigger than the wavelength [20]. We can also say that only single scattering processes need to be considered when the *optical thickness* $\tau \ll 0.1$, where $\exp[-\tau]$ is the attenuation undergone by the intensity in traversing the dispersion [3, 16]. Hereafter, we will term a dispersions possessing the features mentioned above as a *tenuous dispersion*.

If the positions of the particles are randomly distributed, the waves scattered by each one of them into any given direction, with the exception of the forward direction, interfere with random phase difference. As a consequence, the field scattered by the whole dispersion does not depend on the phases of the single waves, and we have to sum their intensities [21]. Using the index ν to label the quantities that refer to the νth particle we can write

$$I = \sum_\nu I_\nu ,$$

and the scattering cross sections too turn out to be additive [21], i.e.,

$$\sigma_S = \sum_\nu \sigma_{S\nu} . \tag{2.20}$$

On the other hand, if forward scattering is considered, it should be borne in mind that any shift of the position of the particles does not alter the phase of the scattered wave. As a consequence, in the forward direction we need to sum the fields and not the intensities. Therefore we write

$$\boldsymbol{E}_\mathrm{S} = \sum_\nu \boldsymbol{E}_{\mathrm{S}\nu} .$$

The optical theorem then decrees that the extinction cross sections are additive

$$\sigma_\mathrm{T} = \sum_\nu \sigma_{\mathrm{T}\nu} . \qquad (2.21)$$

Equations (2.20) and (2.21) state that, although for quite different physical reasons, the extinction and the scattering cross sections are additive, provided the number density of the dispersion is low enough to allow neglecting the multiple scattering processes. From this constatation the additivity of the absorption cross sections follows at once. Therefore, even for these cross sections we can write

$$\sigma_\mathrm{A} = \sum_\nu \sigma_{\mathrm{A}\nu} .$$

2.7 Refractive Index of a Dispersion of Particles

We are now able to show that the propagation of a plane wave through a dispersion of particles can be characterized by a macroscopic *refractive index* that is the analogous of the refractive index of homogeneous media. We consider first the case of a monodispersion, this term being used here in a restricted sense to indicate a dispersion of particles with identical morphology and orientation with respect to the incident field. This restriction, which will be removed later, allows us to use the same scattering amplitude to describe the field scattered by each particle. It is necessary, however, that the dispersion be tenuous, in the sense discussed in the preceding section.

2.7.1 Definition of the Refractive Index

We assume that all the particles be included within a layer limited by a pair of parallel planes a distance l apart and orthogonal to the z axis, N denotes the number density of the particles, so that the volume element $\mathrm{d}V$ includes $N\,\mathrm{d}V$ scatterers (see Fig. 2.2). The plane wave

$$\boldsymbol{E}_\mathrm{I} = E_0 \hat{\boldsymbol{e}}_\mathrm{I} e^{\mathrm{i}(kz-\omega t)}$$

that propagates through the layer undergoes scattering by the particles so that a detector located on the z axis, say at z_0, at a large distance from the layer, records the superposition of the original plane wave and of the

2.7 Refractive Index of a Dispersion of Particles

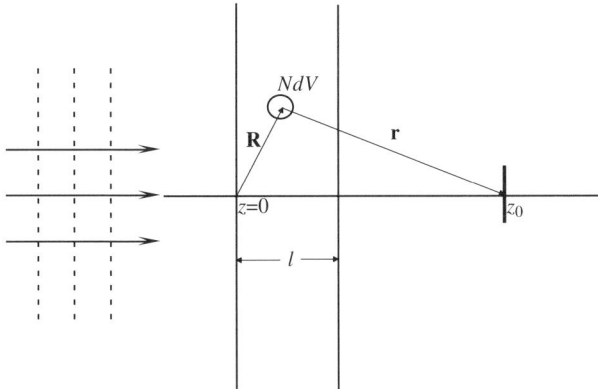

Fig. 2.2. Propagation through a layer of particles

waves scattered by the particles. The waves coming from particles included within the volume of intersection of the layer and of the aperture cone of the observation instrument sum coherently. Provided that this cone is narrow enough, the only waves that will be recorded are those scattered in the forward direction. Now, at z_0 the field scattered by a particle that is located at $\boldsymbol{R}$ is

$$\boldsymbol{E}_\text{S} = E_0 \frac{\mathrm{e}^{\mathrm{i}kR}}{R} \mathrm{e}^{\mathrm{i}kz} \boldsymbol{f}(0) ,$$

where

$$R = \left[x^2 + y^2 + (z_0 - z)^2\right]^{1/2}$$

and the phase factor $\mathrm{e}^{\mathrm{i}kz}$ accounts for the position of the particle within the layer. The total field scattered by the whole dispersion is therefore

$$\boldsymbol{E}_\text{S} = NE_0 \int_V \frac{\mathrm{e}^{\mathrm{i}kR}}{R} \mathrm{e}^{\mathrm{i}kz} \boldsymbol{f}(0) \, \mathrm{d}V , \qquad (2.22)$$

where V is the volume that contains the scatterers that take part in the process. The correctness of (2.22), which contains an integration over the volume instead of a sum over the particles, is granted by the fact that even for a low-density dispersion the volume of integration still contains a large number of scatterers. Since z_0 is large and, however, very much larger than the values of z within the layer, the distance R can be expanded as

$$R \approx (z_0 - z) + (x^2 + y^2)/2(z_0 - z) ,$$

in analogy to the procedure in Sect. 2.5. Accordingly, (2.22) becomes

$$\boldsymbol{E}_\text{S} = NE_0 \mathrm{e}^{\mathrm{i}kz_0} \int_0^l \mathrm{d}z \int_{-\infty}^{\infty} \mathrm{d}x \int_{-\infty}^{\infty} \mathrm{d}y \frac{\exp[\mathrm{i}k(x^2 + y^2)/2(z_0 - z)]}{(z_0 - z)} \boldsymbol{f}(0) .$$

The integrals are of the kind that we already considered in Sect. 2.5, so that the final result is

$$\boldsymbol{E}_\mathrm{S} = N E_0 \mathrm{e}^{\mathrm{i}kz_0} \frac{2\pi\mathrm{i}l}{k} \boldsymbol{f}(0) .$$

Ultimately, the total field recorded by the instrument at z_0 is

$$\boldsymbol{E}_\mathrm{T} = E_0 \mathrm{e}^{\mathrm{i}kz_0} \left[\hat{\boldsymbol{e}}_\mathrm{I} + \frac{2\pi\mathrm{i}Nl}{k} \boldsymbol{f}(0) \right] . \tag{2.23}$$

The preceding result can be interpreted by substituting for the actual matrix of refractive index n containing the assembly a plane parallel plate of homogeneous material with (complex) refractive index $\bar{n}n$ and thickness l. Classical optics states that within the plate the incident plane wave becomes

$$\boldsymbol{E}_\mathrm{I} = E_0 \hat{\boldsymbol{e}}_\mathrm{I} \mathrm{e}^{\mathrm{i}\bar{n}kz} ,$$

so that, after crossing the thickness l, it is attenuated to the ratio

$$|\boldsymbol{E}_\mathrm{I} \cdot \hat{\boldsymbol{e}}_\mathrm{I}|/E_0 = \mathrm{e}^{\mathrm{i}kl(\bar{n}-1)} \approx 1 + \mathrm{i}kl(\bar{n}-1) ,$$

where the last expression is valid provided $|\bar{n} - 1| \ll 1$. Now, by dot multiplying (2.23) by $\hat{\boldsymbol{e}}_\mathrm{I}$ we get

$$\boldsymbol{E}_\mathrm{T} \cdot \hat{\boldsymbol{e}}_\mathrm{I} = E_0 \mathrm{e}^{\mathrm{i}kz_0} \left[1 + \frac{2\pi\mathrm{i}Nl}{k} \boldsymbol{f}(0) \cdot \hat{\boldsymbol{e}}_\mathrm{I} \right]$$

that, when compared with the preceding expression, leads to the conclusion that to a layer of scatterers can be attributed a formal refractive index

$$\bar{n} = 1 + \frac{2\pi}{k^2} N f(0) , \tag{2.24}$$

where $f(0) = \boldsymbol{f}(\hat{\boldsymbol{k}}_\mathrm{I}, \hat{\boldsymbol{k}}_\mathrm{I}) \cdot \hat{\boldsymbol{e}}_\mathrm{I}$ is the component of the forward scattering amplitude in the direction of polarization of the incident field. Equation (2.24) thus relates the scattering properties of the particles with the macroscopic optical properties of a monodispersion of them.

Little modifications make (2.24) applicable to the case of a dispersion that includes more that one kind of particles. In fact, even when the dispersion includes particles with a different morphology and/or orientation with respect to the incident field, the effect of each kind can be considered separately and the results added. This possibility stems from our assumption that the dispersion be tenuous so that the multiple scattering processes can be neglected, and there is no interaction among particles either of the same or of a different kind. Accordingly, denoting with the index ν the quantities that refer to particles of the νth kind, (2.24) can be rewritten as

$$\bar{n} = 1 + \frac{2\pi}{k^2} \sum_{\nu} N_\nu f_\nu(0) \ . \tag{2.25}$$

Of course, the preceding equation should be considered only a definition, because nothing has been said on the procedure to actually perform the sum over the particles of the various kinds. This problem will be addressed in the next chapter.

2.7.2 Features of the Refractive Index

The refractive index of a dispersion of particles that is defined by (2.24) and its extension (2.25) turns out to be a 2×2 matrix with complex elements whose physical meaning can be highlighted by decomposing the incident plane wave into its components parallel and perpendicular to the plane of scattering. We already stressed that in the case of forward scattering the polarization basis vectors of the scattered wave can be chosen to coincide with those of the incident wave. Assuming this choice, the matrix of the refractive index takes on the form

$$\mathcal{N}_{\eta\eta'} = \delta_{\eta\eta'} + \frac{2\pi}{k^2} N f_{\eta\eta'}(0) \ , \tag{2.26}$$

which holds true whether the chosen basis is linear or circular. The diagonal elements of (2.26) give the complex refractive index for the two independent states of polarization. On the contrary the off-diagonal elements give information on the change of the plane of polarization. More precisely, the real part of the diagonal elements

$$\mathrm{Re}[\mathcal{N}_{\eta\eta}] = 1 + \frac{2\pi}{k^2} N \mathrm{Re}[f_{\eta\eta}(0)]$$

describe the change of the phase velocity, or also of the wavelength, for the two states of polarization. This phenomenon is known as *birefringence*. The imaginary part of the diagonal elements

$$\mathrm{Im}[\mathcal{N}_{\eta\eta}] = \frac{2\pi}{k^2} N \mathrm{Im}[f_{\eta\eta}(0)]$$

takes account of the *dichroism*, i.e., the polarization-dependent absorption by the medium. In particular, it is an easy matter to realize that the absorption coefficient of the dispersion for a plane wave polarized along $\hat{u}_\eta$ is

$$\gamma_\eta = 2k\mathrm{Im}[\mathcal{N}_{\eta\eta}] \ ,$$

i.e., by the optical theorem

$$\gamma_\eta = N\sigma_{\mathrm{T}\eta} \ ,$$

where we attached the subscript η to the cross section to recall that, in general, it depends on the polarization of the incident wave.

3. Multipole Expansions and Transition Matrix

The theory in Chap. 2 yields a number of useful relations but gives no prescription for calculating the quantities of interest. In this chapter we use the theory of the multipole fields to establish equations that are suitable for actual calculations. In fact, by expanding both the incident and the scattered field in a series of spherical vector multipole fields, one is lead to introduce the *transition matrix* that proved to be one of the most fruitful concepts in the theory of scattering. Beyond encompassing all the information on the morphology and on the orientation of the scattering particle with respect to the incident field, the transition matrix has well-defined transformation properties under rotation of the coordinate frame. These properties will be shown to be the key tool for the description of the propagation of light through an assembly of nonspherical particles.

3.1 Multipole Expansions

Our starting point is the expansion of the incident and of the scattered field. We assume once for all that the medium through which these fields propagate be homogeneous, and that its refractive index, n, be real.

3.1.1 Expansion of a Homogeneous Plane Wave

Let us consider an electromagnetic plane wave, whose electric and magnetic field vectors are

$$\boldsymbol{E} = E_0 \hat{\boldsymbol{e}} \exp(\mathrm{i} \boldsymbol{k} \cdot \boldsymbol{r}) , \tag{3.1a}$$

$$\mathrm{i} \boldsymbol{B} = \mathrm{i} n E_0 (\hat{\boldsymbol{k}} \times \hat{\boldsymbol{e}}) \exp(\mathrm{i} \boldsymbol{k} \cdot \boldsymbol{r}) , \tag{3.1b}$$

where $\boldsymbol{k} = n k_\mathrm{v} \hat{\boldsymbol{k}}$. When the propagation vector $\boldsymbol{k}$ is real the plane wave is said to be *homogeneous*. We assume that the multipole expansion of the fields (3.1) can be written as

$$\boldsymbol{E} = E_0 \sum_{plm} \boldsymbol{J}_{lm}^{(p)}(\boldsymbol{r}, k) W_{lm}^{(p)}(\hat{\boldsymbol{e}}, \hat{\boldsymbol{k}}) , \tag{3.2a}$$

$$\mathrm{i} \boldsymbol{B} = n E_0 \sum_{plm} \boldsymbol{J}_{lm}^{(p')}(\boldsymbol{r}, k) W_{lm}^{(p)}(\hat{\boldsymbol{e}}, \hat{\boldsymbol{k}}) \text{ with } p' \neq p , \tag{3.2b}$$

where we used J-multipole fields to ensure the finiteness at the origin, and the amplitudes $W_{lm}^{(p)}(\hat{e}, \hat{k})$ will be determined presently. To this end, we dot multiply (3.2) by $\boldsymbol{X}_{lm}^{*}(\hat{r})$ and integrate over the solid angle. The orthogonality relations of the vector spherical harmonics yield

$$E_0 W_{lm}^{(1)} j_l(kr) = \int \boldsymbol{E} \cdot \boldsymbol{X}_{lm}^{*} \, d\Omega \,,$$

$$nE_0 W_{lm}^{(2)} j_l(kr) = \mathrm{i} \int \boldsymbol{B} \cdot \boldsymbol{X}_{lm}^{*} \, d\Omega \,.$$

Substitution in place of the fields of their expressions given by (3.1) leads to

$$W_{lm}^{(1)} j_l(kr) = \int \exp(\mathrm{i}\boldsymbol{k} \cdot \boldsymbol{r}) \hat{e} \cdot \boldsymbol{X}_{lm}^{*} \, d\Omega \,, \qquad (3.3\mathrm{a})$$

$$W_{lm}^{(2)} j_l(kr) = \mathrm{i} \int \exp(\mathrm{i}\boldsymbol{k} \cdot \boldsymbol{r}) (\hat{\boldsymbol{k}} \times \hat{e}) \cdot \boldsymbol{X}_{lm}^{*} \, d\Omega \,. \qquad (3.3\mathrm{b})$$

The integrals in (3.3) can be evaluated by recalling that, according to the definition (1.40),

$$\hat{e} \cdot \boldsymbol{X}_{lm}^{*}(\hat{r}) = [l(l+1)]^{-1/2} \hat{e} \cdot [\mathrm{L} Y_{lm}(\hat{r})]^{*} = [l(l+1)]^{-1/2} [\hat{e}^{*} \cdot \mathrm{L} Y_{lm}(\hat{r})]^{*} \,.$$

Now, in terms of spherical components,

$$\hat{e}^{*} \cdot \mathrm{L} = \sum_{\mu} (-)^{\mu} \hat{e}_{\mu}^{*} \mathrm{L}_{-\mu} \,,$$

where, of course,

$$\mathrm{L}_0 = \mathrm{L}_z \,, \quad \mathrm{L}_{\pm 1} = \mp \tfrac{1}{\sqrt{2}} \mathrm{L}_{\pm} = \mp \tfrac{1}{\sqrt{2}} (\mathrm{L}_x \pm \mathrm{i}\mathrm{L}_y) \,.$$

Therefore, letting $\mathrm{L}_{\pm}$ and L_z operate on Y_{lm} we get [6]

$$\hat{e} \cdot \boldsymbol{X}_{lm}^{*}(\hat{r}) = [l(l+1)]^{-1/2} \{ \hat{e}_z C_z Y_{lm}(\hat{r}) + \tfrac{1}{2}[-\hat{e}_1 C_- Y_{l,\,m-1}(\hat{r}) + \hat{e}_{-1} C_+ Y_{l,\,m+1}(\hat{r})] \}^{*} \,,$$

where

$$C_z = m \,, \qquad C_{\pm} = [l(l+1) - m(m \pm 1)]^{1/2} \,.$$

A quite analogous result is obtained for $(\hat{\boldsymbol{k}} \times \hat{e}) \cdot \boldsymbol{X}_{lm}^{*}(\hat{r})$. At this stage we substitute in place of the exponentials in (3.3) the Bauer expansion

$$\exp(\mathrm{i}\boldsymbol{k} \cdot \boldsymbol{r}) = 4\pi \sum_{lm} \mathrm{i}^l j_l(kr) Y_{lm}^{*}(\hat{\boldsymbol{k}}) Y_{lm}(\hat{r})$$

and perform the integration. The orthogonality properties of the spherical harmonics then yield the result

$$W^{(1)}_{lm}(\hat{e},\hat{k}) = 4\pi i^l \hat{e} \cdot \boldsymbol{X}^*_{lm}(\hat{k}) ,$$
$$W^{(2)}_{lm}(\hat{e},\hat{k}) = 4\pi i^{l+1} (\hat{k} \times \hat{e}) \cdot \boldsymbol{X}^*_{lm}(\hat{k}) ,$$

which, in terms of the transverse vector harmonics [see (1.44)], can be put in the more compact form [11]

$$W^{(p)}_{lm}(\hat{e},\hat{k}) = 4\pi i^{p+l-1} \hat{e} \cdot \boldsymbol{Z}^{(p)*}_{lm}(\hat{k}) . \tag{3.4}$$

3.1.2 Inhomogeneous Plane Wave

When the propagation vector of a plane wave is complex, the wave is said to be *inhomogeneous* [2]. We stress that the wavevector may be complex not only because the medium is absorptive but also because it may be convenient to assume a complex propagation vector in order to perform particular expansions. One such expansion will be used in Chap. 5. When performing the multipole expansion of an inhomogeneous plane wave it should be borne in mind that the orthogonality properties of the vector spherical harmonics in the form stated in Chap. 1 hold true for real argument only. For complex $\hat{k}$ we have

$$(-)^{m+1} \int \boldsymbol{X}_{lm}(\hat{k}) \cdot \boldsymbol{X}_{l,-m}(\hat{k}) \mathrm{d}\Omega_{\hat{k}} = \delta_{ll'}\delta_{mm'} .$$

Note that only for real argument $\boldsymbol{X}^*_{lm} = (-)^{m+1} \boldsymbol{X}_{l,-m}$. With this *proviso*, the procedure described in the preceding section can be applied to the expansion of an inhomogeneous plane wave with the result

$$\hat{e}\exp[i\boldsymbol{k}\cdot\boldsymbol{r}] = 4\pi \sum_{plm} i^{p+l+1}(-)^{m+1} \big[\boldsymbol{Z}^{(p)}_{l,-m}(\hat{k}) \cdot \hat{e}\big] \boldsymbol{J}^{(p)}_{lm}(\boldsymbol{r},k) ,$$

so that

$$W^{(p)}_{lm}(\hat{e},\hat{k}) = 4\pi i^{p+l-1}(-)^{m+1} \hat{e} \cdot \boldsymbol{Z}^{(p)}_{l,-m}(\hat{k}) .$$

3.1.3 Scattered Field

In analogy to the incident field

$$\boldsymbol{E}_{I\eta} = E_0 \sum_{plm} \boldsymbol{J}^{(p)}_{lm}(\boldsymbol{r},k) W^{(p)}_{lm}(\hat{\boldsymbol{u}}_{I\eta}, \hat{\boldsymbol{k}}_I) , \tag{3.5}$$

we expand the scattered field as

$$\boldsymbol{E}_{S\eta} = E_0 \sum_{plm} \boldsymbol{H}^{(p)}_{lm}(\boldsymbol{r},k) A^{(p)}_{\eta lm} , \tag{3.6}$$

where we use $\boldsymbol{H}$-multipole fields because the scattered field should satisfy the radiation condition at infinity [22], and the amplitudes $A^{(p)}_{\eta lm}$ are determined by the boundary conditions across the surface of the particle. The multipole expansion of the normalized scattering amplitude is easily obtained by taking the limit of the $\boldsymbol{H}$-multipole fields for $r \to \infty$. Indeed, with the help of (1.32b), (1.33b), and (1.48) we get

$$\boldsymbol{H}^{(p)}_{lm}(\boldsymbol{r}, k) = (-\mathrm{i})^{l+p} \frac{\mathrm{e}^{\mathrm{i}kr}}{kr} \boldsymbol{Z}^{(p)}_{lm}(\hat{\boldsymbol{k}}_S) \ .$$

Therefore, the asymptotic form of the scattered field is

$$\boldsymbol{E}_{S\eta} = \frac{E_0}{k} \frac{\mathrm{e}^{\mathrm{i}kr}}{r} \sum_{plm} (-\mathrm{i})^{l+p} \boldsymbol{Z}^{(p)}_{lm}(\hat{\boldsymbol{k}}_S) A^{(p)}_{\eta lm} \ .$$

Comparison with (2.4) then yields

$$\boldsymbol{f}_\eta(\hat{\boldsymbol{k}}_S, \hat{\boldsymbol{k}}_I) = \frac{1}{k} \sum_{plm} (-\mathrm{i})^{l+p} \boldsymbol{Z}^{(p)}_{lm}(\hat{\boldsymbol{k}}_S) A^{(p)}_{\eta lm} \ , \qquad (3.7)$$

which is the required expression. The components of the normalized scattering amplitude on the basis for the scattered field are obtained by dot multiplying (3.7) by $\hat{\boldsymbol{u}}^*_{S\eta'}$ with the result

$$f_{\eta'\eta} = -\frac{\mathrm{i}}{4\pi k} \sum_{plm} W^{(p)*}_{lm}(\hat{\boldsymbol{u}}_{S\eta'}, \hat{\boldsymbol{k}}_S) A^{(p)}_{\eta lm} \ , \qquad (3.8)$$

where the definition (3.4) has been used. Equation (3.8) confirms our previous statement in Sect. 2.1 that at a large distance from the particle the scattered wave becomes locally akin to a plane wave.

At this stage we are able to give explicit expressions for the extinction and for the scattering cross section of a particle when the incident plane wave is polarized along $\hat{\boldsymbol{u}}_{I\eta}$. Since the extinction cross section is related to the forward scattering amplitude through the optical theorem (2.19), we get

$$\sigma_{T\eta} = \frac{4\pi}{k} \mathrm{Im}\left[f_{\eta\eta}(\hat{\boldsymbol{k}}_I, \hat{\boldsymbol{k}}_I)\right] = -\frac{1}{k^2} \mathrm{Re}\left[\sum_{plm} W^{(p)*}_{lm}(\hat{\boldsymbol{u}}_{I\eta}, \hat{\boldsymbol{k}}_I) A^{(p)}_{\eta lm}\right] \ . \qquad (3.9)$$

In turn, using the definitions (2.5) and (3.7) and the orthogonality properties of the transverse harmonics (1.42) and (1.43), we obtain

$$\sigma_{S\eta} = \int |\boldsymbol{f}_\eta(\hat{\boldsymbol{k}}_S = \hat{\boldsymbol{r}}, \hat{\boldsymbol{k}}_I)|^2 \mathrm{d}\Omega = \frac{1}{k^2} \sum_{plm} A^{(p)*}_{\eta lm} A^{(p)}_{\eta lm} \ . \qquad (3.10)$$

3.2 Transition Matrix

From a formal point of view, the scattering process can be described by a linear operator S defined by the equation

$$\boldsymbol{E}_{\mathrm{S}\eta} = \mathrm{S}\boldsymbol{E}_{\mathrm{I}\eta} \ .$$

The possibility of introducing such an operator, known as *transition operator*, stems from the linearity of all the equations that describe the field, including those that express the boundary conditions across the surface of the scattering particle. The representation of the operator S on the basis of the spherical multipole fields is known as *transition matrix* and is defined by the equation [23]

$$A_{\eta lm}^{(p)} = \sum_{p'l'm'} \mathcal{S}_{lml'm'}^{(pp')} W_{l'm'}^{(p')}(\hat{\boldsymbol{u}}_{\mathrm{I}\eta}, \hat{\boldsymbol{k}}_{\mathrm{I}}) \ , \qquad (3.11)$$

which relates the multipole amplitudes of the incident field to those of the scattered field. The quantities $\mathcal{S}_{lml'm'}^{(pp')}$, which are the elements of the transition matrix, take into account the morphology of the particle as well as the boundary conditions. Substitution of (3.11) into (3.8) yields

$$f_{\eta\eta'} = -\frac{\mathrm{i}}{4\pi k} \sum_{plm} \sum_{p'l'm'} W_{lm}^{(p)*}(\hat{\boldsymbol{u}}_{\mathrm{S}\eta}, \hat{\boldsymbol{k}}_{\mathrm{S}}) \mathcal{S}_{lml'm'}^{(pp')} W_{l'm'}^{(p')}(\hat{\boldsymbol{u}}_{\mathrm{I}\eta'}, \hat{\boldsymbol{k}}_{\mathrm{I}}) \ , \qquad (3.12)$$

which gives the explicit relation between the scattering amplitude and the transition matrix.

Clearly, the elements of S depend on the multipole basis and on the frame of reference with respect to which the scattering process is described. This dependence will now be determined explicitly with the help of the transformation properties of the multipole fields under rotation [24]. We choose into the laboratory a frame of reference, say Σ, and associate to the scattering particle a local frame of reference, say $\bar{\Sigma}$, whose orientation with respect to Σ is individuated by the Eulerian angles $\Theta \equiv (\alpha, \beta, \gamma)$ that describe the transformation of Σ into $\bar{\Sigma}$. In the laboratory frame the polarized plane wave (3.5) impinges onto the particle and is transformed into the scattered wave (3.6). Of course, the relation among the amplitudes of the incident field and those of the scattered field are given by (3.11). The same scattering process, when described in the local reference frame $\bar{\Sigma}$, implies the expressions

$$\boldsymbol{E}_{\mathrm{I}\eta} = E_0 \sum_{plm} \boldsymbol{J}_{lm}^{(p)}(\boldsymbol{r}', k) W_{lm}^{(p)}(\hat{\boldsymbol{u}}'_{\mathrm{I}\eta}, \hat{\boldsymbol{k}}'_{\mathrm{I}}) \qquad (3.13)$$

for the incident field, and

$$\boldsymbol{E}_{\mathrm{S}\eta} = E_0 \sum_{plm} \boldsymbol{H}_{lm}^{(p)}(\boldsymbol{r}', k) \bar{A}_{\eta lm}^{(p)} \qquad (3.14)$$

for the scattered field, where the variables that refer to the frame $\bar{\Sigma}$ are indicated by a prime. In $\bar{\Sigma}$ the relation among the amplitudes of the incident field and those of the scattered field is given by the analog of (3.11)

$$\bar{A}^{(p)}_{\eta lm} = \sum_{p'l'm'} \bar{S}^{(pp')}_{lml'm'} W^{(p')}_{l'm'}(\hat{\boldsymbol{u}}'_{I\eta}, \hat{\boldsymbol{k}}'_I) \,. \tag{3.15}$$

The spherical multipole fields transform under rotation according to $D^{(l)}$ [5, 12], so that the equations hold

$$\boldsymbol{J}^{(p)}_{lm}(\boldsymbol{r}, k) = \sum_{\mu} \boldsymbol{J}^{(p)}_{l\mu}(\boldsymbol{r}', k) \mathcal{D}^{(l)}_{\mu m}(\Theta) \tag{3.16}$$

and

$$\boldsymbol{H}^{(p)}_{l\mu}(\boldsymbol{r}', k) = \sum_{m} \boldsymbol{H}^{(p)}_{lm}(\boldsymbol{r}, k) \left[D^{(l)}(\Theta) \right]^{-1}_{m\mu} \,. \tag{3.17}$$

Then, we substitute (3.16) into (3.5), (3.17) into (3.14) and compare the resulting equations with (3.13) and (3.6), respectively. Since the field strength cannot depend on the choice of the frame of reference, on account of the mutual independence of the spherical multipole fields, the equalities hold

$$W^{(p)}_{l\mu}(\hat{\boldsymbol{u}}'_{I\eta}, \hat{\boldsymbol{k}}'_I) = \sum_{m} \mathcal{D}^{(l)}_{\mu m}(\Theta) W^{(p)}_{lm}(\hat{\boldsymbol{u}}_{I\eta}, \hat{\boldsymbol{k}}_I) \,, \tag{3.18}$$

which gives the transformation rule for the amplitudes of the incident field, and

$$A^{(p)}_{\eta lm} = \sum_{\mu} \left[D^{(l)}(\Theta) \right]^{-1}_{m\mu} \bar{A}^{(p)}_{\eta lm} = \sum_{\mu} \mathcal{D}^{(l)*}_{\mu m}(\Theta) \bar{A}^{(p)}_{\eta l\mu} \,, \tag{3.19}$$

which gives the transformation rule for the amplitudes of the scattered field under rotation. At this stage, substitution in (3.19) of $\bar{A}$ as given by (3.15) and successive substitution into the resulting equation of (3.18) in place of $W^{(p)}_{l\mu}(\hat{\boldsymbol{u}}'_{I\eta}, \hat{\boldsymbol{k}}'_I)$ yields the result

$$A^{(p)}_{\eta lm} = \sum_{\mu\mu'} \sum_{p'l'm'} \mathcal{D}^{(l)*}_{\mu m}(\Theta) \bar{S}^{(pp')}_{l\mu l'\mu'} \mathcal{D}^{(l')}_{\mu'm'}(\Theta) W^{(p')}_{l'm'}(\hat{\boldsymbol{u}}_{I\eta}, \hat{\boldsymbol{k}}_I) \,. \tag{3.20}$$

Comparison of (3.20) and (3.11) then leads to

$$S^{(pp')}_{lml'm'} = \sum_{\mu\mu'} \mathcal{D}^{(l)*}_{\mu m}(\Theta) \bar{S}^{(pp')}_{l\mu l'\mu'} \mathcal{D}^{(l')}_{\mu'm'}(\Theta) \,, \tag{3.21}$$

which establishes the transformation rule of the transition matrix elements under rotation of the coordinate frame [25]. Note that transformation rules analogous to (3.16) and (3.17) hold true for the transverse harmonics $\boldsymbol{Z}^{(p)}_{lm}$.

3.3 Refractive Index of a Polydispersion of Particles

The results in Sect. 3.2 make us able to extend the definition of the refractive index of a dispersion to the case in which the component particles have the same morphology but are not all oriented alike [24]. As far as we know, using the transformation properties of the transition matrix for performing sums over the orientation of the particles has been proposed by Varadan [26], who also devised a specific formula for the description of light propagation through a dispersion of randomly oriented ellipsoids [27]. Nevertheless, a general study of Khlebtzov [28] proved the occurrence of some drawbacks in Varadan's approach and results, so that the first correct procedure turns out to be the one that we are going to describe [24]. We stress however, that the problem of the orientational average of all the quantities of interest in electromagnetic scattering has been successfully faced later in the framework of the Stokes parameters formalism [29, 30, 31, 32].

Let us assign to each particle a local frame of reference, say $\bar{\Sigma}_\nu$ for the νth particle, whose axes are so chosen that, when two particles are brought into coincidence, their respective local frames of reference also coincide. The Eulerian angles that individuate the orientation of $\bar{\Sigma}_\nu$ with respect to Σ, to which the whole dispersion is referred, will be collectively indicated by the symbol Θ_ν. Now, letting N_ν be the number density of the particles whose orientation is Θ_ν, the refractive index of the dispersion reads

$$\mathcal{N}_{\eta\eta'} = \delta_{\eta\eta'} + \frac{2\pi}{k^2} \sum_\nu N_\nu f_{\nu,\eta\eta'}(\hat{\boldsymbol{k}}_\mathrm{I}, \hat{\boldsymbol{k}}_\mathrm{I}) , \qquad (3.22)$$

where $f_{\nu,\eta\eta'}(\hat{\boldsymbol{k}}_\mathrm{I}, \hat{\boldsymbol{k}}_\mathrm{I})$ is the forward-scattering amplitude of the νth particle. Equation (3.22) seems to imply the need for calculating the scattering amplitudes of all the particles that differ in their orientation with respect to Σ before the sum can be performed. Nevertheless, if we substitute into (3.8) for the scattering amplitude the expression for the amplitudes of the scattered field given by (3.20), we get

$$f_{\nu,\eta\eta'}(\boldsymbol{k}_S, \boldsymbol{k}_I) = -\frac{\mathrm{i}}{4\pi k} \sum_{plm} \sum_{p'l'm'} \sum_{\mu\mu'} W^{(p)*}_{lm}(\hat{\boldsymbol{u}}_{S\eta}, \hat{\boldsymbol{k}}_S)$$

$$\times \mathcal{D}^{(l)*}_{\mu m}(\Theta_\nu) \bar{\mathcal{S}}^{(pp')}_{l\mu l'\mu'} \mathcal{D}^{(l')}_{\mu' m'}(\Theta_\nu) W^{(p')}_{l'm'}(\hat{\boldsymbol{u}}_{I\eta'}, \hat{\boldsymbol{k}}_I) . \quad (3.23)$$

It is important to note that in the preceding equation the amplitudes W are given in the laboratory frame Σ, whereas the elements of the transition matrix are given in the local frame $\bar{\Sigma}_\nu$ and are, therefore, independent of the particle index ν.

At this stage it is reasonable to assume that, although we consider a tenuous dispersion, the number of the particles is still so large that the

56   3. Transition Matrix

orientations can be considered as distributed with continuity. Therefore, the sum over ν can be substituted by an integral so that

$$\mathcal{N}_{\eta\eta'} = \delta_{\eta\eta'} - \frac{i}{2k^3} \sum_{plm} \sum_{p'l'm'} \sum_{\mu\mu'} \int N(\Theta) W_{lm}^{(p)*}(\hat{\boldsymbol{u}}_{\mathrm{I}\eta}, \hat{\boldsymbol{k}}_{\mathrm{I}})$$
$$\times \mathcal{D}_{\mu m}^{(l)*}(\Theta) \bar{\mathcal{S}}_{l\mu l'\mu'}^{(pp')} \mathcal{D}_{\mu'm'}^{(l')}(\Theta) W_{l'm'}^{(p')}(\hat{\boldsymbol{u}}_{\mathrm{I}\eta'}, \hat{\boldsymbol{k}}_{\mathrm{I}}) \, d\Theta \,, \quad (3.24)$$

where $N(\Theta)$ is the number density of the particles with orientation Θ. Ultimately, the integral to be performed has the form

$$\mathcal{I} = \int N(\Theta) \mathcal{D}_{\mu m}^{(l)*}(\Theta) \mathcal{D}_{\mu'm'}^{(l')}(\Theta) \, d\Theta \,,$$

and in several cases can be performed analytically.

3.4 Orientational Averages

The need to perform orientational averages does not arise only when calculating the refractive index of a dispersion. Several quantities of actual interest must be calculated through averages over the orientation of the particles concerned [25]. Therefore, we give explicit formulas for calculating the orientational averages of the most important of them. To this end, let us put $N(\Theta) = N_0 P(\Theta)$, where N_0 is the number density of the particles. Then the orientational average of any quantity of interest, say Q, in a scattering problem reads

$$\langle Q \rangle = \int P(\Theta) Q(\Theta) \, d\Theta \,.$$

Such an average can be performed with the help of (3.21) provided Q is expressed in terms of the transition matrix S and of the amplitudes W, for which we introduce the notation

$$W_{\mathrm{I}\eta lm}^{(p)} = W_{lm}^{(p)}(\hat{\boldsymbol{u}}_{\mathrm{I}\eta}, \hat{\boldsymbol{k}}_{\mathrm{I}}) \quad \text{and} \quad W_{\mathrm{S}\eta lm}^{(p)} = W_{lm}^{(p)}(\hat{\boldsymbol{u}}_{\mathrm{S}\eta}, \hat{\boldsymbol{k}}_{\mathrm{S}}) \,.$$

For instance, the required expressions for $\sigma_{\mathrm{T}\eta}$ and $\sigma_{\mathrm{S}\eta}$, using (3.9) and (3.10), read

$$\sigma_{\mathrm{T}\eta} = -\frac{1}{k^2} \mathrm{Re} \left[\sum_{plm} \sum_{p'l'm'} W_{\mathrm{I}\eta lm}^{(p)*} \mathcal{S}_{lml'm'}^{(pp')} W_{\mathrm{I}\eta l'm'}^{(p')} \right] \quad (3.25)$$

and

$$\sigma_{\mathrm{S}\eta} = \frac{1}{k^2} \sum_{plm} \sum_{p'l'm'} \sum_{p''l''m''} \mathcal{S}_{lml'm'}^{(pp')*} W_{\mathrm{I}\eta l'm'}^{(p')*} \mathcal{S}_{lml''m''}^{(pp'')} W_{\mathrm{I}\eta l''m''}^{(p'')} \,, \quad (3.26)$$

respectively. We remark that, according to their derivation, (3.25) and (3.26) hold true when the incident light is polarized along $\hat{\boldsymbol{u}}_{\mathrm{I}\eta}$. When the polariza-

3.4 Orientational Averages

tion is along a general direction, σ_S and σ_T are obtained by substituting into (3.25) and (3.26) in place of $W^{(p)}_{I\eta lm}$

$$W^{(p)}_{Ilm} = \frac{1}{E_0} \sum_\eta W^{(p)}_{I\eta lm} E_{0\eta} ,$$

with $E_0 = |E_{01}\hat{u}_{I1} + E_{02}\hat{u}_{I2}|$. The resulting expressions are easily seen to coincide with those that are obtained through the more formal procedure that resorts to the Stokes vector and to the Müller matrix for σ_S and to the so-called extinction matrix for σ_T [33, 16]. Moreover, for natural incident light we have

$$\sigma = \frac{\sigma_1 + \sigma_2}{2} .$$

The elements of the Müller transformation matrix of a particle can be expressed in terms of the quantities

$$\mathcal{F}_{\eta\eta'\eta''\eta'''} = f^*_{\eta\eta'}(\hat{\boldsymbol{k}}_S, \hat{\boldsymbol{k}}_I) f_{\eta''\eta'''}(\hat{\boldsymbol{k}}_S, \hat{\boldsymbol{k}}_I) ,$$

from which the co-polarized and the cross-polarized components of the scattered intensity can be obtained by putting $\eta'' = \eta$ and $\eta''' = \eta'$. In fact we have

$$I_{\eta\eta'}(\hat{\boldsymbol{k}}_S, \hat{\boldsymbol{k}}_I) = |\boldsymbol{E}_{S\,\eta'} \cdot \hat{\boldsymbol{u}}^*_{S\,\eta}|^2 = \frac{|E_{0\,\eta'}|^2}{r^2} |f_{\eta\eta'}(\hat{\boldsymbol{k}}_S, \hat{\boldsymbol{k}}_I)|^2 .$$

As a result of the transformation properties of S matrix elements [see (3.21)] and on account of (3.12), (3.25), and (3.26), the quantities to be integrated to get the required averages are actually products of up to four D matrix elements multiplied by the weight function $P(\Theta)$. Of course, the products of D matrix elements can be simplified by a wise use of the Clebsch–Gordan series [see (1.23)]. It is important to note that, when performing averages of the Müller matrix elements and, in particular, of the intensity, it is necessary that the size of the whole dispersion be small in comparison with the distance of observation.

Hereafter, our attention will focus on a few cases of physical interest for which the averages can be performed analytically and thus explicit formulas can be given [25]. The appropriate weight functions have been factorized as

$$P(\Theta) = u(\alpha)v(\beta)w(\gamma)$$

and are listed in Table 3.1.

In Case 1 the particles are randomly oriented and the average is easily performed with the the help of the orthogonality properties (1.13) of the rotation matrix elements $\mathcal{D}^{(l)}_{\mu m}(\Theta)$. The result is

58   3. Transition Matrix

Table 3.1. Orientational weight functions

	$u(\alpha)$	$v(\beta)$	$w(\gamma)$
1	$1/2\pi$	$1/2$	$1/2\pi$
2	$1/2\pi$	$v(\beta)$	$1/2\pi$
3	$1/2\pi$	$\delta(\beta)/\sin\beta$	$1/2\pi$
4	$1/2\pi$	$\delta(\beta)/\sin\beta$	$\delta(\gamma)$
5	$1/2\pi$	$\delta(\beta-\pi/2)/\sin\beta$	$1/2\pi$

$$\langle f_{\eta\eta'}(\hat{\mathbf{k}}_\mathrm{I},\hat{\mathbf{k}}_\mathrm{I})\rangle = c_\mathrm{A} \sum_{pp'}\sum_{l}\sum_{m\mu} \frac{1}{2l+1} W^{(p)*}_{\mathrm{I}\eta lm}\bar{S}^{(pp')}_{l\mu l\mu}W^{(p')}_{\mathrm{I}\eta' lm}, \qquad (3.27\mathrm{a})$$

$$\langle \sigma_{\mathrm{S}\,\eta}\rangle = c_\mathrm{S} \sum_{plm\,p'p''}\sum_{l'}\sum_{m'\mu} \frac{1}{2l'+1}\bar{S}^{(pp')*}_{lml'\mu}W^{(p')*}_{\mathrm{I}\eta l'm'}\bar{S}^{(pp'')}_{lml'\mu}W^{(p'')}_{\mathrm{I}\eta l'm'}, \qquad (3.27\mathrm{b})$$

$$\langle \mathcal{F}_{\eta\eta'\eta''\eta'''}\rangle = c_\mathrm{I} \sum_{L}\sum_{\mu\mu'} \frac{1}{2L+1} F^*_{\eta\eta' L\mu\mu'} F_{\eta''\eta''' L\mu\mu'}, \qquad (3.27\mathrm{c})$$

where

$$c_\mathrm{A} = -\mathrm{i}/(4\pi k), \quad c_\mathrm{S} = 1/k^2, \quad c_\mathrm{I} = 1/(4\pi k)^2.$$

In (3.27c) we define

$$F_{\eta\eta' L\mu\mu'} = \sum_{pp'}\sum_{ll'}\sum_{mm'}\sum_{\bar{m}\bar{m}'} (-)^{m'-\bar{m}'} C(l,l',L;m,-m') C(l,l',L;\bar{m},-\bar{m}')$$

$$\times W^{(p)*}_{\mathrm{S}\eta lm}\bar{S}^{(pp')}_{l\bar{m}l'\bar{m}'}W^{(p')}_{\mathrm{I}\eta'l'm'}\delta_{\mu,m-m'}\delta_{\mu',\bar{m}-\bar{m}'},$$

where, as usual, the C's are Clebsch–Gordan coefficients [5]. Equations (3.27a) and (3.27b) can be simplified by using the sum rule [33, 34]

$$\sum_m W^{(p)*}_{\mathrm{I}\eta lm}W^{(p')}_{\mathrm{I}\eta' lm} = 2\pi(2l+1)[\delta_{\eta\eta'}\delta_{pp'} + \mathrm{i}(-)^\eta(1-\delta_{\eta\eta'})(1-\delta_{pp'})], \quad (3.28)$$

that holds true for linear polarization, whereas for circular polarization the appropriate sum rule is

$$\sum_m W^{(p)*}_{\mathrm{I}\eta lm}W^{(p')}_{\mathrm{I}\eta' lm} = 2\pi(2l+1)\delta_{\eta\eta'}[\delta_{pp'} + (-)^{\eta+1}(1-\delta_{pp'})]. \quad (3.29)$$

In Case 2 the $\bar{z}$ axis of the local reference frame may form an angle β with the z axis of the laboratory frame with probability $v(\beta)$, whereas the particle is randomly oriented around the $\bar{z}$ axis which, in turn, may have any value of α for a given β. This kind of distribution is particularly significant when the $\bar{z}$ axis is chosen to be the highest-symmetry axis of the particle: the set of the orientations that can be assumed by the particle can be visualized by thinking of a spinning symmetrical top precessing around the vertical with its axis at an angle β with probability $v(\beta)$. The averages are

$$\langle f_{\eta\eta'}(\hat{\mathbf{k}}_\text{I}, \hat{\mathbf{k}}_\text{I}) \rangle = c_\text{A} \sum_L F_{\eta\eta' L00} \mathcal{I}_L \,, \tag{3.30a}$$

$$\langle \sigma_{\text{S}\,\eta} \rangle = c_\text{S} \sum_{plm} \sum_{p'p''} \sum_{l'l''} \sum_{m'\mu} \sum_L \bar{\mathcal{S}}_{lml'\mu}^{(pp')*} W_{\text{I}\eta l'm'}^{(p')*} \bar{\mathcal{S}}_{lml''\mu}^{(pp'')} W_{\text{I}\eta l''m'}^{(p'')}$$

$$\times (-)^{m'-\mu} C(l',l'',L;m',-m') C(l',l'',L;\mu,-\mu) \mathcal{I}_L \,, \tag{3.30b}$$

$$\langle \mathcal{F}_{\eta\eta'\eta''\eta'''} \rangle = c_\text{I} \sum_{LL'} \sum_\lambda \sum_{\mu\mu'} F^*_{\eta\eta' L\mu\mu'} F_{\eta''\eta''' L'\mu\mu'} (-)^{\mu-\mu'}$$

$$\times C(L,L',\lambda;\mu,-\mu) C(L,L',\lambda;\mu',-\mu') \mathcal{I}_\lambda \,, \tag{3.30c}$$

where

$$\mathcal{I}_l = \int_0^\pi v(\beta) P_l(\cos\beta) \sin\beta \, d\beta \tag{3.31}$$

in which $P_l(\cos\beta)$ is the Legendre polynomial of degree l. In a sense, the preceding equations are the fundamental ones as regards orientational averages. In fact, by choosing $v(\beta) = 1/2$, (3.31) yields $\mathcal{I}_l = \delta_{l0}$ and one gets back the case of randomly oriented particles. Case 3 too can be obtained by choosing in (3.31) $v(\beta) = \delta(\beta)/\sin\beta$, which yields $\mathcal{I}_l = 1$ for any l. In fact, this choice implies that the $\bar{z}$ axis is parallel to the z axis but the particle is otherwise randomly oriented. Even in this case the set of the orientations can be visualized by thinking of a symmetrical top spinning with its axis vertical. In the next section we will show that this case occurs when prolate particles are oriented by the effect of a strong electrostatic field [34].

Case 4 differs from Case 3 only because the angle γ is held fixed. Nevertheless, a moment's thought will convince the reader that, according to the definition of the Eulerian angles, the actual orientational distribution does not differ from the one considered in Case 3. As a consequence, the formulas for these two cases are equivalent in spite of their formal differences. Since using either formulas is a matter of convenience in actual calculations, we report below the explicit formulas for Case 4. We have

$$\langle f_{\eta\eta'}(\hat{\mathbf{k}}_\text{I}, \hat{\mathbf{k}}_\text{I}) \rangle = c_\text{A} \sum_{pl} \sum_{p'l'} \sum_m W_{\text{I}\eta lm}^{(p)*} \bar{\mathcal{S}}_{lml'm}^{(pp')} W_{\text{I}\eta'l'm}^{(p')} \,, \tag{3.32a}$$

$$\langle \sigma_{\text{S}\eta} \rangle = c_\text{S} \sum_{plm} \sum_{p'l'} \sum_{p''l''} \sum_{m'} \bar{\mathcal{S}}_{lml'm'}^{(pp')*} W_{\text{I}\eta l'm'}^{(p')*} \bar{\mathcal{S}}_{lml''m'}^{(pp'')} W_{\text{I}\eta l''m'}^{(p'')} \,, \tag{3.32b}$$

$$\langle \mathcal{F}_{\eta\eta'\eta''\eta'''} \rangle = c_\text{I} \sum_{mm'} \sum_{m''m'''} T^*_{\eta\eta' mm'} T_{\eta''\eta''' m''m'''} \delta_{m-m'-m''+m''',0} \tag{3.32c}$$

with

$$T_{\eta\eta' mm'} = \sum_{pl} \sum_{p'l'} W_{\text{S}\eta lm}^{(p)*} \bar{\mathcal{S}}_{lml'm'}^{(pp')} W_{\text{I}\eta'l'm'}^{(p')} \,.$$

It may be useful to remark that in (3.32c) the Kronecker delta comes from the integral

$$\mathcal{I}_\mu = \int_0^{2\pi} u(\alpha) e^{i\alpha\mu} \, d\alpha \,.$$

In fact, when $u(\alpha) = 1/2\pi$, i.e., when angle α is randomly distributed, we get $\mathcal{I}_\mu = \delta_{\mu 0}$. When $u(\alpha) = \delta(\alpha)$, i.e., when angle α is fixed, we have $\mathcal{I}_\mu = 1$ and, consequently,

$$\langle \mathcal{F}_{\eta\eta'\eta''\eta'''} \rangle = \mathcal{F}_{\eta\eta'\eta''\eta'''} \,. \tag{3.33}$$

The latter is an obvious result that is due to the fact that no average over angle α has actually been performed. Moreover, when the scatterer has cylindrical symmetry around the $\bar{z}$ axis of $\bar{\Sigma}$, then the transition matrix elements have the property [23]

$$\bar{S}^{(pp')}_{l\mu l'\mu'} = \bar{S}^{(pp')}_{ll';\mu} \delta_{\mu\mu'} \,, \tag{3.34}$$

and again we get (3.33), as expected, because the scatterer is unaffected by any rotation around its cylindrical-symmetry axis.

A case somewhat complementary to Case 4 occurs when a strong electrostatic field acts on oblate particles, the effect being that the $\bar{z}$ axis lies in the xy plane of the laboratory frame where it can have any orientation [34]. At the same time the particle may have any angle γ around $\bar{z}$. This configuration, indicated as Case 5 in Table 3.1, is obtained by choosing $v(\beta) = \delta(\beta - \pi/2)/\sin\beta$ and thus getting $\mathcal{I}_l = P_l(0)$ from (3.31). In this case we can think of a symmetrical top spinning with its axis held horizontal.

Property (3.34) for the particles with cylindrical symmetry around the $\bar{z}$ axis can be used, as we did in Case 4, even in the other cases in Table 3.1 to simplify the reported formulas. The mutual relations among the cases discussed above show that the orientational averaging procedure that we presented in this section is endowed with internal coherence, even in limiting cases. We will return to the subject in the next chapters when discussing the results of specific applications of the theory.

3.4.1 Asymmetry Parameter

The averaging techniques that were developed until now can be applied to the calculation of the so-called *asymmetry parameter* g, a quantity that is in common use in studies on electromagnetic scattering. Although g can be defined for single particles in connection with the calculation of the radiation pressure [3], it is even more useful when defined for a dispersion of particles in random orientation [16]. In fact, by definition

$$g = \frac{1}{k^2 \langle \sigma_\mathrm{S} \rangle} \langle \int |\boldsymbol{f}(\hat{\boldsymbol{k}}_\mathrm{S} = \hat{\boldsymbol{r}}, \hat{\boldsymbol{k}}_\mathrm{I} = \hat{\boldsymbol{e}}_z)|^2 \cos\vartheta \, d\Omega \rangle \,, \tag{3.35}$$

3.4 Orientational Averages

where the angular brackets denote average for random orientation, and ϑ coincides with the angle of scattering. When $g = \pm 1$ the dispersion scatters essentially in the forward $(g = 1)$ or backward direction $(g = -1)$, whereas when $g = 0$ the scattering process is isotropic. According to (3.7), we can rewrite (3.35) in the form

$$g = \frac{1}{k^2 \langle \sigma_S \rangle} \sum_{plm} \sum_{p'l'm'} (-i)^{l'+p'-l+p} \langle A_{lm}^{(p)*} A_{l'm'}^{(p')} \rangle \int \mathbf{Z}_{lm}^{(p)*} \cdot \mathbf{Z}_{l'm'}^{(p')} \cos\vartheta \, d\Omega \; .$$

The average $\langle A_{lm}^{(p)*} A_{l'm'}^{(p')} \rangle$ can be easily calculated with the help of (3.27b) and (3.11). In turn, the integral that involves $\cos\vartheta$ can be calculated using several techniques. For instance, we could operate as we did in Sect. 3.1.1 with $\mathbf{Z}_{lm}^{(p)} = \sum_\eta \hat{\mathbf{u}}_\eta \cdot \mathbf{Z}_{lm}^{(p)} \hat{\mathbf{u}}_\eta$. Nevertheless, we prefer to resort to an approach that is coherent with the one we used earlier in this section to perform configurational averages. In other words, the integration on the directions of observation will be performed by resorting to the transformation properties of the quantities involved. First, we remark that

$$\cos\vartheta = Y_{10} \left(\frac{4\pi}{3}\right)^{1/2} .$$

Therefore, the integral of interest can be written

$$I_{plmp'l'm'} = \frac{(-i)^{l'+p'-l+p}}{k^2} \left(\frac{4\pi}{3}\right)^{1/2} \int \mathbf{Z}_{lm}^{(p)*} \cdot \mathbf{Z}_{l'm'}^{(p')} Y_{10} \, d\Omega$$

$$= \frac{(-i)^{l'+p'-l+p}}{k^2} \left(\frac{4\pi}{3}\right)^{1/2} \sum_{\bar{m}\bar{m}'\mu} \mathbf{Z}_{l\bar{m}}^{(p)*}(\hat{\mathbf{e}}_z) \cdot \mathbf{Z}_{l'\bar{m}'}^{(p')}(\hat{\mathbf{e}}_z) Y_{1\mu}(\hat{\mathbf{e}}_z)$$

$$\times \int \mathcal{D}_{\bar{m}m}^{(l)*} \mathcal{D}_{\bar{m}'m'}^{(l')} \mathcal{D}_{\mu 0}^{(1)} \, d\Omega \; .$$

The integral of the product of three elements of the $\mathcal{D}^{(l)}$ matrices is well known [35]. Consequently,

$$I_{plmp'l'm'} = \left(\frac{4\pi}{3}\right)^{1/2} \frac{8\pi^2}{2l+1} \frac{(-i)^{l'+p'-l+p}}{k^2} C(1, l', l; 0, m')$$

$$\times \sum_{\bar{m}\bar{m}'} C(1, l', l; \bar{m} - \bar{m}', \bar{m}') \mathbf{Z}_{l\bar{m}}^{(p)*}(\hat{\mathbf{e}}_z) \cdot \mathbf{Z}_{l'\bar{m}'}^{(p')}(\hat{\mathbf{e}}_z) Y_{1,\bar{m}-\bar{m}'}(\hat{\mathbf{e}}_z)$$

$$= I_{plp'l';m} \delta_{mm'} \; .$$

Calculation of $I_{plp'l';m}$ is straightforward. In fact, according to (1.44), (1.45), (1.47b), and (1.28), all the Clebsch–Gordan coefficients, including those that are needed to perform the dot product of the transverse harmonics, are of the form $C(1, l, l; \mu, m)$ or $C(1, l, l \pm 1; \mu, m)$ (see Appendix A.3.1), whereas all the spherical harmonics are given by (1.59).

3.5 Effect of an Electrostatic Field

A typical case in which the orientational distribution of a dispersion of particles is nonrandom occurs when the ensemble is subject to the effect of an electrostatic field. In fact, in thermal equilibrium and in the absence of any applied field, the particles of a dispersion are randomly distributed both in space and orientation. Therefore, the dispersion on the whole is isotropic and cannot show either birefringence or linear dichroism. Of course, when an electrostatic field is applied the distribution becomes nonrandom, and the optical properties of the dispersion are bound to change [34]. To give a quantitative content to this intuitive statement, we assume that the particles of interest are uncharged so that an electrostatic field has no effect on their space distribution. We further assume that the particles are identical to each other and that the whole dispersion is in thermal equilibrium at the temperature T. When an electrostatic field is applied the orientational distribution function $P(\Theta)$ can be assumed to be [36]

$$P(\Theta) = \frac{\exp[-U(\Theta)/k_B T]}{Z} ,$$

where $U(\Theta)$ is the interaction energy that, as explicitly indicated, depends on the orientation of the particles, k_B is the Boltzmann constant and

$$Z = \int \exp[-U(\Theta)/k_B T] \, d\Theta .$$

In general, by indicating with x_1, x_2, and x_3 the rectangular coordinates in the laboratory reference frame Σ, the interaction energy $U(\Theta)$ can be written as [2]

$$U(\Theta) = -\tfrac{1}{2} \sum_{ij} \alpha_{ij} E_i E_j ,$$

where α_{ij} are the components of the polarizability tensor of the particle. By choosing the x_3 axis of Σ to be parallel to the electrostatic field, the interaction energy simplifies into

$$U(\Theta) = -\tfrac{1}{2} \alpha_{33} E^2 .$$

We now attach to the particle a local frame of reference $\bar{\Sigma}$ whose coordinates are transformed into those of Σ by the matrix R. Then we can write

$$\alpha_{ij} = \sum_{lk} R_{li} R_{kj} \bar{\alpha}_{lk},$$

where,

$$\bar{\alpha}_{lk} = \bar{\alpha}_k \delta_{lk},$$

provided the axes of $\bar{\Sigma}$ are the principal axes of the polarizability tensor.

Hereafter, we restrict our study to the case in which

$$\bar{\alpha}_1 = \bar{\alpha}_2 = \alpha_\perp, \qquad \bar{\alpha}_3 = \alpha_\parallel, \tag{3.36}$$

which is typical not only of axially symmetric particles but also of particles with an appropriate symmetry group. In terms of the Eulerian angles, the relation between α_{33} and the $\bar{\alpha}_{ij}$ is [5]

$$\alpha_{33} = \sum_k R_{k3}^2 \bar{\alpha}_k = \bar{\alpha}_1 \sin^2 \beta \cos^2 \gamma + \bar{\alpha}_2 \sin^2 \beta \sin^2 \gamma$$
$$+ \bar{\alpha}_3 \cos^2 \beta = \alpha_\perp + (\alpha_\parallel - \alpha_\perp) \cos^2 \beta.$$

Therefore, by defining

$$u = (\alpha_\parallel - \alpha_\perp) E^2 / 2 k_B T \tag{3.37}$$

and

$$I_0 = \int_0^\pi \exp[u \cos^2 \beta] \sin \beta \, d\beta,$$

the orientational distribution function results.

$$P(\Theta) = \frac{1}{4\pi^2} v(\beta) = \frac{1}{4\pi^2 I_0} \exp[u \cos^2 \beta]. \tag{3.38}$$

The latter equation shows that for axially symmetric particles $P(\Theta)$ is a function of β only. Furthermore, according to (3.37) and (3.38), the effect of the electrostatic field will be the more considerable the more $\alpha_\parallel$ and $\alpha_\perp$ are different.

At this stage the orientational average of the forward scattering amplitude can be performed through (3.30a) and (3.31).

3.5.1 Polarizability of a Particle of Arbitrary Shape

The usefulness of the results of the preceding section depends on the knowledge of the polarizability of the particles of interest. The literature reports a few expressions for the polarizability of objects of simple shape, e.g., the Clausius–Mossotti and the Lorentz–Lorenz formulas for spheres [2], and the expression devised by Clark Jones for ellipsoids and disks [37]. On the other

64   3. Transition Matrix

hand, the polarizability has been widely used in the framework of the Rayleigh scattering approximation (RSA) that assumes that the scattered field from a small sphere is well approximated by the field of the dipole moment that is induced by the incident electromagnetic wave [3]. The relation between the incident field and the induced dipole is, indeed, given through the polarizability although the latter, at least in principle, is a static and not a dynamic quantity.

The preceding considerations will presently be extended by working out a general relation between the polarizability tensor and the transition matrix of a particle of arbitrary shape [34]. We start by recalling that, in the long-wavelength limit, the field that is scattered by a particle is well approximated by the field of the electric dipole that is induced by the incident wave and can, therefore, be written as [2]

$$\boldsymbol{E}_S \approx \frac{\exp(ikr)}{r} k^2 (\hat{\boldsymbol{k}}_s \times \boldsymbol{p}) \times \hat{\boldsymbol{k}}_s$$
$$= \frac{\exp(ikr)}{r} k^2 [\boldsymbol{p} - (\boldsymbol{p} \cdot \hat{\boldsymbol{k}}_s) \hat{\boldsymbol{k}}_s] \,, \qquad (3.39)$$

where $\boldsymbol{p}$ is the induced electric dipole moment. Here and hereafter we use the symbol $\approx$ to denote the approximate equalities that become rigorous as $k \to 0$. Thus, if k is sufficiently small, (2.4) and (3.39) become comparable and yield the equality

$$E_0 \boldsymbol{f} \approx k^2 [\boldsymbol{p} - (\boldsymbol{p} \cdot \hat{\boldsymbol{k}}_S) \hat{\boldsymbol{k}}_s] \qquad (3.40)$$

that relates the dynamic and the static features of the scattering particle. At this stage, we substitute into (3.40) the expression of $\boldsymbol{f}$ given by (3.7), dot multiply both sides of the resulting equation by the transverse harmonics $\boldsymbol{Z}_{lm}^{(p)*}(\hat{\boldsymbol{k}}_S)$ and integrate over the solid angle. We get

$$E_0 A_{lm}^{(1)} = \mathrm{i}^{l+1} k E_0 \int \boldsymbol{f} \cdot \boldsymbol{Z}_{lm}^{(1)*}(\hat{\boldsymbol{k}}_S) \, \mathrm{d}\Omega_S$$
$$\approx \mathrm{i}^{l+1} k^3 \boldsymbol{p} \cdot \int \boldsymbol{Z}_{lm}^{(1)*}(\hat{\boldsymbol{k}}_S) \, \mathrm{d}\Omega_S = 0 \qquad (3.41)$$

and

$$E_0 A_{lm}^{(2)} = -\mathrm{i}^l k E_0 \int \boldsymbol{f} \cdot \boldsymbol{Z}_{lm}^{(2)*}(\hat{\boldsymbol{k}}_S) \, \mathrm{d}\Omega_S \approx -\mathrm{i}^l k^3 \boldsymbol{p} \cdot \int \boldsymbol{Z}_{lm}^{(2)*}(\hat{\boldsymbol{k}}_S) \, \mathrm{d}\Omega_S \,. \qquad (3.42)$$

The integral in (3.42) can be evaluated by expressing $\boldsymbol{Z}_{lm}^{(2)}$ in terms of irreducible spherical tensors [see (1.47b)] and writing the dot product in terms of spherical components. The result is

$$E_0 A_{lm}^{(2)} \approx k^3 \sqrt{\frac{8\pi}{3}} (-)^m p_{-m} \delta_{l1} \,. \qquad (3.43)$$

Clearly, (3.41) and (3.43) establish an extension of the RSA to particles of arbitrary shape as they give the amplitudes of the scattered field in terms of the induced dipole moment.

To relate the amplitudes $A_{1m}^{(2)}$ to the polarizability of the particle, we write the relation between the spherical components of the incident field and those of the induced dipole as

$$p_m = \sum_{m'} \alpha_{mm'} E_{m'} .$$

Since the origin of the frame of reference has been chosen within the scattering particle, the incident plane wave reduces to its amplitude and we can write

$$p_m = \sum_{m'} \alpha_{mm'} \hat{e}_{m'} E_0 ,$$

where $\hat{e}_{m'}$ denotes the spherical components of the polarization vector of the incident wave. With the help of the identity

$$\hat{e}_m = \hat{e} \cdot \boldsymbol{\xi}_m = [(\hat{\boldsymbol{k}}_I \times \hat{e}) \times \hat{\boldsymbol{k}}_I] \cdot \boldsymbol{\xi}_m = (\hat{\boldsymbol{k}}_I \times \hat{e}) \cdot (\hat{\boldsymbol{k}}_I \times \boldsymbol{\xi}_m) ,$$

a straightforward calculation gives the final result

$$\begin{aligned} A_{1m}^{(2)} &\approx -\mathrm{i}k^3 \frac{8\pi}{3} \sum_{m'} (-)^{m+m'} \alpha_{-m,-m'} (\hat{\boldsymbol{k}}_I \times \hat{e}) \cdot \boldsymbol{X}_{1m'}^*(\hat{\boldsymbol{k}}_I) \\ &= \frac{2}{3}\mathrm{i}k^3 \sum_{m'} (-)^{m+m'} \alpha_{-m,-m'} W_{1m'}^{(2)}(\hat{e}, \hat{\boldsymbol{k}}_I) . \end{aligned} \quad (3.44)$$

The preceding equation is the required relation between the polarizability of a particle of arbitrary shape and the electric dipole amplitudes of the scattered field. By substituting $A_{1m}^{(2)}$ as given by (3.11) into (3.44), one gets the required relation between the components of the polarizability of a particle of arbitrary shape and the elements of its transition matrix in the form

$$\alpha_{-m,-m'} = -\frac{3}{2}\mathrm{i}(-)^{m+m'} \lim_{k \to 0} \frac{1}{k^3} S_{1m1m'}^{(2\,2)} . \quad (3.45)$$

We stress that it is just the limit for $k \to 0$ that transforms the approximate equality (3.44) into the rigorous equation (3.45).

By choosing the axes of the reference frame so as to diagonalize the Cartesian representation of the polarization tensor the corresponding spherical representation takes on the form [5]

$$\bar{\alpha}_{mm'} = \frac{1}{2} \begin{pmatrix} \bar{\alpha}_1 + \bar{\alpha}_2 & 0 & \bar{\alpha}_2 - \bar{\alpha}_1 \\ 0 & 2\bar{\alpha}_3 & 0 \\ \bar{\alpha}_2 - \bar{\alpha}_1 & 0 & \bar{\alpha}_1 + \bar{\alpha}_2 \end{pmatrix} \quad (3.46)$$

so that (3.45) simplifies into

$$\bar{\alpha}_{-m,-m} = -\frac{3}{2}\mathrm{i}\lim_{k\to 0}\frac{1}{k^3}\bar{S}^{(2\,2)}_{1m1m} \qquad (3.47)$$

for all the particles such that $\bar{\alpha}_1 = \bar{\alpha}_2$. The preceding equation applies, in particular, to axially symmetric particles, provided their high-symmetry axis coincides with the $\bar{z}$ axis. In this case one also has $\bar{\alpha}_{-1,-1} = \bar{\alpha}_{11}$ and, since these latter quantities characterize the polarizability in the plane that is orthogonal to the cylindrical axis, we can put

$$\bar{\alpha}_{-1,-1} = \bar{\alpha}_{11} = \alpha_\perp, \quad \bar{\alpha}_0 = \alpha_\| . \qquad (3.48)$$

A look to (3.46) shows that $\alpha_\|$ and $\alpha_\perp$ as defined above coincide with the corresponding quantities in (3.36).

In the next chapter we will show that, for small spheres, (3.45) reduces to the well known form that is used in the Rayleigh approximation.

3.6 Effect of the Diffusive Motion

The orientational distribution functions that we have considered so far are, in a sense, static as none of them allows for the motion of the particles. Nevertheless, in some cases the diffusion motion may affect the intensity of the field scattered by the whole dispersion. When this happens we have to deal with a dynamic light-scattering process and for its description we have to consider both the electromagnetic and the statistical behavior of the dispersion, including the motion of the particles concerned. The electromagnetic and the statistical aspects of dynamic light scattering are connected by the Wiener–Kinchin theorem that relates the electromagnetic scattering properties and the diffusive motion of the particles. More precisely, the Wiener–Kinchin theorem [38] relates the intensity of the light scattered by a dispersion of particles to the ensemble average of the time autocorrelation function of the field scattered by the individual objects. As a consequence, a reliable interpretation of the observed spectra requires a good description of the scattering properties of the particles involved and a reliable estimate of the dynamical distribution function, both in space and orientation, of the particles themselves. In this section we focus on the averages of the dynamic distribution function by the methods described earlier in this chapter. The calculation of the scattering properties of the particles is deferred to Chap. 6.

Let a polarized plane wave of wavevector $\mathbf{k}_\mathrm{I}$ and (circular) frequency ω_0 propagating in vacuo ($n = 1$) be scattered by an assembly of particles. We refer the scattering process to some plane chosen once for all as the plane of reference, so that the polarization both of the incident and the scattered field

3.6 Effect of the Diffusive Motion

can be analyzed as we did in Sect. 2.2 with respect to the plane of scattering. Accordingly, we denote with

$$\bar{E}_{S\eta\eta'} = \bar{\boldsymbol{E}}_{S\eta'} \cdot \hat{\boldsymbol{u}}_{S\eta}^*$$

the component polarized along $\hat{\boldsymbol{u}}_{S\eta}$ of the field $\bar{\boldsymbol{E}}_{S\eta'}$ scattered by the whole assembly when the incident field is polarized along $\hat{\boldsymbol{u}}_{I\eta'}$. The spectral scattered intensity is then

$$I_{\eta\eta'}(\boldsymbol{r},\omega) = \lim_{T\to\infty} \frac{1}{T} \bar{E}^*_{S\eta\eta'}(\boldsymbol{r},\omega;T)\bar{E}_{S\eta\eta'}(\boldsymbol{r},\omega;T) ,$$

where

$$\bar{E}_{S\eta\eta'}(\boldsymbol{r},\omega;T) = \int_{-\infty}^{\infty} \bar{E}_{S\eta\eta'}(\boldsymbol{r},t;T)\exp[i\omega t]dt$$

with

$$\bar{E}_{S\eta\eta'}(\boldsymbol{r},t;T) = \begin{cases} \bar{E}_{S\eta\eta'}(\boldsymbol{r},t) & \text{when } |t| < T \\ 0 & \text{when } |t| > T \end{cases} .$$

Provided the number density of the dispersion is low enough, the scattered field can be given as the superposition of the fields scattered by the individual particles. For a particle at the origin, we write

$$E_{S\eta\eta'}(\boldsymbol{r},\omega) = E_0 \frac{\exp(ikr)}{r} f_{\eta\eta'}(\hat{\boldsymbol{k}}_S, \hat{\boldsymbol{k}}_I, \omega, \Theta) ,$$

where $f_{\eta\eta'}$ bears the additional arguments ω and Θ to recall that the scattering amplitude depends also on the frequency of the incident wave and on the orientation of the scattering particle. At this stage, the Wiener–Kinchin theorem relates the scattered intensity and the scattering amplitude in the form

$$I_{\eta\eta'}(\boldsymbol{r},\omega) = \frac{N|E_0|^2}{r^2} \int \langle f^*_{\eta\eta'}(\hat{\boldsymbol{k}}_S, \hat{\boldsymbol{k}}_I, \omega, \Theta_0) f_{\eta\eta'}(\hat{\boldsymbol{k}}_S, \hat{\boldsymbol{k}}_I, \omega, \Theta) \rangle \\ \times F(\boldsymbol{k}_I - \boldsymbol{k}_S, \tau)\exp[i(\omega-\omega_0)\tau]d\tau , \qquad (3.49)$$

where N is the number of the particles, $\Theta_0 = \Theta(0)$ is the initial orientation, $\Theta = \Theta(\tau)$ is the orientation at time τ, and $\langle \ldots \rangle$ denotes the average over the orientations on the customary assumption of no correlation existing between the position and orientation of the particles. In turn, $F(\boldsymbol{k}_I - \boldsymbol{k}_S, \tau)$ is the dynamic structure factor of the dispersion that can be defined as

$$F(\boldsymbol{K},\tau) = \frac{1}{N} \int d\boldsymbol{r}' \int d\boldsymbol{R} \exp(i\boldsymbol{K}\cdot\boldsymbol{R})\langle n(\boldsymbol{r}'+\boldsymbol{R},\tau)n(\boldsymbol{r}',0)\rangle ,$$

where

$$n(\boldsymbol{r},t) = \sum_{i=1}^{N} \delta(\boldsymbol{r} - \boldsymbol{R}_i(t))$$

is the number density of a dispersion of identical particles with instantaneous vector position $\boldsymbol{R}_i(t)$.

Equation (3.49) is the most general relation among the quantities of interest in an actual experiment of dynamic light scattering, viz., the spectral intensity and the time-dependent autocorrelation function of the scattered field. Equation (3.49) differs from that reported, e.g. by Reichl [38], because the latter author assumes the particles to be so small that their electromagnetic response can be described by their polarizability tensor. On the contrary, we describe the scattering properties of the particles through their scattering amplitude matrix so that we have no need to assume that the particles concerned are small with respect to the wavelength. Anyway, in order to give a reliable description of the orientational distribution function of the particles, we assume that they are axially symmetric. In fact, such a kind of particles has the dynamical features of a symmetrical top so that the appropriate weight function has the form [39]

$$P(\Theta, \tau; \Theta_0) = \sum_{LKM} \frac{2L+1}{(8\pi^2)^2} \mathcal{D}_{KM}^{(L)}(\Theta_0) \mathcal{D}_{KM}^{(L)*}(\Theta) \exp(-\nu_{LK}\tau) ,$$

with

$$\nu_{LK} = D_\perp L(L+1) + (D_\parallel - D_\perp) K^2 ,$$

where $D_\perp$ and $D_\parallel$ are the rotational diffusion coefficients. As a result, the orientational average

$$\langle f^*_{\eta\eta'}(\Theta_0) f_{\eta\eta'}(\Theta) \rangle = \int d\Theta_0 \, d\Theta \, f^*_{\eta\eta'}(\Theta_0) f_{\eta\eta'}(\Theta) P(\Theta, \tau; \Theta_0) ,$$

can be performed analytically because it implies the well known integrals

$$\int \mathcal{D}_{KM}^{(L)*} \mathcal{D}_{\mu m'}^{(l')*} \mathcal{D}_{\mu m}^{(l)} \, d\Theta = \frac{8\pi^2}{2l+1} C(L, l', l; 0, \mu, \mu) C(L, l', l; M, m', m) \delta_{K0} .$$

Therefore, the final result is [40]

$$\langle f^*_{\eta\eta'}(\Theta_0) f_{\eta\eta'}(\Theta) \rangle = \frac{1}{16\pi^2 k^2} \sum_L R_{\eta\eta' L} \exp\left[-D_\perp L(L+1)\tau\right] , \qquad (3.50)$$

with

$$R_{\eta\eta' L} = (2L+1) \sum_M |F_{\eta\eta' LM}|^2 , \qquad (3.51)$$

where

$$F_{\eta\eta' LM} = \sum_{plm} \sum_{p'l'm'} \sum_\mu \frac{1}{2l+1} C(L, l, l'; 0, \mu, \mu)$$
$$\times C L, l, l'; M, m, m') W_{S\eta lm}^{(p)*} \bar{\mathcal{S}}_{l\mu l' \mu}^{(pp')} W_{I\eta' l' m'}^{(p')} .$$

3.6 Effect of the Diffusive Motion

It may be worth recalling that the configurational average in (3.50), when multiplied by the dynamical structure factor, is just the time-dependent autocorrelation function that is the quantity which is actually measured in photon correlation experiments. We also note that, according to (3.50) the spectral intensity turns out to have an exponential-like behavior as required by statistical physics. More precisely, each of the exponentials has its own amplitude that can be calculated from the knowledge of the transition matrix of the particles. Note that the amplitude of each exponential depends on the polarization and on the angle of scattering, as is evident from the expression of $F_{\eta\eta'LM}$ above. To this angular dependence, one should superpose the contribution from the dynamic structure factor that is customarily assumed to have the form

$$F(\boldsymbol{k}_\mathrm{I} - \boldsymbol{k}_\mathrm{S}, \tau) \propto \exp[-q^2 D\tau] \,,$$

with

$$q = |\boldsymbol{k}_\mathrm{I} - \boldsymbol{k}_\mathrm{S}| \approx 2k\sin(\Phi/2) \,,$$

where Φ is the angle of scattering and D is the translational diffusion coefficient.

We can thus conclude that the averaging procedures that we have described in this chapter are also suitable for dealing with cases in which the motion of the particles need to be taken into account.

4. Transition Matrix of Single and Aggregated Spheres

In Chap. 3 we focused on the relations of the transition matrix of individual particles with the macroscopic optical properties of a dispersion of them. Since the knowledge of the transition matrix of the particles concerned has turned out to be of paramount importance for the interpretation of the observational data, this chapter is devoted to the description of the procedure for calculating the transition matrix of single and aggregated spheres. Limiting our effort to such kinds of scatterers is less restrictive than one may think because, as we outlined in Sect. 1.1, the parameters that individuate the morphology of an aggregate can be chosen so as to fit the properties of a large class of actual particles.

First, we consider homogeneous and radially nonhomogeneous spheres, and, in the latter case, we show the relation with the case of layered spheres. Next, we consider the scatterers composed by aggregated spheres and those composed by a sphere containing either a single eccentric inclusion or more spherical inclusions (often referred to as internal aggregation). Except for the case of single spheres, either homogeneous or concentrically layered, that are isotropic, all other scatterers that we consider in this chapter are anisotropic to a more or less large extent. We will also discuss the onset of optical resonances because these features of the extinction spectrum proved to be of help in studying the morphology of the scattering particles.

In the last section we will report a brief survey of a few other methods that are commonly used to calculate the field scattered by nonspherical particles. In fact, we feel that this survey will be useful for a better understanding of our conclusions when we compare our results with those of other researchers in the field.

4.1 Homogeneous Spheres

The complete solution to the problem of scattering from homogeneous spheres of arbitrary size and refractive index was given by Mie as early as 1908 [41]. For this reason it is customarily referred to as the Mie theory. No generality is lost by assuming that the center of the sphere, of radius ρ and (possibly complex) refractive index n_0, coincides with the origin of the coordinate frame. The incident electric field is assumed to be that of a plane polarized wave, whose multipole expansion is given by (3.5). In turn, the electric field of the scattered wave has, at any point outside the sphere, the multipole expansion (3.6). Finally, since the field within the sphere must be regular at the origin, its expansion can be taken in the form

$$\boldsymbol{E}_{\text{T}\eta} = E_0 \sum_{plm} \boldsymbol{J}_{lm}^{(p)}(\boldsymbol{r}, k_0) C_{\eta lm}^{(p)} ,$$

with $k_0 = n_0 k_{\text{v}}$. The magnetic field, that is also needed for imposing the boundary conditions, is given by the equation

$$\mathrm{i}\boldsymbol{B} = \frac{1}{k_{\text{v}}} \nabla \times \boldsymbol{E} , \qquad (4.1)$$

both within and outside the sphere. In fact, the amplitudes of the field within the sphere as well as those of the field in the external region, where there is the superposition of the incident and of the scattered fields, must satisfy the appropriate boundary conditions across the surface of the sphere.

We assume that the material of the sphere and that of the surrounding medium are nonmagnetic, so that the boundary conditions reduce to the requirement of continuity of the tangential components both of the electric and of the magnetic field. These conditions are often stated by taking the components of the fields along the unit vectors $\hat{\vartheta}$ and $\hat{\varphi}$ of a spherical coordinate system with origin at the center of the sphere [3, 16]. In our approach, all the quantities are referred to the plane of scattering so that a more general procedure is in order. In fact, we equate the dot products of the internal and of the external fields by the transverse harmonics $\boldsymbol{Z}_{lm}^{(p)}$ and integrate the resulting equations over the solid angle. Accordingly, we write the equalities

$$\int (\boldsymbol{E}_{\text{I}\eta} + \boldsymbol{E}_{\text{S}\eta}) \cdot \boldsymbol{Z}_{lm}^{(p)} \, \mathrm{d}\Omega = \int \boldsymbol{E}_{\text{T}\eta} \cdot \boldsymbol{Z}_{lm}^{(p)} \, \mathrm{d}\Omega , \qquad (4.2\text{a})$$

$$\int (\boldsymbol{B}_{\text{I}\eta} + \boldsymbol{B}_{\text{S}\eta}) \cdot \boldsymbol{Z}_{lm}^{(p)} \, \mathrm{d}\Omega = \int \boldsymbol{B}_{\text{T}\eta} \cdot \boldsymbol{Z}_{lm}^{(p)} \, \mathrm{d}\Omega , \qquad (4.2\text{b})$$

that, due to the orthogonality of the multipole fields, yield a system of equations among which the the amplitudes of the internal field can be easily eliminated. In practice, by substituting into (4.2) the expansions for the fields and performing the integrations we get, for each l and m, the four equations

$$\left[h_l(x) A_{\eta lm}^{(1)} + j_l(x) W_{\text{I}\eta lm}^{(1)} \right] = j_l(x_0) C_{\eta lm}^{(1)} , \qquad (4.3\text{a})$$

$$\frac{1}{x} \left[(xh_l(x))' A_{\eta lm}^{(2)} + (xj_l(x))' W_{\text{I}\eta lm}^{(2)} \right] = \frac{1}{x_0} (x_0 j_l(x_0))' C_{\eta lm}^{(2)} , \qquad (4.3\text{b})$$

$$\frac{n}{x} \left[(xh_l(x))' A_{\eta lm}^{(1)} + (xj_l(x))' W_{\text{I}\eta lm}^{(1)} \right] = \frac{n_0}{x_0} (x_0 j_l(x_0))' C_{\eta lm}^{(1)} , \qquad (4.3\text{c})$$

$$n \left[h_l(x) A_{\eta lm}^{(2)} + j_l(x) W_{\text{I}\eta lm}^{(2)} \right] = n_0 j_l(x_0) C_{\eta lm}^{(2)} , \qquad (4.3\text{d})$$

where the prime denotes differentiation with respect to the argument and we put $x = k\rho$, $x_0 = k_0\rho$. Then, by eliminating the amplitudes of the internal field $C^{(p)}_{\eta lm}$ among (4.3) and solving the resulting equations for the amplitudes $A^{(p)}_{\eta lm}$, we get

$$A^{(1)}_{\eta lm} = -\frac{n_0 u'_l(x_0) u_l(x) - n u_l(x_0) u'_l(x)}{n_0 u'_l(x_0) w_l(x) - n u_l(x_0) w'_l(x)} W^{(1)}_{\eta lm} = -R^{(1)}_l W^{(1)}_{I\eta lm}, \qquad (4.4a)$$

$$A^{(2)}_{\eta lm} = -\frac{n u'_l(x_0) u_l(x) - n_0 u_l(x_0) u'_l(x)}{n u'_l(x_0) w_l(x) - n_0 u_l(x_0) w'_l(x)} W^{(2)}_{\eta lm} = -R^{(2)}_l W^{(2)}_{I\eta lm}, \qquad (4.4b)$$

where $u_l(x) = x j_l(x)$ and $w_l(x) = x h_l(x)$ are Riccati–Bessel and Riccati–Hankel functions [7], respectively. The quantities $R^{(1)}_l$ and $R^{(2)}_l$ coincide with the well known Mie scattering coefficients, b_l and a_l, respectively, for a homogeneous sphere of refractive index n_0 embedded into a medium of refractive index n [3]. Equations (4.4) can be written in more compact form as

$$A^{(p)}_{\eta lm} = -R^{(p)}_l W^{(p)}_{I\eta lm}, \qquad (4.5)$$

where

$$R^{(p)}_l = \frac{(1 + \bar{n}\delta_{p1}) u'_l(x_0) u_l(x) - (1 + \bar{n}\delta_{p2}) u_l(x_0) u'_l(x)}{(1 + \bar{n}\delta_{p1}) u'_l(x_0) w_l(x) - (1 + \bar{n}\delta_{p2}) u_l(x_0) w'_l(x)} \qquad (4.6)$$

with $\bar{n} = (n_0/n) - 1$. Equation (4.5) shows that the quantities $R^{(p)}_l$ can also be interpreted as the elements of the transition matrix for a homogeneous sphere according to the equation

$$S^{(pp')}_{lml'm'} = -R^{(p)}_l \delta_{pp'} \delta_{ll'} \delta_{mm'}. \qquad (4.7)$$

Their independence of m and their diagonality in l are easily understood on the basis of trivial symmetry considerations. Moreover, although according to (3.8) and (4.5), the scattering amplitude has the form

$$f_{\eta\eta'} = \frac{i}{4\pi k} \sum_{plm} W^{(p)*}_{S\eta lm} R^{(p)}_l W^{(p)}_{I\eta' lm}, \qquad (4.8)$$

it is actually diagonal in η on account of the reciprocity theorem [3, 42]. It should be borne in mind, however, that the diagonal elements $f_{\eta\eta}$ are complex numbers with a different phase, so that the scattered wave turns out to be elliptically polarized even when the incident wave is linearly polarized.

To get an accurate representation of the scattered field, the sum in (4.8) must be extended to a sufficiently high value of l, say l_M. In other words, the convergence of the calculation must always be checked. A detailed discussion of this point has been given, e.g. by Stratton [4], who shows that, to get a fair convergence for a sphere of *size parameter* x, it is necessary to include into (4.8) terms up to $l_M > x$. In practice, when $x \leq 0.1$ one needs to consider the $l = 1$ or, at most, the terms up $l = 2$ only and, for smaller values of x, one can expand the elements of the transition matrix in powers of x thus obtaining the well known RSA (see Sect. 3.5.1) [3]. It should be borne in mind, however, that the elements of the transition matrix as well as the convergence of the scattered field depends not only on the *size parameter* but also on the refractive index n_0 (that is contained in x_0). Therefore, as long as the refractive indexes are frequency independent, the response of a spherical scatterer does not depend separately on ρ and λ, but rather on their ratio. This is the principle of *optical scaling* that allows the experimentalists to test the reliability of the theoretical predictions using microwave devices and large scale scatterers [43]. We will return on this point in Chap. 6, just when discussing the comparison of theoretical predictions with the experimental data.

We conclude this section by showing that (3.45) for the polarizability of a particle of arbitrary shape reduces to the well known Lorentz–Lorenz form when the particle at hand is a small homogeneous sphere of radius ρ and refractive index n_0. Due to spherical symmetry, the transition matrix elements are independent of m so that from (4.7) we get

$$\alpha = \frac{3}{2} i \lim_{k \to 0} \frac{1}{k^3} R_1^{(2)} \tag{4.9}$$

and thus the polarizability reduces to a scalar. It is then an easy matter to see that, by expanding $R_1^{(2)}$ in terms of the size parameter, one gets by (4.9) the Lorentz–Lorenz formula [3], which, for a homogeneous sphere, gives α in terms of the refractive index.

4.2 Radially Nonhomogeneous Spheres

The Mie theory has been extended by Wyatt [44] to radially nonhomogeneous spheres, viz., to spheres of radius ρ whose refractive index, $n_0 = n_0(r)$, is a regular function of the distance from the center. The external medium is again assumed to be homogeneous and nonabsorbing so that both the incident and the scattered field still have the multipole expansions (3.5) and (3.6), respectively. On the contrary, the expansion of the field within the sphere requires some further consideration, because, when the refractive index is space-dependent, the electric and the magnetic field do not satisfy the Helmholtz equation but rather the system

$$\nabla \times \nabla \times \boldsymbol{E} - n_0^2 k_v^2 \boldsymbol{E} = 0 , \tag{4.10a}$$
$$\nabla \times \nabla \times \boldsymbol{B} - n_0^2 k_v^2 \boldsymbol{B} = -i k_v \nabla n_0^2 \times \boldsymbol{E} , \tag{4.10b}$$

that is coupled because of the assumed nonhomogeneity of the medium. Due to the spherical symmetry of the scatterer, however, the internal field can still be expanded in a series of vector spherical harmonics, whereas the inhomogeneity of the refractive index suggests that a single radial function is not sufficient for the description of the field. Therefore, we assume that the electric and the magnetic field can be expanded in the form

$$\boldsymbol{E}_{T\eta} = \sum_{lm}\left[C^{(1)}_{\eta lm}\Phi_l(r)\boldsymbol{X}_{lm}(\hat{\boldsymbol{r}}) + C^{(2)}_{\eta lm}\frac{1}{n_0^2}\frac{1}{k_v}\nabla\times\Psi_l(r)\boldsymbol{X}_{lm}(\hat{\boldsymbol{r}})\right], \quad (4.11\text{a})$$

$$i\boldsymbol{B}_{T\eta} = \sum_{lm}\left[C^{(1)}_{\eta lm}\frac{1}{k_v}\nabla\times\Phi_l(r)\boldsymbol{X}_{lm}(\hat{\boldsymbol{r}}) + C^{(2)}_{\eta lm}\Psi_l(r)\boldsymbol{X}_{lm}(\hat{\boldsymbol{r}})\right], \quad (4.11\text{b})$$

where the radial functions $\Phi_l(r)$ and $\Psi_l(r)$ must satisfy appropriate equations. The expansions (4.11) are suggested by the fact that, for whatever choice of the radial functions $\nabla\cdot\boldsymbol{B} = 0$ and, as it is easily verified, $\nabla\cdot n_0^2\boldsymbol{E} = 0$ for any radial dependence of the refractive index. We now substitute the expansions (4.11) into (1.2a) with the result

$$\sum_{lm}\left[C^{(1)}_{\eta lm}\nabla\times\Phi_l(r)\boldsymbol{X}_{lm}(\hat{\boldsymbol{r}}) + C^{(2)}_{\eta lm}\frac{1}{n_0^2}\frac{1}{k_v}\nabla\times\nabla\times\Psi_l(r)\boldsymbol{X}_{lm}(\hat{\boldsymbol{r}})\right.$$
$$\left. + C^{(2)}_{\eta lm}\left(\frac{d}{dr}\frac{1}{n_0^2}\right)\frac{1}{k_v}\hat{\boldsymbol{r}}\times\nabla\times\Psi_l(r)\boldsymbol{X}_{lm}(\hat{\boldsymbol{r}})\right]$$
$$= \sum_{lm}\left[C^{(1)}_{\eta lm}\nabla\times\Phi_l(r)\boldsymbol{X}_{lm}(\hat{\boldsymbol{r}}) + k_v C^{(2)}_{\eta lm}\Psi_l(r)\boldsymbol{X}_{lm}(\hat{\boldsymbol{r}})\right],$$

from which, due to the independence of the vector spherical harmonics, it follows

$$\frac{1}{n_0^2}\frac{1}{k_v}\nabla\times\nabla\times\Psi_l(r)\boldsymbol{X}_{lm}(\hat{\boldsymbol{r}}) - \frac{2}{n_0^3}\frac{dn_0}{dr}\frac{1}{k_v}\hat{\boldsymbol{r}}\times\nabla\times\Psi_l(r)\boldsymbol{X}_{lm}(\hat{\boldsymbol{r}})$$
$$= k_v\Psi_l(r)\boldsymbol{X}_{lm}(\hat{\boldsymbol{r}})\ .$$

Now, according to (1.35b), $\nabla\cdot\Psi_l(r)\boldsymbol{X}_{lm}(\hat{\boldsymbol{r}}) = 0$, for whatever choice of the radial function. Therefore, with the help of the identity (1.4) we get

$$-\nabla^2\Psi_l(r)\boldsymbol{X}_{lm}(\hat{\boldsymbol{r}}) + \frac{2}{n_0}\frac{dn_0}{dr}\frac{1}{r}\frac{d}{dr}\left[r\Psi_l(r)\right]\boldsymbol{X}_{lm}(\hat{\boldsymbol{r}}) - k_v^2 n_0^2\Psi_l(r)\boldsymbol{X}_{lm}(\hat{\boldsymbol{r}}) = 0\ .$$

Since the vector spherical harmonics are eigenvectors of the angular momentum, we conclude that the radial function $\Psi_l(r)$ should be a solution, regular at $r = 0$, of the equation

$$\left[\frac{d^2}{dr^2} - \frac{l(l+1)}{r^2} + k_v^2 n_0^2 - \frac{2}{n_0}\frac{dn_0}{dr}\frac{d}{dr}\right]\left[r\Psi_l(r)\right] = 0\ . \quad (4.12)$$

4. Single and Aggregated Spheres

Analogously, by substituting the expansions (4.11) into (1.2b), we find that the radial function $\Phi_l(r)$ must be a solution, regular at $r = 0$, of the equation

$$\left[\frac{d^2}{dr^2} - \frac{l(l+1)}{r^2} + k_v^2 n_0^2\right][r\Phi_l(r)] = 0 . \tag{4.13}$$

It may be useful to stress that the radial equations (4.12) and (4.13) can also be obtained by substituting the expansions (4.11) into (4.10) instead than into the Maxwell equations (1.2). Anyway, by introducing the variable $\xi = k_v r$ and the functions

$$G_l^{(1)} = \xi \Phi_l , \quad G_l^{(2)} = \xi \Psi_l ,$$

equations (4.12) and (4.13) can be rewritten in more compact form as

$$\frac{d^2 G_l^{(p)}}{d\xi^2} - \frac{2}{n_0}\frac{dn_0}{d\xi}\frac{dG_l^{(2)}}{d\xi}\delta_{p2} + \left[n_0^2 - \frac{l(l+1)}{\xi^2}\right] G_l^{(p)} = 0 , \tag{4.14}$$

that can be integrated numerically. Needless to say that, if n_0 becomes a constant, (4.13) becomes identical to (4.12), and one can choose the normalization so that $\Phi_l = \Psi_l = j_l(n_0 k_v r)$.

We are now able to impose the boundary conditions by the same procedure that we used for homogeneous spheres. The result is

$$A_{\eta l m}^{(p)} = -R_l^{(p)} W_{I\eta l m}^{(p)} ,$$

where [45]

$$R_l^{(p)} = \frac{G_l^{(p)\prime}(x_v) u_l(x) - (1 + \bar{n}\delta_{p2})^2 G_l^{(p)}(x_v) u_l'(x)}{G_l^{(p)\prime}(x_v) w_l(x) - (1 + \bar{n}\delta_{p2})^2 G_l^{(p)}(x_v) w_l'(x)} \tag{4.15}$$

with $x_v = k_v \rho$. We stress again that, due to the spherical symmetry, the transition matrix is diagonal and its elements are independent of m.

An interesting application of the theory expounded so far is the calculation of the field scattered by spheres with a soft surface [46, 47], i.e., by spherical scatterers for which the transition from the material of the sphere to that of the surrounding medium is not sharp but occurs with a continuous variation both of the refractive index and of its radial derivative. Accordingly, we assume that in the interval $r_- \leq r \leq r_+$ the refractive index varies from n_- to n_+. By defining the quantities

$$\Delta = n_+^2 - n_-^2 , \quad s = \frac{r - r_-}{r_+ - r_-} ,$$

we can assume that, in the layer between r_- and r_+, the refractive index varies according to the formula

$$n_0^2(r) = n_-^2 + (3s^2 - 2s^3)\Delta . \tag{4.16}$$

The preceding law of variation ensures the continuity both of n_0^2 and of its radial derivative and can therefore be inserted into (4.14) for calculating the radial functions. The same technique can be used to describe the scattering from spheres composed of several concentric layers. To this end it suffices to interpose between any pair of contiguous layers a thin transition layer within which the refractive index varies according to (4.16). Our experience proved that, provided the transition layer is thin enough, the scattering properties of the spheres are not altered within numerical accuracy [46].

Finally, we note that (4.9) for the polarizability of a small sphere applies also to radially nonhomogeneous spheres provided that $R_1^{(2)}$ is given by (4.15).

4.3 Resonances

When the extinction cross section of spheres is plotted versus the size parameter, more or less sharp peaks may occur, and we say that the scattering process has undergone resonances. The origin of this phenomenon for homogeneous spheres is easily understood on the ground of the theory in Sect. 4.1. Equations (4.4), in fact, show that, when either

$$n_0 u_l'(x_0) w_l(x) - n u_l(x_0) w_l'(x) = 0 , \quad (4.17a)$$

or

$$n u_l'(x_0) w_l(x) - n_0 u_l(x_0) w_l'(x) = 0 , \quad (4.17b)$$

the amplitude of the scattered wave may become large even for small amplitude of the incident wave. In other words, spheres of radius ρ satisfying either of (4.17) support (almost) self-sustaining scattering modes even when the strength of the incident field is vanishingly small. Equations (4.17) can be satisfied by complex values of x_0 and, since the radius ρ is real, the wavevector must be complex thus yielding a damping of the field. Provided the imaginary part of x_0 is small, the damping itself will be small, and a resonance peak of finite height and width could occur. The interested reader can find the details of the analysis referred to above in the books of van de Hulst [3] and of Newton [21].

An alternative fruitful formulation of the theory of resonances both of homogeneous and radially nonhomogeneous spheres has been given by Johnson [48]. The starting point of this formulation is the remark that (4.12) becomes identical to the radial Schrödinger equation for a spherically symmetric system whose potential is

$$v_l(r) = \begin{cases} k_v^2[n^2 - n_0^2(r)] + l(l+1)/r^2 , & \text{for } 0 \leq r \leq \rho \\ l(l+1)/r^2 , & \text{for } \rho \leq r \leq \infty \end{cases} . \quad (4.18)$$

The eigenvalue is $E = k_v^2$ and, as usual, is determined by the condition of continuity of the radial function and of its radial derivative. The connection with the theory of the resonances stems from the remark that the potential function (4.18) may have a well in the vicinity of the surface of the particle for some ranges of k_v^2. In this respect, it should be borne in mind that the potential function itself is a function of the eigenvalue. Any eigenvalue that is located in the range between the top and the bottom of the well then belongs to a state in which the field can tunnel out of the potential well. This is just a resonance state in the same sense in which this term is used in quantum mechanics. Therefore, in Johnson's theory a resonance is to be interpreted as due to the shape of the potential. At this stage, using techniques that are well known in quantum mechanics, Johnson is able to give formulas for the width of the resonance for all the cases of interest for the refractive index of the sphere. In spite of its applicability to spherically symmetric scatterers only, Johnson's approach is, in our opinion, the method of choice for the interpretation of the observed resonances.

4.4 Aggregates of Spheres

The scatterers considered so far (homogeneous and radially nonhomogeneous spheres), on account of the ease of computation of their optical properties, have been widely used to attempt the interpretation of the observational data. These attempts have often been unsuccessful, however, because the particles that are most commonly met in actual observations are nonspherical to a more or less large extent and, even when their orientations are randomly distributed, their collective optical behavior differs from that of *average* spherical particles. This is not surprising because all the properties of the scattering particles are contained in their transition matrix that, according to the results of Chap. 3, is unaffected by any orientational averaging. Therefore, the effects that stem from their lack of sphericity may be attenuated but never fully cancelled by the averaging procedure. The correctness of this statement will be proved in Chap. 6 through the discussion of actual calculations.

Several attempts have been made to devise model nonspherical particles whose optical properties could be calculated as exactly as possible, i.e., without resorting to any approximation. The first real progress was marked by Bruning and Lo [49] who devised a technique to calculate the optical properties of linear chains of identical spherical scatterers. The properties of this model were investigated by Peterson and Ström [50] for general geometry of the aggregation, whereas, as far as we know, the first application of the cluster model to the description of real particles is due to Gérardy and Ausloos [51]. Here we present the procedure devised for the calculation of the transition matrix of a group of N not necessarily equal spheres whose mutual distances are so small that they must be dealt with as one object [52]. The geometry of such kind of scatterer is arbitrary to a large extent, so that aggregates can be built to model particles of various shapes, in the sense of Sect. 1.1. The emphasis is on the transition matrix on account of the usefulness of the latter for performing orientational averages. The surrounding medium is assumed to be a homogeneous dielectric so that the incident field still has the form of a polarized plane wave whose multipole expansion is given by (3.5).

We number the spheres by an index α and indicate with $\boldsymbol{R}_\alpha$ the vector position of the center of the αth sphere of radius ρ_α and refractive index n_α (see Fig. 4.1). The following theory refers to aggregates of homogeneous spheres but this assumption is hardly necessary. To deal with radially non-homogeneous spheres the theory requires only minor modifications that will be expounded later. We write the field scattered by the whole aggregate as the superposition of the fields scattered by each of the spheres in the form

$$\boldsymbol{E}_{\mathrm{S}\eta} = \sum_\alpha \sum_{plm} \boldsymbol{H}_{lm}^{(p)}(\boldsymbol{r}_\alpha, k) \mathcal{A}_{\eta\alpha lm}^{(p)} , \qquad (4.19)$$

where the amplitudes $\mathcal{A}_{\eta\alpha lm}^{(p)}$ should be calculated so that $\boldsymbol{E}_{\mathrm{S}\eta}$ satisfy the appropriate boundary conditions at the surface of each of the spheres. The radiation condition at infinity is automatically satisfied by the field (4.19), because the expansion includes $\boldsymbol{H}$-multipole fields only. The field within each sphere is in turn taken in the form

$$\boldsymbol{E}_{\mathrm{T}\eta\alpha} = \sum_{plm} \boldsymbol{J}_{lm}^{(p)}(\boldsymbol{r}_\alpha, k_\alpha) \mathcal{C}_{\eta\alpha lm}^{(p)} , \qquad (4.20)$$

with $k_\alpha = n_\alpha k_\mathrm{v}$, so that it is regular everywhere within the sphere. The magnetic field is given by (4.1).

We now remark that the scattered field is given by a linear combination of multipole fields that have different origins, whereas the incident field is given by a combination of multipole fields centered at the origin of the coordinates.

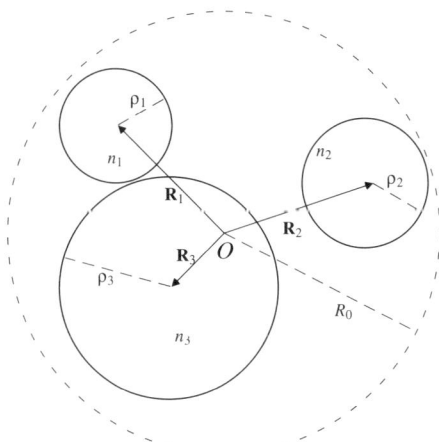

Fig. 4.1. Geometry for a cluster of three spheres. The dashed sphere separates the near-field from the far-field region

Since the boundary conditions must be imposed at the surface of each of the spheres, e.g. of the αth sphere, we resort to the addition theorem of Sect. 1.8 to rewrite the whole field in terms of multipole fields centered at $\boldsymbol{R}_\alpha$. In fact, with the help of this theorem the scattered field at the surface of the αth sphere turns out to be

$$\boldsymbol{E}_{S\eta} = \sum_{plm} \left[\boldsymbol{H}_{lm}^{(p)}(\boldsymbol{r}_\alpha, k) \mathcal{A}_{\eta\alpha lm}^{(p)} + \sum_{\alpha'} \sum_{p'l'm'} \boldsymbol{J}_{lm}^{(p)}(\boldsymbol{r}_\alpha, k) \mathcal{H}_{\alpha l m \alpha' l' m'}^{(pp')} \mathcal{A}_{\eta\alpha' l' m'}^{(p')} \right],$$

where the quantities $\mathcal{H}$ are given by (1.58a) in the notation (1.61). Analogously, the incident field at the surface of the αth sphere is

$$\boldsymbol{E}_{I\eta} = \sum_{plm} \sum_{p'l'm'} \boldsymbol{J}_{lm}^{(p)}(\boldsymbol{r}_\alpha, k) \mathcal{J}_{\alpha lm 0 l' m'}^{(pp')} W_{I\eta l' m'}^{(p')}, \qquad (4.21)$$

where $\boldsymbol{R}_0 = 0$ is the vector position of the origin, and the quantities $\mathcal{J}$ are defined in (1.58b) with the notation (1.61). At this stage, it is an easy matter to impose the boundary conditions by the same technique that we used for homogeneous spheres. Accordingly, we get for each α four equations among which the eliminations of the amplitudes of the internal field is trivial. Once this elimination is effected, we remain with the system of linear nonhomogeneous equations

$$\sum_{\alpha'} \sum_{p'l'm'} \mathcal{M}_{\alpha lm\alpha'l'm'}^{(pp')} \mathcal{A}_{\eta\alpha'l'm'}^{(p')} = -\mathcal{W}_{I\eta\alpha lm}^{(p)}, \qquad (4.22)$$

where we define

$$\mathcal{W}_{I\eta\alpha lm}^{(p)} = \sum_{p'l'm'} \mathcal{J}_{\alpha lm 0 l' m'}^{(pp')} W_{I\eta l' m'}^{(p')}, \qquad (4.23)$$

and

$$\mathcal{M}_{\alpha lm\alpha'l'm'}^{(pp')} = \left(R_{\alpha l}^{(p)}\right)^{-1} \delta_{\alpha\alpha'} \delta_{pp'} \delta_{ll'} \delta_{mm'} + \mathcal{H}_{\alpha lm\alpha'l'm'}^{(pp')}. \qquad (4.24)$$

We notice that the quantities $R_{\alpha l}^{(p)}$ are the Mie coefficients for the scattering from the αth sphere. These quantities are the only ones that must be modified in case the spheres in the cluster are radially nonhomogeneous. In this case, in fact, they must be substituted by the quantities given by (4.15) with the index α inserted where appropriate. In turn the matrix H describes the multiple scattering processes that, in view of the small mutual distance, occur with noticeable strength among the spheres of the aggregate. The occurrence of these processes is a manifestation of the coupling between spheres. As a consequence, the amplitudes of the scattered field cannot be calculated one at a time, as is instead the case for isolated spheres, but must be calculated by solving the system (4.22). We remark that the elements $\mathcal{H}_{\alpha lm\alpha'l'm'}^{(pp')}$ of the transfer matrix couple multipole fields both of the same and of different

4.4 Aggregates of Spheres

parity with origin on different spheres. This is an immediate consequence of the addition theorem for the VHHs.

We now consider the formal solution to system (4.22)

$$\mathcal{A}^{(p)}_{\eta\alpha lm} = - \sum_{p'l'm'} [\mathrm{M}^{-1}]^{(pp')}_{\alpha lm\alpha'l'm'} \mathcal{W}^{(p')}_{\mathrm{I}\eta\alpha'l'm'}. \qquad (4.25)$$

This equation may lead to the conclusion that matrix M^{-1} be the transition matrix of the aggregate. This conclusion is wrong because, according to (3.11), the transition matrix relates the multipole amplitudes of the incident field to those of the field scattered by the whole object. On the contrary, (4.25) relates the amplitudes of the incident field to those of the fields scattered by each sphere in the aggregate. In order to define the transition matrix for the whole aggregate it is necessary to express the scattered field in terms of multipole fields with the same origin. Actually, with the help of the addition theorem for the VHH, the scattered field can be cast into the form

$$\boldsymbol{E}_{\mathrm{S}\eta} = \sum_{plm}\sum_{\alpha'}\sum_{p'l'm'} \boldsymbol{H}^{(p)}_{lm}(\boldsymbol{r},k) \mathcal{J}^{(pp')}_{0lm\alpha'l'm'} \mathcal{A}^{(p')}_{\eta\alpha'l'm'} = \sum_{plm} \boldsymbol{H}^{(p)}_{lm}(\boldsymbol{r},k) A^{(p)}_{\eta lm},$$

which is valid at a large distance from the aggregate or, at least, outside the smallest sphere with center at O that includes the whole aggregate. The preceding equation shows that the field scattered by the whole cluster can be expanded as a series of vector multipole fields with a single origin (monocentered expansion) provided that the amplitudes are

$$A^{(p)}_{\eta lm} = \sum_{\alpha'}\sum_{p'l'm'} \mathcal{J}^{(pp')}_{0lm\alpha'l'm'} \mathcal{A}^{(p')}_{\eta\alpha'l'm'}. \qquad (4.26)$$

Substituting for the amplitudes $\mathcal{A}$ their expression (4.25) one is led to define the transition matrix of the aggregate as

$$\mathcal{S}^{(pp')}_{lml'm'} = -\sum_{\alpha\alpha'}\sum_{qLM}\sum_{q'L'M'} \mathcal{J}^{(pq)}_{0lm\alpha LM} [\mathrm{M}^{-1}]^{(qq')}_{\alpha LM\alpha'L'M'} \mathcal{J}^{(q'p')}_{\alpha'L'M'0l'm'}. \qquad (4.27)$$

The transition matrix defined in the preceding equation has the correct transformation properties under rotation, although it is non diagonal as a consequence of the lack of spherical symmetry of the aggregate as a whole.

The principle of electromagnetic scaling holds true also for aggregates. In fact, the elements (4.27) depend only on n, n_α, x_α and on the adimensional parameters $x_{\alpha\beta} = kR_{\alpha\beta}$ [see (1.58)]. As a result, provided all the refractive indices (i.e., those of the component spheres, n_α, and that of the external medium, n) are independent of wavelength, the scattered field turns out to depend on the parameters mentioned above, but it is independent of the absolute size of the aggregate.

Calculating the transition matrix of a cluster requires the inversion of matrix M whose order is, in principle, infinite. Naturally, the system (4.22) must be truncated to some finite order by including into the multipole expansions of the fields all the terms up to the multipole order l_I chosen so as to ensure the numerical stability of the results. l_I is the maximum value which l can take in expansions (4.19)–(4.21), so that it depends, in general, on the sphere index α. In fact, according to their scattering power, the spherical components may give considerably different contribution to the scattering of the cluster as a whole. The refinement of assigning a specific l_I for each sphere is unnecessary, however, when the scattering powers of the individual spheres are similar and one deals with compact structures. Anyway, to streamline our argument, we will assign henceforth only one value to l_I to be applied to all the spheres composing the cluster. For a cluster of N spheres this implies inverting a matrix of order $d_\mathrm{I} = 2Nl_\mathrm{I}(l_\mathrm{I} + 2)$ that may grow rather big. Actually, the inversion of matrix M is a $O(d_\mathrm{I}^3)$ process that accounts for a large share of all the calculations one is confronted with to solve this kind of scattering problem. Once matrix M^{-1} has been obtained, one has still to insert its elements into (4.27) to get the transition matrix of the aggregate. In this respect, we remark that the translation matrix J need not be a square matrix. In fact, according to (4.27), the *internal* indices L, M and L', M' are actually summed up so that $L, L' \leq l_\mathrm{I}$. In turn, the *external* indices l, m and l', m' are summed up in (3.6) and in (4.23) to get the field scattered by the whole cluster. Clearly, there is no reason to perform these sums for $l, l' \leq l_\mathrm{I}$. We could rather choose a different upper limit, say l_E, and sum over $l, l' \leq l_\mathrm{E}$. In order to reduce the computational effort required by the inversion of M, l_I must be kept as low as possible, whereas it has a substantially lower cost to be generous in the choice of l_E taking $l_\mathrm{E} \geq l_\mathrm{I}$ in order to have a better description of the scattered field.

A case in which the geometry reduces the computational effort is that of a linear cluster. Provided the centers of the spheres are chosen to lie on the $\bar{z}$ axis, matrix M turns out to be diagonal with respect to m and m' because this property pertains to matrix H according to (1.60). As a result, the inversion of M factorizes into the inversion of the matrices M_m that one obtains for each value of m. The order of M_m is $d_m = 2N(l_\mathrm{I} - |m| + 1 - \delta_{m0})$ and the whole inversion process speeds up dramatically.

As regards the choice of l_I, it should also be borne in mind that n, n_α, x_α, and $x_{\alpha\beta}$ all contribute in determining the rate of convergence. As a result, even when the choice of l_I would ensure convergent calculations for all the component spheres, the field scattered by the whole cluster may be not well convergent. Therefore, no general criterion of convergence can be given and the actual convergence should be checked for any specific application.

At last, let us remark that, when we substitute (4.27) into (3.12), the resulting expression for elements of the scattering amplitude matrix $f_{\eta\eta'}$ turns out to contain the quantities

$$\sum_{p'l'm'} \mathcal{J}^{(pp')}_{\alpha lm 0 l' m'} W^{(p')}_{\mathrm{I}\eta l' m'} \quad \text{and} \quad \sum_{plm} W^{(p)*}_{\mathrm{S}\eta lm} \mathcal{J}^{(pp')}_{0 l m \alpha' l' m'} .$$

4.4 Aggregates of Spheres

Now, on account that

$$\exp[i\boldsymbol{k}_\mathrm{I} \cdot \boldsymbol{r}] = \exp[i\boldsymbol{k}_\mathrm{I} \cdot \boldsymbol{R}_\alpha]\exp[i\boldsymbol{k}_\mathrm{I} \cdot \boldsymbol{r}_\alpha]\,,$$

by expanding $\hat{\boldsymbol{u}}_{\mathrm{I}\eta}\exp[i\boldsymbol{k}_\mathrm{I}\cdot\boldsymbol{r}]$ and $\hat{\boldsymbol{u}}_{\mathrm{I}\eta}\exp[i\boldsymbol{k}_\mathrm{I}\cdot\boldsymbol{r}_\alpha]$ we get

$$\mathcal{W}^{(p)}_{\mathrm{I}\eta\alpha lm} = \exp[i\boldsymbol{k}_\mathrm{I}\cdot\boldsymbol{R}_\alpha]W^{(p)}_{\mathrm{I}\eta lm}\,. \tag{4.28}$$

Comparison of last equation with (4.21) then yields

$$\sum_{p'l'm'} \mathcal{J}^{(pp')}_{\alpha lm 0 l'm'} W^{(p')}_{\mathrm{I}\eta l'm'} = \exp[i\boldsymbol{k}_\mathrm{I}\cdot\boldsymbol{R}_\alpha]W^{(p)}_{\mathrm{I}\eta lm}\,. \tag{4.29}$$

Substitution in (4.29) of $\hat{\boldsymbol{u}}_{\mathrm{S}\eta}$ in place of $\hat{\boldsymbol{u}}_{\mathrm{I}\eta}$ and $\boldsymbol{k}_\mathrm{S}$ in place of $\boldsymbol{k}_\mathrm{I}$ and complex conjugation then yields

$$\sum_{p'l'm'} \mathcal{J}^{(pp')*}_{\alpha lm 0 l'm'} W^{(p')*}_{\mathrm{S}\eta l'm'} = \exp[-i\boldsymbol{k}_\mathrm{S}\cdot\boldsymbol{R}_\alpha]W^{(p)*}_{\mathrm{S}\eta lm}\,.$$

Provided the refractive index n is real, the definition of J in Sect. 1.8 implies

$$\mathcal{J}^{(pp')*}_{\alpha lm 0 l'm'} = \mathcal{J}^{(p'p)}_{0 l'm'\alpha lm}\,, \tag{4.30}$$

so that

$$\sum_{plm} W^{(p)*}_{\mathrm{S}\eta lm}\mathcal{J}^{(pp')}_{0 lm\alpha' l'm'} = W^{(p')*}_{\mathrm{S}\eta l'm'}\exp[-i\boldsymbol{k}_\mathrm{S}\cdot\boldsymbol{R}_{\alpha'}]\,. \tag{4.31}$$

Using (4.30) to get $f_{\eta\eta'}$ slightly reduces the amount of calculation. Using (4.29) and (4.31) is somewhat more advantageous but at a cost: their use prevents, indeed, obtaining the transition matrix as defined by (4.27). As a result, we cannot use the averaging procedure of Sects. 3.4 and 3.6. In fact, the true transition matrix cannot be reminiscent of information either on the excitation (through the factors $\exp[i\boldsymbol{k}_\mathrm{I}\cdot\boldsymbol{R}_\alpha]$) or on observation (through the factors $\exp[-i\boldsymbol{k}_\mathrm{S}\cdot\boldsymbol{R}_{\alpha'}]$). Therefore, use of (4.29) and (4.31) is appropriate only when orientational averages are not required.

Hereafter, if different l_E and l_I are not used or when a value of l_E cannot be defined for the problem at hand, we will everywhere indicate the convergence value as l_M.

On account of the preceding considerations on convergence, one is led to conclude that, to get fairly convergent results, say to three or four decimal places, the computational effort may grow rather heavy. Nevertheless, there are several cases in which pursuing such an accuracy is pointless: in fact, as we outlined in Sect. 1.1, when one deals with a dispersion of particles, the distribution may be such that the scattering features of the single objects are smoothed out. When this is the case, achieving an accuracy smaller than the one that is mathematically possible may be sufficient to all practical effects. Actually, one can save a safe control of the degree of convergence by starting the calculations with a comparatively low value of l_M and then increasing this value until necessary to achieve the required accuracy.

4.5 Spheres Containing Spherical Inclusions

An interesting case that is often met in practice is that of a homogeneous sphere containing one or more spherical inclusions. Clearly, in case there exist a single centered inclusion the problem reduces to that of a layered sphere.

The procedure we are going to describe [53, 54] is similar to that devised by Fikioris and Uzunoglu [55]. These authors, however, limited themselves to approximate calculations appropriate for the case of a small inclusion whose refractive index differs little from that of the host sphere. A different approach has been used by Skaropoulos, Ioannidou, and Chrissoulidis [56] and by Ioannidou, and Chrissoulidis [57], but the final equations are essentially similar to those we are going to present and give identical results.

We partition the space into three regions (Fig. 4.2): the external region that is filled by a homogeneous nondispersive medium of refractive index n; the interstitial region, of radius ρ_0, filled by a homogeneous, possibly absorptive, medium of refractive index n_0; and the region within the inclusions, each of radius ρ_α, whose material has a refractive index n_α that may be radially nonhomogeneous and thus depend on r_α. The propagation constants in the regions defined above are

$$k = nk_v, \quad k_0 = n_0 k_v, \quad k_\alpha = n_\alpha k_v,$$

respectively. The field in the external region is, as usual, the superposition of the incident field and of the field scattered by the whole object. The

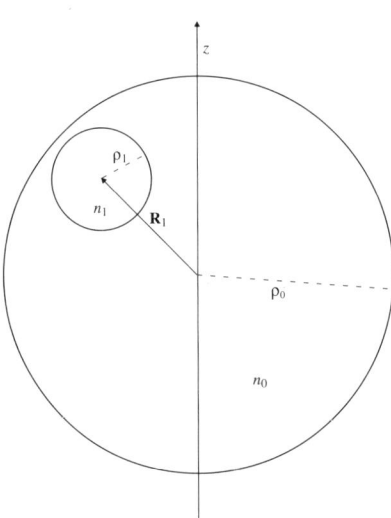

Fig. 4.2. Geometry for a sphere containing inclusions. Only the αth inclusion is shown for the sake of clarity

4.5 Spheres Containing Spherical Inclusions

respective multipole expansions are given by (3.5) and (3.6). In turn, the field within the inclusions can be expanded in the form of (4.20) provided that they are homogeneous, whereas, when they are radially nonhomogeneous, the appropriate expansion is that in (4.11) as it reads for the αth sphere. Finally, we assume that the field in the interstitial region is the superposition of the field that would exist in the absence of any inclusion and of the field scattered by the inclusions themselves. Accordingly, within the interstitial region we expand the field in the form

$$\boldsymbol{E}_{\mathrm{T}\eta 0} = E_0 \sum_{plm} \left[\sum_{\alpha} \boldsymbol{H}_{lm}^{(p)}(\boldsymbol{r}_\alpha, k_0) \mathcal{P}_{\eta\alpha lm}^{(p)} + \boldsymbol{J}_{lm}^{(p)}(\boldsymbol{r}_0, k_0) \mathcal{P}_{\eta 0 lm}^{(p)} \right] . \quad (4.32)$$

The calculation of the multipole amplitudes of the field scattered by the whole object is performed in two steps. First, we impose the boundary conditions to the internal field and to the interstitial field across the surface of each one of the inclusions. This procedure leads us to eliminating the amplitudes of the field internal to each of the inclusions and to determine the amplitudes of the interstitial field. Second, we impose the boundary conditions to the interstitial field and to the external field across the surface of the host sphere to get the amplitudes of the scattered field. Of course, both steps require using the addition theorem for multipole fields in order to get the expansions in terms of multipole fields centered at the appropriate origin. Anyway, as a result of the first step, we get a system of linear nonhomogeneous equations that we write in matrix form as

$$\begin{vmatrix} (\mathrm{R})^{-1} + \mathrm{H} & \mathrm{J}_{\leftarrow 0} \\ \mathrm{R}_W \mathrm{J}_{0\leftarrow} & (\mathrm{R}_0)^{-1} \end{vmatrix} \begin{vmatrix} \mathrm{P} \\ \mathrm{P}_0 \end{vmatrix} = \begin{vmatrix} 0 \\ \mathrm{W} \end{vmatrix} , \quad (4.33)$$

or, more compactly

$$\mathcal{MP} = \mathcal{W} .$$

To solve the system (4.33) we have to invert the matrix $\mathcal{M}$, but the actual calculation of the P's, on account of the particular form of the vector $\mathcal{W}$, that contains a subvector of zeros, involves only the rightmost columns of the inverted matrix. So, if we write $\mathcal{M}^{-1}$ in the partitioned form

$$\mathcal{M}^{-1} = \begin{vmatrix} \mathrm{Z}_1 & \mathrm{Z}_{10} \\ \mathrm{Z}_{01} & \mathrm{Z}_0 \end{vmatrix} ,$$

where all the submatrices are of the same order as the corresponding ones appearing in $\mathcal{M}$, and define the rectangular submatrix

$$\mathcal{Z} = \begin{vmatrix} \mathrm{Z}_{10} \\ \mathrm{Z}_0 \end{vmatrix} ,$$

the P's are given by the equation

$$\mathcal{P} = \mathcal{Z}\,\mathrm{W}\,.$$

Once the amplitudes P have been calculated, we proceed to the second step which yields

$$\mathrm{A} = \mathcal{T}\mathcal{P} = \mathrm{SW}\,, \tag{4.34}$$

where $\mathcal{T}$ is the rectangular matrix

$$\mathcal{T} = \begin{vmatrix} \mathrm{M_W J_{0\leftarrow}} & \mathrm{M_0} \end{vmatrix}\,.$$

In the preceding equations, matrix R characterizes the inclusions and its matrix elements are given by the equation

$$\mathcal{R}^{(pp')}_{\alpha l m \alpha' l' m'} = \delta_{pp'}\delta_{ll'}\delta_{mm'}\delta_{\alpha\alpha'}R^{(p)}_{\alpha l}\,,$$

where, in the case of homogeneous inclusions, the quantities $R^{(p)}_{\alpha l}$ are identical to those defined in Sect. 4.4 for aggregated spheres. We stress again that they are the only quantities that must be modified when the inclusions are radially nonhomogeneous. In the latter case, set aside the index α, the appropriate expression for $R^{(p)}_{\alpha l}$ is that given by (4.15). As the inclusions are embedded into the interstitial medium, in calculating $R^{(p)}_{\alpha l}$ we must use $\bar{n}_\alpha = (n_\alpha/n_0)-1$.

Matrix R_0 characterizes the external sphere and its elements are given by

$$\mathcal{R}^{(pp')}_{0\,lml'm'} = \delta_{pp'}\delta_{ll'}\delta_{mm'}\,R^{(p)}_{0l}$$

with

$$R^{(p)}_{0l} = \mathrm{i}[(1+\bar{n}_0\delta_{p1})u_l(k_0\rho_0)w'_l(k\rho_0) - (1+\bar{n}_0\delta_{p2})u'_l(k_0\rho_0)w_l(k\rho_0)]^{-1}\,.$$

In turn, matrix H describes the multiple scattering processes that occur among the inclusions, whereas the matrices $\mathrm{J}_{\leftarrow 0}$ and $\mathrm{J}_{0\leftarrow}$ describe multiple scattering processes that occur among the inclusions and the external sphere. The matrix elements of H, $\mathrm{J}_{\leftarrow 0}$, and $\mathrm{J}_{0\leftarrow}$ are $\mathcal{H}^{(pp')}_{\alpha l m \alpha' l' m'}$, $\mathcal{J}^{(pp')}_{\alpha l m 0 l' m'}$, and $\mathcal{J}^{(pp')}_{0 l m \alpha' l' m'}$, respectively. They are just the ones that we already used in Sect. 4.4 for aggregated spheres, with n_0 substituted for n, however.

Matrix $\mathrm{M_0}$ is diagonal and its elements are given by the equation

$$\mathcal{M}^{(pp')}_{0lml'm'} = \delta_{pp'}\delta_{ll'}\delta_{mm'}\,M^{(p)}_{0l}\,,$$

where

$$M^{(p)}_{0l} = \mathrm{i}[(1+\bar{n}_0\delta_{p1})u_l(k_0\rho_0)u'_l(k\rho_0) - (1+\bar{n}_0\delta_{p2})u'_l(k_0\rho_0)u_l(k\rho_0)]\,.$$

4.5 Spheres Containing Spherical Inclusions

Finally, the elements of the matrices R_W and M_W are

$$\mathcal{R}^{(pp')}_{Wlml'm'} = \delta_{pp'}\delta_{ll'}\delta_{mm'} R^{(p)}_{Wl}$$

and

$$\mathcal{M}^{(pp')}_{Wlml'm'} = \delta_{pp'}\delta_{ll'}\delta_{mm'} M^{(p)}_{Wl},$$

where

$$R^{(p)}_{Wl} = -\mathrm{i}[(1+\bar{n}_0\delta_{p1})w_l(k_0\rho_0)w'_l(k\rho_0) - (1+\bar{n}_0\delta_{p2})w'_l(k_0\rho_0)w_l(k\rho_0)]$$

and

$$M^{(p)}_{Wl} = \mathrm{i}[(1+\bar{n}_0\delta_{p1})w_l(k_0\rho_0)u'_l(k\rho_0) - (1+\bar{n}_0\delta_{p2})w'_l(k_0\rho_0)u_l(k\rho_0)].$$

Note that, in defining $R^{(p)}_{0l}$, $M^{(p)}_{0l}$, $R^{(p)}_{Wl}$, and $M^{(p)}_{Wl}$, we use $\bar{n}_0 = (n/n_0) - 1$. The preceding equations solve the problem at hand. In particular, (4.34) gives the definition of the transition matrix S and allows us to perform, when necessary, orientational averages.

When the host sphere contains a single eccentric inclusion, (4.33) and (4.34) can be formally solved to get the transition matrix. In fact, (4.33) does not contain H, that appears only when more than one inclusion is present, so that

$$S = \left(M_0 - M_W J_{0\leftarrow 1} R_1 J_{1\leftarrow 0}\right)\left(R_0^{-1} - R_W J_{0\leftarrow 1} R_1 J_{1\leftarrow 0}\right)^{-1}. \quad (4.35)$$

It may be useful to also consider the case in which there exist a single spherical inclusion concentric to the host sphere. In this case, both the matrices $J_{\leftarrow 0}$ and $J_{0\leftarrow}$ in (4.33) tend to **1** whereas the matrix H is not present. As a consequence, the system for the amplitudes P and P_0 becomes

$$\begin{vmatrix} (R)^{-1} & \mathbf{1} \\ R_W & (R_0)^{-1} \end{vmatrix} \begin{vmatrix} P \\ P_0 \end{vmatrix} = \begin{vmatrix} 0 \\ W \end{vmatrix}, \quad (4.36)$$

Equation (4.36) is easily identified with the system reported by Kerker [45] for the solution of the problem of the layered sphere.

Considerations on convergence can be made along the lines of Sect. 4.4. Although one could introduce l_E and different values for l_I, here we consider a single l_M only. Once a convenient value of l_M is determined so as to ensure the convergence of the scattered field, the order d_M of the matrix $\mathcal{M}$ to be inverted is $d_M = 2(N+1)l_M(l_M+2)$ for a sphere with N inclusions. Furthermore, when the inclusions form a linear chain lying on the diameter of the host sphere, i.e., when the scatterer as a whole has axial symmetry, the transition matrix turns out to be diagonal in m and matrix $\mathcal{M}$ factorizes into submatrices $\mathcal{M}_m$ of order $d_m = 2(N+1)(l_M - |m| + 1 - \delta_{m0})$.

The two-step procedure that was used to calculate the scattered field is a direct matching procedure. In fact, it directly uses the boundary conditions to match the multipole expansions of the field across the surfaces of the inclusions and of the host sphere. This procedure is similar to the one used by Fikioris and Uzunoglu [55]. However, we mentioned above that the scattered field has been calculated by Skaropoulos, Ioannidou, and Chrissoulidis [56] and by Ioannidou and Chrissoulidis [57] through a one-step indirect matching procedure. These authors, indeed, use multipole expansions to describe the field within each region and match them across the surfaces present by resorting to Green's second vector identity. As a result, they obtain a single matrix equation whose solution yields all the expansion coefficients, including those of the scattered field. This procedure is quite correct but requires the inversion of a bigger matrix than that implied in our two-step procedure. In fact, the one-step procedure in [56] and [57] implies the inversion of a matrix of order $d'_M = 2(N+2)l_M(l_M + 2) > d_M$. Even the one-step procedure yields factorization of the matrix to be inverted into matrices of order $d'_m = 2(N+2)(l_M - |m| + 1 - \delta_{m0}) > d_m$, when the inclusions form a linear chain along the diameter of the host sphere. Since matrix inversion is an $O(N^3)$ process, the increase in computational effort is noticeable and is not compensated by the further calculations required by (4.34) to get the scattered field in the two-step procedure. In any case, we can safely state that the variety of contributions either in $\mathcal{M}$ and $\mathcal{T}$ or in the matrix of the two-step procedure prevents establishing any rule of thumb for determining l_M.

4.6 Finite Elements Methods

In the preceding sections we discussed a set of prescriptions for calculating the field scattered by some kinds of nonspherical model particles. Of course, the transition matrix approach on which our formulas are based is only one of the methods that were devised to study the scattering properties of nonspherical particles. Moreover, when one is interested in effects that depend on the precise shape of the particles concerned, using more sophisticated methods is in order. In this respect, we recall that the transition matrix method was derived by Waterman [23] starting from the integral equation formulation of electromagnetic scattering from particles. This formulation, also called extended boundary conditions method, relies on integrals over the volume or, more often, over the surface of the particle. Provided these integrals can be performed with a sufficient accuracy, the extended boundary conditions method accounts very well for the actual shape of the particle. We do not enter into the details that are fully discussed by Waterman in his original paper. A survey of all the methods that have been developed to deal with scattering from particles of general shape has been given by Mishchenko and Travis [58]. Here, we report a brief description of two of the so-called finite elements methods, which are in common use for studying electromagnetic scattering from particles of complicated shape. Our description does not pretend to be exhaustive and is meant to help the reader to better understand both the similarities and the differences of our results in comparison with those of other researchers.

4.6.1 Discrete Dipole Approximation

The discrete dipole approximation (DDA), that is also called the coupled dipole method, is a finite elements method originally devised by Purcell and Pennypacker [59] and later improved by Draine and Flatau [60]. The DDA simulates the actual particle by an array of N polarizable points, each characterized by its own polarizability tensor whose elements may be complex. The medium among the dipoles is always the vacuum. The dipole moment acquired by the point at $\boldsymbol{r}_j$ with polarizability $\boldsymbol{\alpha}_j$ is

$$\boldsymbol{P}_j = \boldsymbol{\alpha}_j \boldsymbol{E}_j$$

where $\boldsymbol{E}_j$ is the total electric field at $\boldsymbol{r}_j$. In fact, the field $\boldsymbol{E}_j$ is the superposition of the incident wave field and of the fields due to all dipoles other than the jth. Therefore, we can write

$$\boldsymbol{P}_j = \boldsymbol{\alpha}_j \left(\boldsymbol{E}_{\mathrm{I}j} - \sum_{k \neq j} \mathrm{A}_{jk} \boldsymbol{P}_k \right),$$

where $\boldsymbol{E}_{\mathrm{I}j}$ is the incident field at $\boldsymbol{r}_j$ and $-\mathrm{A}_{jk}\boldsymbol{P}_k$ is the contribution to the field at $\boldsymbol{r}_j$ of the dipole at $\boldsymbol{r}_k$. Matrices A_{jk}, $j \neq k$, that couple to each other all the dipoles present, are given by

$$\mathrm{A}_{jk}\boldsymbol{P}_k = \frac{\exp(\mathrm{i}k_{\mathrm{v}} r_{jk})}{r_{jk}^3} \left\{ k_{\mathrm{v}}^2 \boldsymbol{r}_{jk} \times (\boldsymbol{r}_{jk} \times \boldsymbol{P}_k) \right. \\ \left. + \frac{1 - \mathrm{i}k_{\mathrm{v}} r_{jk}}{r_{jk}^2} \left[r_{jk}^2 \boldsymbol{P}_k - 3\boldsymbol{r}_{jk}(\boldsymbol{r}_{jk} \cdot \boldsymbol{P}_k) \right] \right\},$$

with $\boldsymbol{r}_{jk} = \boldsymbol{r}_j - \boldsymbol{r}_k$. Draine found it convenient to also define the matrices

$$\mathrm{A}_{jj} = \boldsymbol{\alpha}_j^{-1},$$

so that the dipole moments are the solution of the system of linear nonhomogeneous equations

$$\sum_{k=1}^{N} \mathrm{A}_{jk} \boldsymbol{P}_k = \boldsymbol{E}_{\mathrm{I}j}.$$

The scattered field, once the $\boldsymbol{P}_k$ are known, can be calculated through the well known formulas reported, e.g. by Jackson [2].

4. Single and Aggregated Spheres

In itself, the DDA is a powerful method that allows one to approximate particles of any shape by an appropriate choice of the array on which the point dipoles are located. The possible lack of homogeneity of the particles is not a problem because the polarizabilities need not be identical to each other. The absorption by the particle is accounted for by the imaginary part of the polarizability tensor. The serious problem with the DDA comes instead from the number of the dipoles that are needed to approximate adequately the particle concerned. In fact, since A_{jk} are 3×3 matrices and the dipole moments are three-dimensional vectors, the system to be solved is of order $3N$, and N itself may be of order 10^4–10^5 to approximate particles of not too complex shape. These problems have been faced by locating the dipoles on a regular cubic array, with lattice spacing d, so as to take advantage of numerical techniques based on the fast Fourier transform and on the conjugate gradient method in order to reduce the computational effort. However, the problem remains of choosing the polarizability tensor of each of the N points, the most obvious choice being the well known Lorentz–Lorenz formula [3]

$$\alpha_j^{(0)} = \frac{3d^3}{4\pi} \left(\frac{n_j^2 - 1}{n_j + 2} \right),$$

where n_j is the refractive index at the position of the jth point. Nevertheless, Draine [61] has shown that, since the scattered field must satisfy the optical theorem, the polarizability should include a radiative reaction correction of order $(k_\mathrm{v} d)^3$ so that

$$\alpha_j = \frac{\alpha^{(\mathrm{nr})}}{1 - (2/3)\mathrm{i}(\alpha^{(\mathrm{nr})}/d^3)(k_\mathrm{v} d)^3} \mathbf{1},$$

where $\alpha^{(\mathrm{nr})}$ denotes the polarizability with no radiative correction included. In his earlier work Draine chose

$$\alpha_j^{(\mathrm{nr})} = \alpha_j^{(0)},$$

thus assuming a Lorentz–Lorenz polarizability including the radiative correction. This subject was further investigated by Draine and Goodman [62] by a thorough study of the propagation of a plane wave through a cubic array of point dipoles. Their main result can be summarized by stating that the array can sustain the propagation of a plane wave provided the polarizability satisfies a well defined dispersion relation that holds true for any value of $k_\mathrm{v} d$. Of course, if the lattice is meant to mimic the dielectric behavior of a continuous medium, as is the case with the DDA, one can consider the series expansion of dispersion relation in the long wavelength limit. As a result, to zero order one recovers the Lorentz–Lorenz formula and to first order the equation that includes the radiative correction.

Once a reliable choice of the polarizability has been made, the next problem with the DDA is to asses its range of validity. To this end the predictions of the DDA for homogeneous spheres have been compared with the analytic Mie theory solutions. The scattering from tetrahedra, for which no analytic solution is known, has instead been compared with the results of FDTD (see Sect. 4.6.2 below). The conclusions that stem from all these comparisons suggest that, provided the number of the dipoles N is sufficiently big to ensure good fitting of the shape of the particle, the interdipole distance must satisfy the condition

$$|n_0| k_v d \leq 1$$

where n_0 is the refractive index of the particle. In particular, the scattered field is described with best accuracy when $|n_0 - 1| \leq 1$. In practice, when the refractive index is large, using the DDA requires great care in checking the reliability of the results.

The above considerations might convince the reader that the DDA should be the method of choice when dealing with scattering by irregularly shaped particles. Nevertheless, unlike the transition matrix approach, the DDA does not imply any simple relation between the particle orientation and the scattered field. Therefore, when orientational averages, such as those discussed in Chap. 3, are needed, it is necessary to calculate anew the scattered field for each of a sufficiently large number of orientations of the particle. As a result, the computational effort may soon become unsustainable.

4.6.2 Finite Difference Time Domain Method

Most of the methods that are in common use for calculating the scattered field from particles seek for solutions of the Maxwell equations in the frequency domain. From a mathematical point of view, Maxwell equations in the frequency domain are elliptic, so that the imposition of the boundary conditions is essential for finding solutions that describe actual wave propagation. On the contrary, the finite difference time domain method (FDTD) is a finite-elements method based on the numerical integration of the Maxwell equations in the time domain [63]. This means that the solution to the problem of scattering is dealt with as an initial value problem and the incident field need not be a monochromatic wave. In principle, boundary conditions are not necessary because in the time domain the Maxwell equations are hyperbolic [64].

The region of space containing the scattering particle is discretized by defining a regular array of cells whose vertices, commonly referred to as the nodes, are individuated by three integers such that the vector position of node (i, j, k) is

$$\boldsymbol{r}_{ijk} = i \Delta x \, \hat{\boldsymbol{e}}_x + j \Delta y \, \hat{\boldsymbol{e}}_y + k \Delta z \, \hat{\boldsymbol{e}}_z \,.$$

The particle is then characterized by assigning the value of the refractive index (or of the dielectric function) at each node within the volume occupied by the particle itself. This procedure, which is similar to the one used in the DDA, does not require the particle to be homogeneous or to have a regular shape. Each space- and time-dependent variable, such as the components of the fields, is discretized as

$$F(\boldsymbol{r},t) \to F_\mathrm{d}(i,j,k;n) = F(i\Delta x, j\Delta y, k\Delta z; n\Delta t) ,$$

and this procedure is used to write the discretized version of the Maxwell curl equations to be integrated numerically.

The actual situation is more complicated than stated above. First, in order to ensure the stability of the numerical integration of the Mawxell equations the space and time increments must satisfy the condition [65]

$$c\Delta t \leq \left[1/(\Delta x)^2 + 1/(\Delta y)^2 + 1/(\Delta z)^2\right]^{-1/2}$$

that sets a lower limit to the number of nodes that are necessary to get reliable results.

Second, any scattering process occurs in infinite space, whereas, for evident computational reasons, the integration of the discretized Maxwell equations must be performed within a finite region. It is therefore necessary to introduce appropriate boundary conditions across the surface which limits the region of integration to prevent unphysical reflections of the field that would otherwise modify, even severely, the near field [65, 66, 67].

Third, the integration procedure must preserve the continuity of the appropriate components of the fields when going from node to node. To this end a cubic cell is considered around each node with vertices at $(i\pm\frac{1}{2}, j\pm\frac{1}{2}, k\pm\frac{1}{2})$. Then, the components of the field are selected so that the electric field is calculated at the center of the edges of the cell, whereas the magnetic field is calculated at the center of the faces of the cell [63]. This choice ensures that in going from cell to cell the orthogonal components of the magnetic field and the tangential components of the electric field are continuous. It is only at this stage that one can proceed to the integration of the discretized Maxwell equations. In practice, given the field at $t = 0$, one is confronted with the numerical integration of a system of six coupled finite-difference equations for the space-time propagation of the field. Six schemes of integrations have been devised, each of which is applicable to one of the six rules that are used to approximate the derivatives of the field components. Once a scheme is selected, one gets the field within the whole region of integration as a function of time.

Of course, the field calculated as outlined above is the near field, whereas to calculate the quantities of interest, such as the cross sections, one needs the far field. This task is faced by means of Green's second vector identity that ensures the matching of the near and the far field through an integration on any suitable surface containing the particle [68, 69].

Finally, let us recall that, since the incident field is, in general, nonmonochromatic, the far field also is nonmonochromatic. Therefore, in order to get the frequency response of the particle one has to resort to a discrete Fourier transform, so that the results of the FDTD method can be compared with those of more conventional methods that yield the field in the frequency domain. The results of such comparisons prove that the FDTD method achieves an extremely good precision in the calculation of the fields [70]. Nevertheless, the FDTD requires a computational effort even bigger than required by the DDA method. Moreover, even in this case there is no simple relation between the orientation of the particle and the scattered field so that the computational demand may soon become unsustainable.

5. Scattering from Particles on a Plane Surface

In this chapter, we study the scattering of electromagnetic waves from particles near to or deposited on the plane surface that separates two homogeneous media of different optical properties. The plane surface will be termed *the interface*, whereas we will call *accessible half-space* the one in which the particles and the detecting instrumentation are located. The medium that fills the accessible half-space is assumed to be nonabsorptive and to have, therefore, a *real* refractive index.

The presence of the interface has, in general, a striking effect on the scattering pattern from the particles. Indeed, unlike the case of particles in free space, the exciting field does not coincide with the incident plane wave and the observed field does not coincide with the field scattered by the particle. The field that illuminates the particles is partly or totally reflected by the interface and the reflected field contributes both to the exciting and to the observed field. Moreover, the field scattered by the particles is reflected by the interface and thus contributes to the exciting field. In other words there are multiple scattering processes between the particles and the interface. As a result, the field in the accessible half-space includes the incident field $\boldsymbol{E}_\mathrm{I}$, the reflected field $\boldsymbol{E}_\mathrm{R}$, the scattered field $\boldsymbol{E}_\mathrm{S}$ and, finally, the field $\boldsymbol{E}_\mathrm{SR}$ that after scattering by the particles is reflected by the surface. Therefore, a detecting instrument in the far zone within the accessible half-space records, in general, the superposition of $\boldsymbol{E}_\mathrm{S}$ and $\boldsymbol{E}_\mathrm{SR}$, whereas the reflected field will be recorded in the direction of reflection only.

The mathematical difficulties that are met in calculating the scattering pattern are due to the need that the field in the accessible half-space satisfy the boundary conditions both across the (closed) surface of the particles and across the (infinite) interface. In other words, even by assuming that we are able to impose the boundary conditions across the surface of the particle, the problem still remains of imposing the boundary conditions across the interface. In this respect, we remark that the particles may be in contact with the interface, so that the boundary conditions must be imposed to the near field, that, as remarked in Sect. 2.1, has a complex space dependence. Moreover, the form of the boundary conditions across the interface depends on whether the medium that fills the nonaccessible half-space is a dielectric or a metal. For this reason we will deal separately with the case of a metallic surface [71] and with that of a dielectric substrate [72, 73, 74]. Nevertheless, in both cases, it is possible to define the transition matrix for particles in the presence of the interface. Of course, this amounts to the possibility of performing averages over the orientation of the particles with a reduced computational effort.

5.1 Incident and Reflected Fields

The reflection of a plane wave on a plane surface can be dealt with in general terms, i.e., without specifying whether the medium that fills the nonaccessible half-space is a dielectric or a metal. This information can, indeed, be supplied at the end of the algebraic manipulations. Let us thus assume that the interface is the plane $z = 0$ of a Cartesian frame of reference and that the half-space $z < 0$, which we take as the accessible half-space, is filled by a homogeneous medium of (real) refractive index n'. The half-space $z > 0$ is assumed to be filled by a homogeneous medium with (possibly complex) refractive index n''. The plane wave field

$$\boldsymbol{E}_\mathrm{I} = E_0 \hat{\boldsymbol{e}}_\mathrm{I} \exp(\mathrm{i}\boldsymbol{k}_\mathrm{I} \cdot \boldsymbol{r}) ,$$

which propagates within the accessible half-space, is reflected by the interface into the plane wave

$$\boldsymbol{E}_\mathrm{R} = E_0' \hat{\boldsymbol{e}}_\mathrm{R} \exp(\mathrm{i}\boldsymbol{k}_\mathrm{R} \cdot \boldsymbol{r}) ,$$

where $\boldsymbol{k}_\mathrm{I} = k'\hat{\boldsymbol{k}}_\mathrm{I}$ and $\boldsymbol{k}_\mathrm{R} = k'\hat{\boldsymbol{k}}_\mathrm{R}$ are the propagation vectors of the incident and of the reflected wave, respectively, $k' = n'k_\mathrm{v}$ and $\hat{\boldsymbol{e}}_\mathrm{I}$ and $\hat{\boldsymbol{e}}_\mathrm{R}$ are the respective unit polarization vectors. The polarization is analyzed with respect to the two pairs of unit vectors $\hat{\boldsymbol{u}}_{\mathrm{I}\eta}$ and $\hat{\boldsymbol{u}}_{\mathrm{R}\eta}$ that are parallel ($\eta = 1$) and perpendicular ($\eta = 2$) to the plane of incidence that, as usual, is defined as the plane that contains $\boldsymbol{k}_\mathrm{I}$, $\boldsymbol{k}_\mathrm{R}$ and the z axis. Our choice of the orientation is defined by the equations

$$\hat{\boldsymbol{u}}_{\mathrm{I}1} \times \hat{\boldsymbol{u}}_{\mathrm{I}2} = \hat{\boldsymbol{k}}_\mathrm{I} , \qquad \hat{\boldsymbol{u}}_{\mathrm{R}1} \times \hat{\boldsymbol{u}}_{\mathrm{R}2} = \hat{\boldsymbol{k}}_\mathrm{R} ,$$

with $\hat{\boldsymbol{u}}_{\mathrm{I}2} \equiv \hat{\boldsymbol{u}}_{\mathrm{R}2}$. In terms of the projections on the polarization basis, the incident and the reflected field can be written

$$\boldsymbol{E}_\mathrm{I} = E_0 \sum_\eta (\hat{\boldsymbol{e}}_\mathrm{I} \cdot \hat{\boldsymbol{u}}_{\mathrm{I}\eta}) \hat{\boldsymbol{u}}_{\mathrm{I}\eta} \exp(\mathrm{i}\boldsymbol{k}_\mathrm{I} \cdot \boldsymbol{r}) ,$$

and

$$\boldsymbol{E}_\mathrm{R} = E_0' \sum_\eta (\hat{\boldsymbol{e}}_\mathrm{R} \cdot \hat{\boldsymbol{u}}_{\mathrm{R}\eta}) \hat{\boldsymbol{u}}_{\mathrm{R}\eta} \exp(\mathrm{i}\boldsymbol{k}_\mathrm{R} \cdot \boldsymbol{r}) .$$

In the preceding equations the incident field $\boldsymbol{E}_\mathrm{I}$ and the reflected field $\boldsymbol{E}_\mathrm{R}$ are decomposed into their components parallel and orthogonal to the plane of incidence and can be referred to each other by means of the Fresnel coefficients F_η for the reflection of a plane wave with polarization along $\hat{\boldsymbol{u}}_\eta$. The reflection condition is imposed as usual [2] and yields the equation

$$E_0'(\hat{\boldsymbol{e}}_\mathrm{R} \cdot \hat{\boldsymbol{u}}_{\mathrm{R}\eta}) = E_0 F_\eta(\vartheta_\mathrm{I})(\hat{\boldsymbol{e}}_\mathrm{I} \cdot \hat{\boldsymbol{u}}_{\mathrm{I}\eta}) \,, \tag{5.1}$$

where the Fresnel coefficients are defined as

$$F_1(\vartheta_\mathrm{I}) = \frac{\bar{n}^2 \cos\vartheta_\mathrm{I} - \beta}{\bar{n}^2 \cos\vartheta_\mathrm{I} + \beta}\,, \quad F_2(\vartheta_\mathrm{I}) = \frac{\cos\vartheta_\mathrm{I} - \beta}{\cos\vartheta_\mathrm{I} + \beta}\,, \tag{5.2}$$

in which ϑ_I is the angle between $\hat{\boldsymbol{k}}_\mathrm{I}$ and the z axis, $\bar{n} = n''/n'$, and

$$\beta = \left[(\bar{n}^2 - 1) + \cos^2\vartheta_\mathrm{I}\right]^{1/2} \,.$$

Using (5.1), the reflected wave can be rewritten as

$$\boldsymbol{E}_\mathrm{R} = E_0 \sum_\eta F_\eta(\vartheta_\mathrm{I})(\hat{\boldsymbol{e}}_\mathrm{I} \cdot \hat{\boldsymbol{u}}_{\mathrm{I}\eta}) \hat{\boldsymbol{u}}_{\mathrm{R}\eta} \exp(\mathrm{i}\boldsymbol{k}_\mathrm{R} \cdot \boldsymbol{r}) \,.$$

We expand both the incident and the reflected field in terms of spherical vector multipole fields along the lines of Sect. 3.1. The result is

$$\boldsymbol{E}_\mathrm{I} = E_0 \sum_\eta (\hat{\boldsymbol{e}}_\mathrm{I} \cdot \hat{\boldsymbol{u}}_{\mathrm{I}\eta}) \sum_{plm} \boldsymbol{J}_{lm}^{(p)}(\boldsymbol{r},k') W_{\mathrm{I}\eta lm}^{(p)}\,,$$

$$\boldsymbol{E}_\mathrm{R} = E_0 \sum_\eta F_\eta(\vartheta_\mathrm{I})(\hat{\boldsymbol{e}}_\mathrm{I} \cdot \hat{\boldsymbol{u}}_{\mathrm{I}\eta}) \sum_{plm} \boldsymbol{J}_{lm}^{(p)}(\boldsymbol{r},k') W_{\mathrm{R}\eta lm}^{(p)}\,,$$

where

$$W_{\mathrm{I}\eta lm}^{(p)} = W_{lm}^{(p)}(\hat{\boldsymbol{u}}_{\mathrm{I}\eta}, \hat{\boldsymbol{k}}_\mathrm{I})\,, \quad W_{\mathrm{R}\eta lm}^{(p)} = W_{lm}^{(p)}(\hat{\boldsymbol{u}}_{\mathrm{R}\eta}, \hat{\boldsymbol{k}}_\mathrm{R})\,,$$

and the amplitudes $W_{lm}^{(p)}(\hat{\boldsymbol{e}}, \hat{\boldsymbol{k}})$ are given by (3.4). Because of the reflection condition, the amplitudes $W_{\mathrm{I}\eta lm}^{(p)}$ and $W_{\mathrm{R}\eta lm}^{(p)}$ cannot be mutually independent. Indeed, if ϑ_I and φ_I are the polar angles of $\hat{\boldsymbol{k}}_\mathrm{I}$, we have

$$\hat{\boldsymbol{u}}_{\mathrm{I}1} \equiv \left(\vartheta_\mathrm{I} + \frac{\pi}{2}, \varphi_\mathrm{I}\right)\,, \quad \hat{\boldsymbol{u}}_{\mathrm{I}2} \equiv \left(\frac{\pi}{2}, \varphi_\mathrm{I} + \frac{\pi}{2}\right)\,,$$

and

$$\hat{\boldsymbol{k}}_\mathrm{R} \equiv (\pi - \vartheta_\mathrm{I}, \varphi_\mathrm{I})\,, \quad \hat{\boldsymbol{u}}_{\mathrm{R}1} \equiv \left(\vartheta_\mathrm{I} + \frac{\pi}{2}, \varphi_\mathrm{I} + \pi\right)\,, \quad \hat{\boldsymbol{u}}_{\mathrm{R}2} \equiv \left(\frac{\pi}{2}, \varphi_\mathrm{I} + \frac{\pi}{2}\right)\,.$$

Substituting these relations into the expression for $W^{(p)}_{R\eta lm}$ and using the transformation rules of the spherical harmonics under change of their argument, we get

$$W^{(p)}_{R\eta lm} = (-)^{\eta+p+l+m} W^{(p)}_{I\eta lm}, \tag{5.3}$$

so that the amplitudes of the reflected field never need to be explicitly considered.

In practice, for further developments, it is convenient to define the exciting field as the superposition

$$\boldsymbol{E}_{\mathrm{E}} = \boldsymbol{E}_{\mathrm{I}} + \boldsymbol{E}_{\mathrm{R}}, \tag{5.4}$$

whose multipole expansion can be written as

$$\boldsymbol{E}_{\mathrm{E}\eta} = E_0 \sum_{plm} \boldsymbol{J}^{(p)}_{lm}(\boldsymbol{r}, k') W^{(p)}_{\mathrm{E}\eta lm} \tag{5.5}$$

with

$$W^{(p)}_{\mathrm{E}\eta lm} = [1 + F_\eta(\vartheta_\mathrm{I})(-)^{\eta+p+l+m}] W^{(p)}_{I\eta lm}. \tag{5.6}$$

5.2 Perfectly Reflecting Surface

A well-polished metallic surface is a good approximation of a perfectly reflecting surface. Owing to the high conductivity of the metal, the tangential component of the electric field vanishes on the surface. This is, indeed, the actual boundary condition for perfect reflection. In this case the problem of scattering can be solved by resorting to *image theory*, according to which the field scattered into the accessible half-space by a particle in the presence of the reflecting surface coincides with the field scattered by the compound object that includes the particle and its image (hereafter referred to as the *effective scatterer*), provided the *exciting field* is the superposition of the actual incident field and of the field that comes from the image source. Since the latter field coincides with field reflected by the surface, the exciting field is the superposition (5.4) with multipole expansion given by (5.5) whose amplitude is obtained by substituting into (5.6) the limiting values of the Fresnel coefficients

$$F_\eta(\vartheta_\mathrm{I}) = (-)^{\eta-1},$$

which are appropriate for the case of a perfectly reflecting surface, i.e., for $n'' \to \infty$. Therefore, the resulting amplitudes are

$$W^{(p)}_{\mathrm{E}\eta lm} = [1 - (-)^{p+l+m}] W^{(p)}_{I\eta lm}. \tag{5.7}$$

Of course, image theory reduces the complexity of the original problem provided one is able to solve the problem of scattering by the effective scatterer that must be considered as a single particle. Therefore, in no case can the scattered field be considered as the superposition of the fields that the actual particle and its image scatter independently. The process of scattering must be considered as a dependent scattering problem. Assuming, for instance, that the actual particle is a sphere, the problem becomes that of scattering by an aggregate of two identical spheres. This problem can be solved by the techniques expounded in Chap. 4. In case the actual object were itself an aggregate of spheres, the problem to be solved would be that for a cluster including twice the spheres as the original aggregate.

When the interface is perfectly reflecting, a further problem can be solved, namely, that of a hemisphere with its flat face on the surface. The effective scatterer is, in fact, a sphere whose scattering properties can be calculated through the well known Mie theory. The fact that the exciting field is not a simple plane wave but the superposition defined in (5.4) adds no further difficulty. Of course, nothing prevents the actual hemisphere from being radially nonhomogeneous, or from containing spherical inclusions or hemispherical inclusions whose flat face lies on the reflecting surface. The effective scatterer is, indeed, a sphere radially nonhomogeneous or containing one or more spherical inclusions (see Fig. 5.1).

With this in mind, we expand the field scattered by the effective scatterer in a series of $\boldsymbol{H}$-multipole fields, so that the radiation condition at infinity is fulfilled. The required expansion is

$$\boldsymbol{E}_{\mathrm{S}\eta} = \sum_{plm} \boldsymbol{H}_{lm}^{(p)}(\boldsymbol{r}, k') A_{\eta lm}^{(p)} ,$$

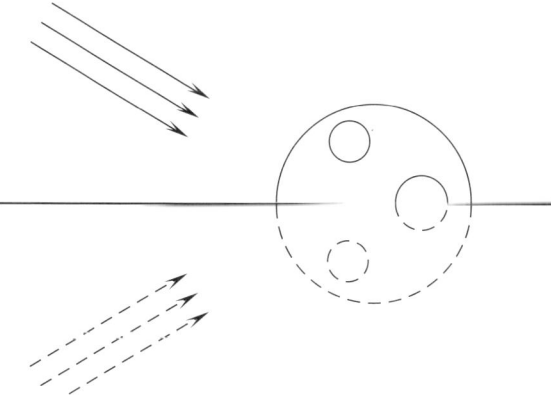

Fig. 5.1. Effective scatterer for the case of a hemisphere containing a hemispheric inclusion and a spherical eccentric inclusion. The images are characterized by *dashed lines*

and, according to our earlier remark, is valid only within the accessible half-space. In the spirit of the transition matrix approach, the amplitudes of the scattered field are related to those of the exciting field by the equation

$$A^{(p)}_{\eta lm} = \sum_{p'l'm'} S^{(pp')}_{lml'm'} W^{(p')}_{E\eta l'm'},$$

where the transition matrix is that of the effective scatterer.

Once the scattered field has been calculated, the scattered intensity is given, as usual, by the matrix

$$I_{\eta\eta'} = \left| \frac{E_0}{r} f_{\eta\eta'} \right|^2.$$

Of course, the scattering amplitude that we consider here is that of the effective scatterer. The elements $f_{\eta\eta'}$ are related to the elements of the transition matrix through the equation

$$f_{\eta\eta'} = -\frac{i}{4\pi k'} \sum_{plm} \sum_{p'l'm'} W^{(p)*}_{S\eta lm} S^{(pp')}_{lml'm'} W^{(p')}_{E\eta'l'm'}. \tag{5.8}$$

It may be worth noticing that the direction of observation need not lie in the plane of incidence and that (5.8), except for the presence of the amplitudes of the exciting field in place of those of the incident field, is identical to (3.12) for particles in free space. We also notice that the convergence considerations that we made in Sect. 4.4 still apply, provided the number of the spheres that we consider be that appropriate for the effective scatterer.

5.2.1 Orientational Averages

If one has to calculate the scattering pattern from a single particle in the vicinity of the reflecting surface, (5.8) is quite efficient as its form shows that, once the transition matrix is known, the scattered intensity can be calculated without undue computational effort for any $\hat{k}_I$ and $\hat{k}_S$ and for whatever choice of the polarization both of the incident and the scattered field.

When one is interested in the pattern of the scattered radiation from an assembly of identical scatterers in front of the surface, the distribution of their orientations must be taken into account. This can be done, as explained in Sect. 3.3, by associating to each particle a local system of axes chosen so that, when two particles are brought into coincidence, their respective local axes coincide. Then, the components of the scattering amplitude of the νth particle of orientation Θ_ν are given by

$$f_{\nu,\eta\eta'} = -\frac{i}{4\pi n'k} \sum_{plm} \sum_{p'l'm'} \sum_{\mu\mu'} W^{(p)*}_{S\eta lm} \mathcal{D}^{(l)*}_{\mu m}(\Theta_\nu) \bar{S}^{(pp')}_{l\mu l'\mu'} \mathcal{D}^{(l')}_{\mu' m'}(\Theta_\nu) W^{(p')}_{E\eta'l'm'}, \tag{5.9}$$

where $\Theta_\nu \equiv (\alpha_\nu, \beta_\nu, \gamma_\nu)$ is a shorthand for the three Euler angles that characterize the orientation of the local frame of the νth particle with respect to the laboratory frame. Again we stress that (5.9) is identical to (3.23) except for the substitution of the amplitudes of the exciting field in place of those of the incident field. In fact, even in the presence of the interface, both $W^{(p)*}_{S\eta lm}$ and $W^{(p')}_{E\eta' l'm'}$ in (5.9) need to be calculated in the laboratory frame, whereas the elements of the transition matrix $\bar{S}^{(pp')}_{l\mu l'\mu'}$ are calculated in the local frame and are thus independent of the orientation of the scatterer. Ultimately, (5.9) shows that, even when dealing with an assembly of identical scatterers in the presence of a perfectly reflecting surface, we need to calculate the transition matrix only once.

We have now all the ingredients that are necessary to calculate the scattered intensity from an assembly of identical particles in the vicinity of the interface. It is to be stressed, however, that the applicability of the method of the images requires that we consider a monolayer of particles that are so arranged that all the effective scatterers are identical to each other and that the orientations of any two of them differ for a rotation around an axis that is perpendicular to the surface. A moment's thought and a look to Fig. 5.2 will convince the reader of the need for this restriction on the orientation of the actual particles in spite of their identity. With this in mind, we can write the total scattered intensity as

$$\bar{I}_{\eta\eta'} = \left(\sum_\nu E^*_{\nu, \eta\eta'}\right)\left(\sum_{\nu'} E_{\nu', \eta\eta'}\right)$$
$$= N\langle|E_{\eta\eta'}(\boldsymbol{R}, \Theta)|^2\rangle + (N^2 - N)\langle E^*_{\eta\eta'}(\boldsymbol{R}, \Theta)E_{\eta\eta'}(\boldsymbol{R}', \Theta')\rangle,$$

where ν and ν' are, as usual, particle indexes, N is the number of the particles on the reflecting surface, $E_{\eta\eta'}(\boldsymbol{R}, \Theta)$ is the scattered field from a particle at

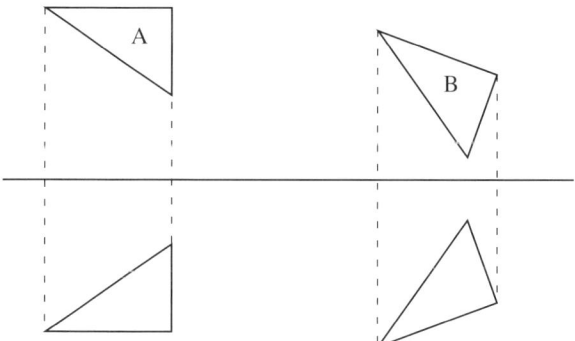

Fig. 5.2. Particles A and B are identical, but the respective effective scatterers they form are not because of the different distance and tilt of A and B with respect to the interface

the position $\boldsymbol{R}$ with orientation Θ, and the brackets denote the ensemble average over both the position and the orientation of the particles. The first term on the right-hand side, the so-called self term, contributes to the scattered intensity for any direction of observation. Even the second term, that is a two-body term, may become important for nonrandom distributions of the scatterers. On the contrary, when they are randomly distributed upon the surface, it does contribute in the direction of reflection only [36]. Ultimately, as long as we deal just with monolayers of randomly distributed scatterers, we will disregard the two-body term so that $\bar{I}_{\eta\eta'}$ reduces to

$$\bar{I}_{\eta\eta'} = N\langle I_{\eta\eta'}\rangle = N \int I_{\eta\eta'}(\boldsymbol{R},\Theta) P(\boldsymbol{R},\Theta)\,\mathrm{d}\boldsymbol{R}\,\mathrm{d}\Theta\;,$$

where $P(\boldsymbol{R},\Theta)$ is the probability that a particle has position within $\boldsymbol{R}$ and $\boldsymbol{R} + \mathrm{d}\boldsymbol{R}$ and orientation within Θ and $\Theta + \mathrm{d}\Theta$. Since we assume that there is no correlation between the position and the orientation of the scatterers, P can be factorized as $P(\boldsymbol{R},\Theta) = P_{\boldsymbol{R}}(\boldsymbol{R})P_{\Theta}(\Theta)$. Furthermore, the distance of observation is assumed, as usual, to be very large with respect to the size of the reflecting surface, so that we can put $\boldsymbol{R} = 0$ and write

$$\langle I_{\eta\eta'}\rangle = \int I_{\eta\eta'}(\boldsymbol{R},\Theta) P_{\boldsymbol{R}}(\boldsymbol{R}) P_{\Theta}(\Theta)\,\mathrm{d}\boldsymbol{R}\,\mathrm{d}\Theta$$
$$\approx \int P_{\boldsymbol{R}}(\boldsymbol{R})\,\mathrm{d}\boldsymbol{R} \int I_{\eta\eta'}(0,\Theta) P_{\Theta}(\Theta)\,\mathrm{d}\Theta = \int I_{\eta\eta'}(0,\Theta) P_{\Theta}(\Theta)\,\mathrm{d}\Theta$$

on account that

$$\int P_{\boldsymbol{R}}(\boldsymbol{R})\,\mathrm{d}\boldsymbol{R} = 1\;.$$

We stated above that the orientation of any two particles may differ only by a rotation around an axis that is orthogonal to the surface. Thus, by choosing the z axis of the local reference frame that we attached to each particle to coincide with that axis, $\beta = \gamma = 0$, and the average involves the Euler angle α only. This kind of orientational distribution has already been considered in Sect. 3.4 as Case 4 in Table 3.1. Accordingly, the total scattered intensity is

$$\bar{I}_{\eta\eta'} = N\langle I_{\eta\eta'}\rangle\;,$$

where the average $\langle I_{\eta\eta'}\rangle$ is given by (3.32c) provided k' is substituted in place of k and $W_{\mathrm{E}\eta}$ in place of $W_{\mathrm{I}\eta}$. Let us also recall that $\bar{S}$ is not the transition matrix of the actual scatterer but must be that of the effective scatterer.

The average over the orientations is, of course, not necessary and is not performed in actual calculations, in two important cases: when all the effective scatterers are oriented alike or when they have cylindrical symmetry around an axis perpendicular to the surface. However, even in these two special cases our orientational averaging procedure yields the correct results as we already remarked at the end of Sect. 3.4.

5.3 Dielectric Half-Space

When the half-space $z > 0$ is filled by a homogeneous dielectric material, the interface is not perfectly reflecting, and the field can propagate also within this half-space. When this occurs, the reflection condition for the vector spherical multipole fields must be reformulated. Obviously no problem exists for the reflection of the plane waves and thus neither for the $\boldsymbol{J}$-multipole fields that enter their expansion. This has already been shown in Sect. 5.1. On the contrary, the problem arises when we have to reflect the field scattered by a particle located within the accessible half-space. This field is indeed a superposition of $\boldsymbol{H}$-multipole fields whose reflection rule we are able to establish in general terms [11].

5.3.1 Reflection Rule for $\boldsymbol{H}$-Multipole Fields

We start by extending to the vector case the formula

$$h_l(kr)Y_{lm}(\vartheta,\varphi) = \frac{(-\mathrm{i})^l}{2\pi} \int_0^{2\pi} \mathrm{d}\varphi_\mathrm{k} \int_0^{\pi/2-\mathrm{i}\infty} \mathrm{d}\vartheta_\mathrm{k} \sin\vartheta_\mathrm{k}\, Y_{lm}(\vartheta_\mathrm{k},\varphi_\mathrm{k})\, \mathrm{e}^{\mathrm{i}\boldsymbol{k}\cdot\boldsymbol{r}} , \tag{5.10}$$

which Bobbert and Vlieger [75] obtained by extending to $l > 0$ an integral expression originally devised by Weyl [76]. Equation (5.10) expresses in the half-space $z > 0$ the scalar multipole field $h_l(kr)Y_{lm}(\vartheta,\varphi)$ as a superposition of plane waves with complex propagation vectors (inhomogeneous plane waves); the limits of integration are such that the real part of $\boldsymbol{k}$ points towards the half-space $z > 0$, so that the multipole field result from the superposition of plane waves coming from a point source at the origin and propagating in the half-space $z > 0$. Should the need arise, it would be possible to write an analogous expansion valid in the half-space $z < 0$ by considering the superposition of plane waves the real part of whose propagation vector points towards this half-space.

The geometry that we adopt is depicted in Fig. 5.3. Accordingly, we define a Cartesian frame of reference whose origin O lies on the plane interface between two media. Without loss of generality, we can assume that the interface coincides with the plane $z = 0$ and that a homogeneous medium of refractive index n' fills the half-space $z < 0$, whereas a different homogeneous medium of refractive index n'' fills the half-space $z > 0$ (see Fig. 5.3). We assume that a source of $\boldsymbol{H}$-multipole fields lies entirely within the half-space $z < 0$ and define a further frame of reference that is translated with respect to O and whose origin, O', lies within the source at a distance d from the interface. For our purposes, it is also convenient to define a third frame of reference that is also translated with respect to O and whose origin, O'', is the mirror image of O' with respect to the interface. We will denote with $\boldsymbol{R}'$ and $\boldsymbol{R}''$ the vector position of the origins O' and O'' in the frame of

104   5. Particles on a Plane Surface

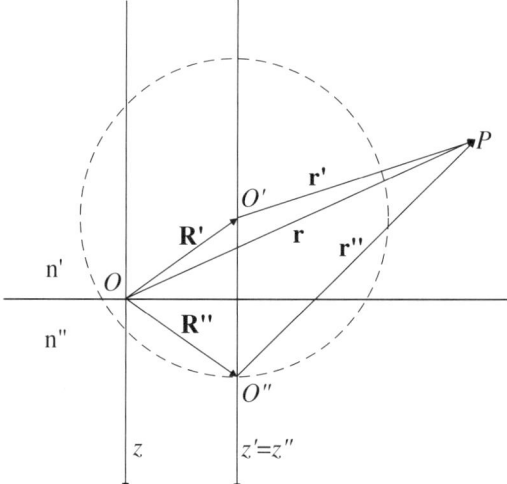

Fig. 5.3. Geometry for the reflection of multipole fields

reference with origin at O, respectively, whereas the vector position of the point of observation P in the three frames defined above will be denoted with r, r', and r'', respectively. Then, we recall the definitions of the H-multipole fields in terms of irreducible spherical tensors [see (1.45) and (1.46)] and the definition of the spherical tensors themselves, (1.28), and apply (5.10) to each of the relevant scalar multipole fields. As a result, we obtain for the expansion of the vector multipole fields $\boldsymbol{H}^{(p)}_{lm}(\boldsymbol{r}', k')$ with origin at O'

$$\boldsymbol{H}^{(p)}_{lm}(\boldsymbol{r}', k') = \frac{(-\mathrm{i})^{p+l-1}}{2\pi} \int_{\mathcal{D}} \boldsymbol{Z}^{(p)}_{lm}(\hat{\boldsymbol{k}}) \exp(\mathrm{i}\boldsymbol{k}\cdot\boldsymbol{r}')\, d\hat{\boldsymbol{k}}, \qquad (5.11)$$

where the domain of integration $\mathcal{D}$ is the same as in (5.10). Equation (5.11) expresses the multipole field $\boldsymbol{H}$ as a superposition of vector plane waves with complex propagation vector (inhomogeneous plane waves) whose amplitudes are the transverse vector harmonics $\boldsymbol{Z}^{(p)}_{lm}$. The components parallel and orthogonal to the plane of incidence of the waves concerned can be reflected on the interface with the help the Fresnel coefficients (5.2). Before we do this, however, it is convenient to refer the waves in (5.11) to the origin O on the interface. Since $\boldsymbol{r}' = \boldsymbol{r} - \boldsymbol{R}'$, the result of this translation of origin is

$$\boldsymbol{H}^{(p)}_{lm}(\boldsymbol{r}', k') = \frac{(-\mathrm{i})^{p+l-1}}{2\pi} \int_{\mathcal{D}} \sum_{\eta} [\hat{\boldsymbol{u}}_\eta \cdot \boldsymbol{Z}^{(p)}_{lm}(\hat{\boldsymbol{k}})]$$
$$\times \hat{\boldsymbol{u}}_\eta \exp(\mathrm{i}\boldsymbol{k}\cdot\boldsymbol{r}) \exp(-\mathrm{i}\boldsymbol{k}\cdot\boldsymbol{R}')\, d\hat{\boldsymbol{k}}.$$

In fact, the term $\hat{\boldsymbol{u}}_\eta \exp(\mathrm{i}\boldsymbol{k}\cdot\boldsymbol{r})$ is a vector plane wave referred to the origin at O that can thus be reflected by the Fresnel reflection rule. In this respect we recall that the Fresnel coefficients hold true both for complex indices of refraction and for complex angle of incidence [4, 22]. As a result of the reflection we get the equation

$$\boldsymbol{H}_{\mathrm{R}lm}^{(p)} = \frac{(-\mathrm{i})^{p+l-1}}{2\pi} \int_{\mathcal{D}} \sum_\eta F_\eta(\vartheta_{\mathrm{k}}) \bigl[\hat{\boldsymbol{u}}_\eta \cdot \boldsymbol{Z}_{lm}^{(p)}(\hat{\boldsymbol{k}}) \bigr]$$
$$\times \hat{\boldsymbol{u}}_{\mathrm{R}\eta} \exp(\mathrm{i}\boldsymbol{k}_{\mathrm{R}} \cdot \boldsymbol{r}) \exp(-\mathrm{i}\boldsymbol{k} \cdot \boldsymbol{R}') \, \mathrm{d}\hat{\boldsymbol{k}} \,, \qquad (5.12)$$

where $\boldsymbol{k}_{\mathrm{R}}$ and $\hat{\boldsymbol{u}}_{\mathrm{R}\eta}$ are, respectively, the wavevector and the polarization vector of the reflected plane wave. The integrand in (5.12) can be referred back to the origin at O' by the phase factor $\exp(\mathrm{i}\boldsymbol{k}_{\mathrm{R}} \cdot \boldsymbol{R}')$ with the result

$$\boldsymbol{H}_{\mathrm{R}lm}^{(p)} = \frac{(-i)^{p+l-1}}{2\pi} \int_{\mathcal{D}} \sum_\eta F_\eta(\vartheta_{\mathrm{k}}) \bigl[\hat{\boldsymbol{u}}_\eta \cdot \boldsymbol{Z}_{lm}^{(p)}(\hat{\boldsymbol{k}}) \bigr]$$
$$\times \hat{\boldsymbol{u}}_{\mathrm{R}\eta} \exp(\mathrm{i}\boldsymbol{k}_{\mathrm{R}} \cdot \boldsymbol{r}') \exp[\mathrm{i}(\boldsymbol{k}_{\mathrm{R}} - \boldsymbol{k}) \cdot \boldsymbol{R}'] \, \mathrm{d}\hat{\boldsymbol{k}} \,. \qquad (5.13)$$

Let us now recall that according to Sect. 3.1.2 the multipole expansion of the inhomogeneous plane wave in (5.13) is

$$\hat{\boldsymbol{u}}_{\mathrm{R}\eta} \exp[\mathrm{i}\boldsymbol{k}_{\mathrm{R}} \cdot \boldsymbol{r}'] = 4\pi \sum_{plm} \mathrm{i}^{p+l+1} (-)^{m+1} \bigl[\boldsymbol{Z}_{l,-m}^{(p)}(\hat{\boldsymbol{k}}_{\mathrm{R}}) \cdot \hat{\boldsymbol{u}}_{\mathrm{R}\eta} \bigr] \boldsymbol{J}_{lm}^{(p)}(\boldsymbol{r}', k')$$

that, when substituted into (5.13), leads us to write the reflected field as

$$\boldsymbol{H}_{\mathrm{R}lm}^{(p)} = \sum_{p'l'm'} \boldsymbol{J}_{l'm'}^{(p')}(\boldsymbol{r}', k') \mathcal{F}_{l'm'lm}^{(p'p)} \,, \qquad (5.14)$$

where we define

$$\mathcal{F}_{l'm'lm}^{(p'p)} = 2\mathrm{i}^{p'-p+l'-l}(-)^{m+1} \int_{\mathcal{D}} \sum_\eta F_\eta(\vartheta_{\mathrm{k}}) \bigl[\hat{\boldsymbol{u}}_\eta \cdot \boldsymbol{Z}_{lm}^{(p)}(\hat{\boldsymbol{k}}) \bigr]$$
$$\times \bigl[\hat{\boldsymbol{u}}_{\mathrm{R}\eta} \cdot \boldsymbol{Z}_{l',-m'}^{(p')}(\hat{\boldsymbol{k}}_{\mathrm{R}}) \bigr] \exp(2\mathrm{i}k'a \cos\vartheta_{\mathrm{k}}) \, \mathrm{d}\hat{\boldsymbol{k}} \,, \qquad (5.15)$$

a being the distance of O' from the interface. The quantities $\mathcal{F}_{l'm'lm}^{(p'p)}$ can be understood as the elements of the matrix F that effects the reflection of the $\boldsymbol{H}$-multipole fields on the plane interface and thus gives a formal solution to the problem at hand. It may be surprising, however, that the reflected field is expressed in terms of $\boldsymbol{J}$-multipole fields that do not satisfy the radiation condition at infinity. It must be stressed, however, that this is a deceiving appearance because the region of validity of the series expansion in (5.14) is still to be determined. The best procedure to clarify this point is to rewrite (5.12) in terms of multipole fields with origin at O'', the mirror image of the source of the original $\boldsymbol{H}$ fields. The required expression is

$$\boldsymbol{H}_{\text{R}lm}^{(p)} = \frac{(-\mathrm{i})^{p+l-1}}{2\pi} \int_{\mathcal{D}} \sum_{\eta} F_{\eta}(\vartheta_{\text{k}}) \left[\hat{\boldsymbol{u}}_{\eta} \cdot \boldsymbol{Z}_{lm}^{(p)}(\hat{\boldsymbol{k}})\right] \hat{\boldsymbol{u}}_{\text{R}\eta} \exp(\mathrm{i}\boldsymbol{k}_{\text{R}} \cdot \boldsymbol{r}'') \, \mathrm{d}\hat{\boldsymbol{k}} \,,$$
(5.16)

whose integrand is expressed in terms of reflected plane waves referred to the origin at O'': the phase factor $\exp(-\mathrm{i}\boldsymbol{k} \cdot \boldsymbol{R}')$ in (5.12) effects, indeed, the translation of origin from O to O''. Now, as a consequence of (5.3) that relates the amplitudes of the incident and reflected field, we have

$$\hat{\boldsymbol{u}}_{\text{R}\eta} \cdot \boldsymbol{Z}_{lm}^{(p)}(\hat{\boldsymbol{k}}_{\text{R}}) = (-)^{\eta+p+l+m} \hat{\boldsymbol{u}}_{\eta} \cdot \boldsymbol{Z}_{lm}^{(p)}(\hat{\boldsymbol{k}}).$$
(5.17)

Therefore, (5.16) can be rewritten as

$$\boldsymbol{H}_{\text{R}lm} = \frac{(-)^{p+l+m-1}(-\mathrm{i})^{p+l-1}}{2\pi}$$
$$\times \int_{\mathcal{D}} \sum_{\eta} (-)^{\eta+1} F_{\eta}(\vartheta_{\text{k}}) \left[\hat{\boldsymbol{u}}_{\text{R}\eta} \cdot \boldsymbol{Z}_{lm}^{(p)}(\hat{\boldsymbol{k}}_{\text{R}})\right] \hat{\boldsymbol{u}}_{\text{R}\eta} \exp(\mathrm{i}\boldsymbol{k}_{\text{R}} \cdot \boldsymbol{r}'') \, \mathrm{d}\hat{\boldsymbol{k}} \,,$$
(5.18)

whose interpretation is straightforward when the interface is perfectly reflecting. In fact, the Fresnel coefficients for a perfectly reflecting surface take on the limiting values

$$F_{\eta} = (-)^{\eta-1},$$

so that, in this case (5.18) represents, except for a sign, the multipole field $\boldsymbol{H}_{lm}^{(p)}(\boldsymbol{r}'', k')$ with origin at O''. This representation is valid in the half-space $z'' < 0$. This suggests that, in the general case, (5.18) represents in the half-space $z'' < 0$ a linear combination of $\boldsymbol{H}$-multipole fields with origin at O''. We write this combination as

$$\boldsymbol{H}_{\text{R}lm}^{(p)} = \sum_{p''l''m''} \boldsymbol{H}_{l''m''}^{(p'')}(\boldsymbol{r}'', k') a_{l''m''lm}^{(p''p)},$$
(5.19)

whose coefficients $a_{l''m''lm}^{(p''p)}$ can be determined by expressing each of the $\boldsymbol{H}$-multipole fields in (5.19) as a combination of multipoles centered at O' by means of the addition theorem of Sect. 1.8. Indeed, for the region inside the sphere of radius $|\boldsymbol{R}' - \boldsymbol{R}''|$ (see Fig. 5.3) we get

$$\boldsymbol{H}_{\text{R}lm}^{(p)} = \sum_{p''l''m''} \sum_{p'l'm'} \boldsymbol{\mathcal{J}}_{l'm'}^{(p')}(\boldsymbol{r}', k') \mathcal{H}_{l'm'l''m''}^{(p'p'')}(\boldsymbol{R}' - \boldsymbol{R}'', k') a_{l''m''lm}^{(p''p)}.$$
(5.20)

In turn, in the region outside this sphere, i.e., for $r' > |\boldsymbol{R}' - \boldsymbol{R}''|$,

$$\boldsymbol{H}_{\text{R}lm}^{(p)} = \sum_{p''l''m''} \sum_{p'l'm'} \boldsymbol{H}_{l'm'}^{(p')}(\boldsymbol{r}', k') \mathcal{J}_{l'm'l''m''}^{(p'p'')}(\boldsymbol{R}' - \boldsymbol{R}'', k') a_{l''m''lm}^{(p''p)}.$$
(5.21)

Comparison of (5.20) with (5.14) leads to the following expression for the coefficients $a_{l''m''lm}^{(p''p)}$:

$$a_{l''m''lm}^{(p''p)} = \sum_{p'l'm'} [H^{-1}]_{l''m''l'm'}^{(p''p')} \mathcal{F}_{l'm'lm}^{(p'p)} . \tag{5.22}$$

When (5.22) is substituted into (5.21) we get the reflected field in a form that is valid at a large distance from the source and satisfies the radiation condition at infinity as it contains $\boldsymbol{H}$-multipole fields only.

5.3.2 Calculation of the Reflected Field

Equation (5.15), which defines the elements of the multipole reflection matrix, may seem to give only a formal solution to the problem of reflection. Nevertheless, in this section we use the properties of the spherical multipole fields to put (5.15) into a form that is suitable for actual calculations. Indeed, in our opinion, showing the details of this kind of calculation may be a valuable tool to make the reader better acquainted with the usage of the machinery expounded so far.

First, by substituting (5.17) into (5.15) we eliminate the reflected direction $\hat{\boldsymbol{k}}_R$ and the polarization vectors $\hat{\boldsymbol{u}}_{R\eta}$ so that the only dot product within the integrand is $\hat{\boldsymbol{u}}_\eta \cdot \boldsymbol{Z}_{lm}^{(p)}(\hat{\boldsymbol{k}})$. The latter can be calculated by representing the polarization unit vectors on the spherical basis as

$$\hat{\boldsymbol{u}}_1 = \frac{\cos \vartheta_k \exp(i\varphi_k)}{\sqrt{2}} \boldsymbol{\xi}_{-1} - \sin \vartheta_k \boldsymbol{\xi}_0 - \frac{\cos \vartheta_k \exp(-i\varphi_k)}{\sqrt{2}} \boldsymbol{\xi}_1 ,$$

$$\hat{\boldsymbol{u}}_2 = \frac{i}{\sqrt{2}} \exp(i\varphi_k) \boldsymbol{\xi}_{-1} + \frac{i}{\sqrt{2}} \exp(-i\varphi_k) \boldsymbol{\xi}_1 .$$

We get

$$\hat{\boldsymbol{u}}_1 \cdot \boldsymbol{Z}_{lm}^{(1)} = \left[-\frac{1}{\sqrt{2}} c_{lm}^{(1)} \bar{P}_{l,m+1}(\cos \vartheta_k) \cos \vartheta_k + c_{lm}^{(0)} \bar{P}_{lm}(\cos \vartheta_k) \sqrt{1 - \cos^2 \vartheta_k} \right.$$
$$\left. + \frac{1}{\sqrt{2}} c_{lm}^{(-1)} \bar{P}_{l,m-1}(\cos \vartheta_k) \cos \vartheta_k \right] \exp(im\varphi_k) ,$$

$$\hat{\boldsymbol{u}}_2 \cdot \boldsymbol{Z}_{lm}^{(1)} = \left[\frac{i}{\sqrt{2}} c_{lm}^{(1)} \bar{P}_{l,m+1}(\cos \vartheta_k) + \frac{i}{\sqrt{2}} c_{lm}^{(-1)} \bar{P}_{l,m-1}(\cos \vartheta_k) \right] \exp(im\varphi_k) ,$$

where we define the functions

$$\bar{P}_{lm}(z) = \sqrt{(2l+1)\frac{(l-m)!}{(l+m)!}} P_{lm}(z) . \tag{5.23}$$

$P_{lm}(z)$ are the Legendre functions of complex argument that are defined by (1.15). The quantities $c_{lm}^{(0)}$ and $c_{lm}^{(\mp)}$ are the Clebsch–Gordan coefficients

$$c_{lm}^{(0)} = C(1,l,l;0,m) = -\frac{m}{\sqrt{l(l+1)}},$$

$$c_{lm}^{(\mp 1)} = C(1,l,l;\pm 1, m\mp 1) = \pm\sqrt{\frac{(l\pm m)(l\mp m+1)}{2l(l+1)}}.$$

Then, it can be easily verified that the relation $\hat{\boldsymbol{u}}_2 = \hat{\boldsymbol{k}} \times \hat{\boldsymbol{u}}_1$ yields the rules

$$\hat{\boldsymbol{u}}_2 \cdot \boldsymbol{Z}_{lm}^{(2)} = -\hat{\boldsymbol{u}}_1 \cdot \boldsymbol{Z}_{lm}^{(1)} = -\hat{\boldsymbol{u}}_1 \cdot \boldsymbol{X}_{lm},$$

$$\hat{\boldsymbol{u}}_1 \cdot \boldsymbol{Z}_{lm}^{(2)} = \hat{\boldsymbol{u}}_2 \cdot \boldsymbol{Z}_{lm}^{(1)} = \hat{\boldsymbol{u}}_2 \cdot \boldsymbol{X}_{lm},$$

whereas the properties of the vector spherical harmonics yield the relation

$$\hat{\boldsymbol{u}}_\eta \cdot \boldsymbol{X}_{l,-m} = (-)^{\eta+m} \hat{\boldsymbol{u}}_\eta \cdot \boldsymbol{X}_{lm} \exp(-2im\varphi_k).$$

We are now able to perform at once the integration over the angle φ_k with the result

$$\int_0^{2\pi} \exp[i(m-m')\varphi_k]\,d\varphi_k = 2\pi\delta_{mm'},$$

so that the elements of the multipole reflection matrix $\mathcal{F}_{l'm'lm}^{(p'p)}$ with $m \neq m'$ do vanish. This property is conveniently expressed by the equation

$$\mathcal{F}_{l'm'lm}^{(p'p)} = \mathcal{F}_{l'l;m}^{(p'p)} \delta_{mm'}, \tag{5.24}$$

that, when introduced into (5.15) yield the useful relation

$$\mathcal{F}_{l'l;m}^{(p'p)} = \mathcal{F}_{ll';-m}^{(pp')}. \tag{5.25}$$

At this stage, each element of matrix F is given by an integral of the form

$$\int_0^{\frac{\pi}{2}-i\infty} f(\cos\vartheta_k) \exp(2ik'a\cos\vartheta_k) \sin\vartheta_k\,d\vartheta_k$$

that, through the substitution $x = 2ik'a(1-\cos\vartheta_k)$ takes on the form

$$\frac{\exp(2ik'a)}{2ik'a} \int_0^{+\infty} f\left(1 - \frac{x}{2ik'a}\right) \exp(-x)\,dx,$$

and is thus suitable for numerical integration through the Gauss–Laguerre method [77]. In practice, the integration formula is

$$\int_0^\infty f(x)\exp(-x)\,dx \approx \sum_{i=1}^N w_i f(x_i),$$

where the weights w_i are

$$w_i = \frac{(N!)^2 x_i}{[(N+1)L_N(x_{i+1})]^2} ,$$

x_i being the ith zero of the Laguerre polynomial L_N. Thus, to get reliable values of the elements of F, the order N must be wisely chosen so as to ensure a fair convergence of the integrals.

Let us now recall that, according to (5.22), the calculation of the reflected field in the far zone requires the elements of the reflection matrix F as well as the quantities $\mathcal{H}^{(p'p)}_{l'm'lm}(\boldsymbol{R}' - \boldsymbol{R}'', k')$. Since in the present case the translation vector is parallel to the z axis, the involved quantities are $\mathcal{H}^{(p'p)}_{l'm'lm}(-2a\hat{\boldsymbol{e}}_z, k')$ that, as well as the quantities $[\mathrm{H}^{-1}]^{(p'p)}_{l'm'lm}(-2a\hat{\boldsymbol{e}}_z, k')$, have the property expressed by (1.60). On account of similar property of the elements of F [see (5.24)], also the amplitudes $a^{(p'p)}_{l'm'lm}$ do vanish unless $m' = m$, i.e.,

$$a^{(p'p)}_{l'm'lm} = a^{(p'p)}_{l'l;m}\delta_{mm'} . \tag{5.26}$$

5.4 Scattering from a Sphere on a Dielectric Substrate

The theory of the preceding section is the starting point for calculating the scattering pattern from a sphere in the vicinity of a dielectric substrate. In fact, the field scattered by any particle can always be expanded in a series of $\boldsymbol{H}$-multipole fields with origin within the particle itself. The reflection of these multipole fields on the plane interface can, of course, be dealt with through the theory of the preceding section.

We assume that a spherical scatterer lies entirely within the accessible half-space and is illuminated by a plane wave. We already pointed out that outside the scatterer the total field is

$$\boldsymbol{E}_{\mathrm{Ext}} = \boldsymbol{E}_{\mathrm{I}} + \boldsymbol{E}_{\mathrm{R}} + \boldsymbol{E}_{\mathrm{S}} + \boldsymbol{E}_{\mathrm{SR}} , \tag{5.27}$$

where $\boldsymbol{E}_{\mathrm{R}}$ and $\boldsymbol{E}_{\mathrm{I}}$ are the same as we would have if no particle were present. $\boldsymbol{E}_{\mathrm{SR}}$ and $\boldsymbol{E}_{\mathrm{S}}$ are related to each other by the reflection condition and $\boldsymbol{F}_{\mathrm{S}}$ is determined by imposing to $\boldsymbol{E}_{\mathrm{Ext}}$ the appropriate boundary conditions across the surface of the scattering particle. The formalism that we are going to expound is appropriate to a homogeneous sphere, but we will show later that its application to radially nonhomogeneous spheres implies minor changes only.

The geometry that we adopt for our study is sketched in Fig 5.4 and differs slightly from that depicted in Fig. 5.3. In fact, we found it convenient to take the center of the spherical scatterer O' and its image O'' on the z axis, a minor difference that allows us to make full use of the symmetry of the problem while preserving the generality of our approach.

110   5. Particles on a Plane Surface

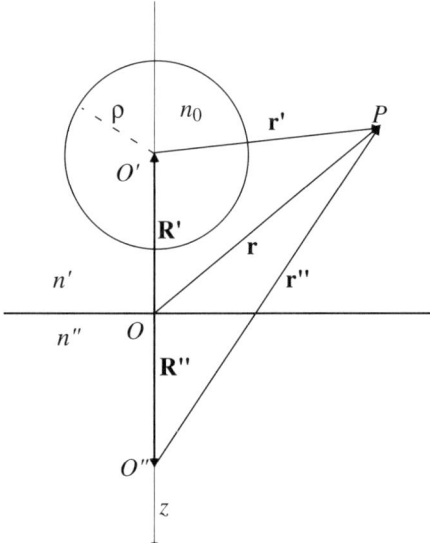

Fig. 5.4. Geometry for scattering from a sphere in the vicinity of a dielectric surface

5.4.1 Reflection of the Incident and Scattered Wave

The incident and the reflected fields have the same multipole expansion that has been established in Sect. 5.1 for the case of a perfectly reflecting surface. The respective multipole amplitudes are related to each other by (5.3) that holds true for general values of n' and n''. Nevertheless, since $\boldsymbol{E}_\mathrm{I}$ and $\boldsymbol{E}_\mathrm{R}$ must also satisfy the boundary conditions on the surface of the scattering sphere, it is convenient to expand both fields in terms of multipole fields with origin at the center of the sphere, O'. The simplest way to do this is by resorting to the appropriate phase factors with the results

$$\boldsymbol{E}_\mathrm{I} = \exp(\mathrm{i}\boldsymbol{k}_\mathrm{I} \cdot \boldsymbol{R}') \sum_\eta E_{0\eta} \sum_{plm} \boldsymbol{J}^{(p)}_{lm}(\boldsymbol{r}', k') W^{(p)}_{\mathrm{I}\eta lm} \;, \tag{5.28}$$

$$\boldsymbol{E}_\mathrm{R} = \exp(\mathrm{i}\boldsymbol{k}_\mathrm{R} \cdot \boldsymbol{R}') \sum_\eta E_{0\eta} F_\eta(\vartheta_\mathrm{I}) \sum_{plm} \boldsymbol{J}^{(p)}_{lm}(\boldsymbol{r}', k') W^{(p)}_{\mathrm{R}\eta lm} \;.$$

Of course, $W^{(p)}_{\mathrm{R}\eta lm} = W^{(p)}_{lm}(\hat{\boldsymbol{u}}_{\mathrm{R}\eta}, \hat{\boldsymbol{k}}_\mathrm{R})$. The field that is scattered by a sphere that lies entirely in the accessible half-space can be expanded in a series of vector multipole fields that satisfy the radiation condition at infinity. The component of the incident field that is polarized along $\hat{\boldsymbol{u}}_{\mathrm{I}\eta}$ yields, indeed, a scattered field that, with respect to an origin O' within the particle, can be written as

$$\boldsymbol{E}_{\mathrm{S}\eta} = E_{0\eta} \sum_{plm} \boldsymbol{H}^{(p)}_{lm}(\boldsymbol{r}', k') \mathcal{A}^{(p)}_{\eta lm} \;.$$

The amplitudes $\mathcal{A}$, which are as yet unknown, are determined by the boundary conditions at the surface of the particle. The scattered field impinges on the plane surface and, by reflection, yields a field that can be obtained through the technique that was described in Sect. 5.2. Accordingly, the reflected-scattered field in the vicinity of the surface of the particle can be written as the superposition of $\boldsymbol{J}$-multipole fields with origin at O'

$$\boldsymbol{E}_{\mathrm{SR}\eta} = E_{0\eta} \sum_{plm} \sum_{p'l'} \boldsymbol{J}_{lm}^{(p)}(\boldsymbol{r}',k') \mathcal{F}_{ll';m}^{(pp')} \mathcal{A}_{\eta l'm}^{(p')} , \qquad (5.29)$$

where the quantities $\mathcal{F}$ are defined by (5.15). We remark that in (5.29) we used the property (5.25).

5.4.2 Amplitudes of the Scattered Field

The results in Sect. 5.4.1 show that the fields that contribute to $\boldsymbol{E}_{\mathrm{Ext}}$ in the vicinity of the surface of the particle, according to (5.28)–(5.29), are all expressed in terms of vector multipole fields with origin at O'. $\boldsymbol{E}_{\mathrm{Ext}}$ is thus ready for the imposition of the boundary conditions.

We now explicitly assume that the scattering particle is a homogeneous sphere with (possibly complex) refractive index n_0 and radius ρ. Since the field within the sphere must be regular at O', its multipole expansion can be taken in the form

$$\boldsymbol{E}_{\mathrm{T}\eta} = E_{0\eta} \sum_{plm} \boldsymbol{J}_{lm}^{(p)}(\boldsymbol{r}',k_0) \mathcal{C}_{\eta lm}^{(p)} .$$

The multipole expansion of the magnetic field $\boldsymbol{B}$, that is also needed to impose the boundary conditions, is readily obtained through the Maxwell equation

$$\mathrm{i}\boldsymbol{B}_\eta = \frac{1}{k} \nabla \times \boldsymbol{E}_\eta .$$

In order to impose the boundary conditions across the surface of the scattering sphere, we apply to $\boldsymbol{E}$ and $\boldsymbol{B}$ the procedure that we described in Chap. 4. This procedure yields, for each p, l, and m, four equations among which the amplitudes of the internal field $\mathcal{C}$ can be easily eliminated. As a result, we get, for each m, a system of linear nonhomogeneous equations for the amplitudes $\mathcal{A}_{\eta lm}^{(p)}$, namely

$$\sum_{p'l'} \mathcal{M}_{ll';m}^{(pp')} \mathcal{A}_{\eta l'm}^{(p')} = -\mathcal{W}_{\eta lm}^{(p)} , \qquad (5.30)$$

where

$$\mathcal{M}_{ll';m}^{(pp')} = \left(R_l^{(p)}\right)^{-1} \delta_{pp'} \delta_{ll'} + \mathcal{F}_{ll';m}^{(pp')} , \qquad (5.31)$$

and

$$\mathcal{W}_{\eta lm}^{(p)} = \exp(i\boldsymbol{k}_{\mathrm{I}} \cdot \boldsymbol{R}')W_{\mathrm{I}\eta lm}^{(p)} + \exp(i\boldsymbol{k}_{\mathrm{R}} \cdot \boldsymbol{R}')F_\eta W_{\mathrm{R}\eta lm}^{(p)}. \qquad (5.32)$$

The quantities $R_l^{(1)}$ and $R_l^{(2)}$, as usual, coincide with the Mie coefficients b_l and a_l, respectively, for a homogeneous sphere of refractive index n_0 embedded into a homogeneous medium of refractive index n'. We remark that our theory can easily deal also with radially nonhomogeneous spheres. Even in this case, in fact, one gets (5.31), but the quantities $R_l^{(p)}$ are given by (4.15).

5.4.3 Scattering Amplitude and Transition Matrix

Once the amplitudes $\mathcal{A}_{\eta lm}^{(p)}$ of $\boldsymbol{E}_{\mathrm{S}\eta}$ have been calculated by solving (5.30), the reflected-scattered field $\boldsymbol{E}_{\mathrm{SR}\eta}$ is also determined by (5.29). The latter equation, however, gives an expression of $\boldsymbol{E}_{\mathrm{SR}\eta}$ that is valid only in the vicinity of the surface of the sphere as it includes multipole fields that do not satisfy the radiation condition at infinity. Nevertheless, the considerations in Sect. 5.2. led us to conclude that, at any point of the accessible half-space, $\boldsymbol{E}_{\mathrm{SR}\eta}$ is given by the equation

$$\boldsymbol{E}_{\mathrm{SR}\eta} = E_{0\eta} \sum_{plm} \boldsymbol{H}_{lm}^{(p)}(\boldsymbol{r}'', k') \mathcal{A}_{\mathrm{R}\eta lm}^{(p)}, \qquad (5.33)$$

where

$$\mathcal{A}_{\mathrm{R}\eta lm}^{(p)} = \sum_{p'l'} a_{ll';m}^{(pp')} \mathcal{A}_{\eta l'm}^{(p')}, \qquad (5.34)$$

and the coefficients $a_{ll';m}^{(pp')}$ are given by (5.22) and (5.26). The superposition of $\boldsymbol{E}_{\mathrm{SR}\eta}$, (5.33), and of $\boldsymbol{E}_{\mathrm{S}\eta}$ yields the field that would be observed by an optical instrument in the far zone. The reflected field $\boldsymbol{E}_{\mathrm{R}\eta}$ would be observed in the direction of reflection only. Since $\boldsymbol{E}_{\mathrm{S}\eta}$ and $\boldsymbol{E}_{\mathrm{SR}\eta}$ are given as expansions in terms of $\boldsymbol{H}$-multipole fields with origin at O' and O'', respectively, it is convenient to again use the addition theorem of Sect. 1.8 [14] to refer both fields to a common origin that, for symmetry reasons, we choose to be the point O that is defined in Fig. 5.4. The observed field then becomes

$$\boldsymbol{E}_{\mathrm{Obs}\,\eta} = \sum_{plm} \boldsymbol{H}_{lm}^{(p)}(\boldsymbol{r}, k') A_{\eta lm}^{(p)},$$

where

$$A_{\eta lm}^{(p)} = \sum_{p'l'} \left[\mathcal{J}_{ll';m}^{(pp')}(a\hat{\boldsymbol{e}}_z, k') \mathcal{A}_{\eta l'm}^{(p')} + \mathcal{J}_{ll';m}^{(pp')}(-a\hat{\boldsymbol{e}}_z, k') \mathcal{A}_{\mathrm{R}\eta l'm}^{(p')} \right]. \qquad (5.35)$$

5.4 Scattering from a Sphere on a Dielectric Substrate

We notice that in (5.35) we used the property (1.60) of the $\mathcal{J}$ translation matrix elements that holds true when, as in the present case, the translation takes place along the z axis. As usual, we have

$$f_{\eta\eta'} = -\frac{i}{4\pi k'} \sum_{plm} W^{(p)*}_{S\eta lm} \mathcal{A}^{(p)}_{\eta' lm} . \tag{5.36}$$

The efficiency of the preceding equations can be improved because, as we assumed throughout, the refractive index n' is real. In this case, indeed, with the help of (4.29) and (4.31), (5.36) can be rewritten as

$$f_{\eta\eta'} = -\frac{i}{4\pi k'} \sum_{plm} W^{(p)*}_{S\eta lm}$$
$$\times \left[\exp(i\boldsymbol{k}_{\mathrm{S}} \cdot \hat{\boldsymbol{e}}_z a) \mathcal{A}^{(p)}_{\eta' lm} + \exp(-i\boldsymbol{k}_{\mathrm{S}} \cdot \hat{\boldsymbol{e}}_z a) \mathcal{A}^{(p)}_{\mathrm{R}\eta' lm} \right] . \tag{5.37}$$

Equation (5.37) is the one that we actually used for our calculations.

We proceed now to a further modification of (5.37) that leads to the definition of the *transition matrix for a particle in the presence of a plane interface*. To this end, let us recall that the formal solution to (5.30) is

$$\mathcal{A}^{(p)}_{\eta lm} = -\sum_{p'l'} [\mathrm{M}^{-1}]^{(pp')}_{ll';m} \mathcal{W}^{(p')}_{\eta l'm} , \tag{5.38}$$

where M^{-1} is the inverse to the matrix M in (5.31). In this respect we notice that, on account of (5.25), the inversion of matrix M needs to be performed for $m \geq 0$ only. With the help of (4.28) and (5.3) we transform the amplitudes $\mathcal{W}$, (5.32), into the form

$$\mathcal{W}^{(p)}_{\eta lm} = \sum_{p'l'} \mathcal{J}^{(pp')}_{ll';m}(-a\hat{\boldsymbol{e}}_z, k') W^{(p')}_{E\eta l'm} , \tag{5.39}$$

where the multipole amplitudes of the exciting field are given by (5.6). Substitution of (5.39) into (5.38) and of (5.38) into (5.35) allows us to write the scattering amplitude as

$$f_{\eta\eta'} = -\frac{i}{4\pi k'} \sum_{pp'} \sum_{ll'} \sum_{m} W^{(p)*}_{S\eta lm} \mathcal{S}^{(pp')}_{ll';m} W^{(p')}_{E\eta' l'm} , \tag{5.40}$$

so that the quantities

$$\mathcal{S}^{(pp')}_{ll';m} = -\sum_{qL} \sum_{q'L'} \left[\mathcal{J}^{(pq)}_{lL;m}(a\hat{\boldsymbol{e}}_z, k') [\mathrm{M}^{-1}]^{(qq')}_{LL';m} \right.$$
$$\left. + \mathcal{J}^{(pq)}_{lL;m}(-a\hat{\boldsymbol{e}}_z, k') \sum_{q''L''} a^{(qq'')}_{LL'';m} [\mathrm{M}^{-1}]^{(q''q')}_{L''L';m} \right] \mathcal{J}^{(q'p')}_{L'l';m}(-a\hat{\boldsymbol{e}}_z, k')$$

can be interpreted as the elements of the transition matrix for the spherical scatterer in the presence of the plane surface [23]. We remark that the multiple scattering processes that occur between the sphere and the substrate are accounted for by the elements of the transfer matrix J.

Using the transition matrix to get the observed field does not yield any advantage in the case that we deal with in this section; in fact, as we stated above, (5.37) is computationally more effective. However, our purpose was to show that the transition matrix can be defined even when a plane interface is present. The advantages yielded by the use of the transition matrix will become evident later when we will deal with the problem of the orientational average over an assembly of nonspherical particles deposited on a plane substrate.

5.5 Aggregated Spheres on a Dielectric Substrate

We go now to consider the scattering pattern from an aggregate of spherical scatterers in the vicinity of or deposited on a dielectric substrate. More precisely, our aim is to calculate the scattering pattern from an aggregate as a function of its orientation with respect to an incident plane wave field. This information is, in fact, the basis for the calculation of the scattering pattern from an assembly of identical aggregates with a known or assumed distribution of their orientations. $\boldsymbol{E}_{\text{Ext}}$ is still given by (5.27), provided $\boldsymbol{E}_\text{S}$ is the field scattered by the whole aggregate. Even in this case the scattered field $\boldsymbol{E}_\text{S}$ is determined by imposing to $\boldsymbol{E}_{\text{Ext}}$ the appropriate boundary conditions across the surface of the particle, i.e., across the surface of each one of the spheres in the aggregate. It will become apparent later that $\boldsymbol{E}_\text{S}$ and $\boldsymbol{E}_{\text{SR}}$ include all the multiple scattering processes that occur both among the spheres of the aggregate and between the aggregate and the substrate.

The geometry that we adopt for our study of the scattering from an aggregate of spheres is depicted in Fig. 5.5.

5.5.1 Multipole Expansion of the Fields

The expansion of $\boldsymbol{E}_{\text{I}\eta}$ and $\boldsymbol{E}_{\text{R}\eta}$ in terms of vector multipole fields with origin at O are

$$\begin{aligned} \boldsymbol{E}_{\text{I}\eta} &= E_{0\eta} \sum_{plm} \boldsymbol{J}_{lm}^{(p)}(\boldsymbol{r}, k') W_{\text{I}\eta lm}^{(p)} \;, \\ \boldsymbol{E}_{\text{R}\eta} &= E_{0\eta} F_\eta(\vartheta_\text{I}) \sum_{plm} \boldsymbol{J}_{lm}^{(p)}(\boldsymbol{r}, k') W_{\text{R}\eta lm}^{(p)} \;. \end{aligned} \quad (5.41)$$

We recall that $\boldsymbol{E}_\text{I}$ and $\boldsymbol{E}_\text{R}$ must also satisfy the boundary conditions on the surface of each of the scattering spheres. Therefore we need to rewrite the expansions (5.41) in terms of spherical vector multipole fields with origin

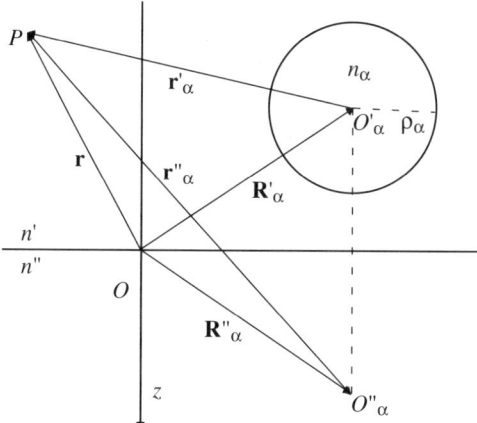

Fig. 5.5. Geometry for the calculation of the scattering pattern from aggregates on a dielectric surface. Only the αth sphere is shown for the sake of clarity

at the center of the αth sphere, O'_α. This can be done in the simplest way by resorting to the appropriate phase factors [72]. Actually, we used this technique in the case of a single sphere in the vicinity of the interface. We stress, however, that this procedure, though exact, yields expressions that are not suitable to effect averages over the orientation of the scatterers. Therefore we start from (5.41) and apply the addition theorem of Sect. 1.8 to express each of the $\boldsymbol{J}$-multipole fields with origin at O as a series of $\boldsymbol{J}$-multipole fields with origin at O'_α. The result is

$$\boldsymbol{E}_{\mathrm{I}\eta} = E_{0\eta} \sum_{plm} \sum_{p'l'm'} \boldsymbol{J}^{(p)}_{lm}(\boldsymbol{r}'_\alpha, k') \mathcal{J}^{(pp')}_{lml'm'}(\boldsymbol{R}'_\alpha, k') W^{(p')}_{\mathrm{I}\eta l'm'} ,$$

$$\boldsymbol{E}_{\mathrm{R}\eta} = E_{0\eta} F_\eta(\vartheta_{\mathrm{I}}) \sum_{plm} \sum_{p'l'm'} \boldsymbol{J}^{(p)}_{lm}(\boldsymbol{r}'_\alpha, k') \mathcal{J}^{(pp')}_{lml'm'}(\boldsymbol{R}'_\alpha, k') W^{(p')}_{\mathrm{R}\eta l'm'} .$$

The field that is scattered by an aggregate of spheres embedded within a homogeneous medium can be described as the superposition of the fields that are scattered by each one of the spheres. Since the latter fields must satisfy the radiation condition at infinity, we write

$$\boldsymbol{E}_{\mathrm{S}\eta} = E_{0\eta} \sum_\alpha \sum_{plm} \boldsymbol{H}^{(p)}_{lm}(\boldsymbol{r}'_\alpha, k') \mathcal{A}^{(p)}_{\eta\alpha lm} ,$$

where the amplitudes $\mathcal{A}^{(p)}_{\eta\alpha lm}$, which are as yet unknown, are determined by the boundary conditions at the surface of each of the component spheres.

We still need to consider the field $\boldsymbol{E}_{\mathrm{RS}}$ that after scattering by the aggregate is reflected by the surface. According to [72, 78] and in analogy to what

116   5. Particles on a Plane Surface

we already did in the case of a single sphere, at any point of the accessible half-space, $\boldsymbol{E}_{\mathrm{SR}\eta}$ can be written as

$$\boldsymbol{E}_{\mathrm{SR}\eta} = E_{0\eta} \sum_{\alpha} \sum_{plm} \boldsymbol{H}^{(p)}_{lm}(\boldsymbol{r}''_\alpha, k') \mathcal{A}^{(p)}_{\mathrm{R}\eta\alpha lm} , \qquad (5.42)$$

i.e., as a superposition of $\boldsymbol{H}$-multipole fields with origin at the image points O''_α. The amplitudes $\mathcal{A}^{(p)}_{\mathrm{R}\eta\alpha lm}$ in (5.42) are related to the amplitudes of the scattered field $\boldsymbol{E}_\mathrm{S}$ by the equation

$$\mathcal{A}^{(p)}_{\mathrm{R}\eta\alpha lm} = \sum_{p'l'} a^{(pp')}_{\alpha ll';m} \mathcal{A}^{(p')}_{\alpha\eta l'm} , \qquad (5.43)$$

with the amplitudes $a^{(pp')}_{\alpha ll';m}$ explicitly given by

$$a^{(pp')}_{\alpha ll';m} = \sum_{p''l''} [\mathrm{H}_\alpha^{-1}]^{(pp'')}_{ll'';m} \mathcal{F}^{(p''p')}_{\alpha l''l';m} , \qquad (5.44)$$

where H_α^{-1} is the inverse to the matrix $\mathrm{H}(\bar{\boldsymbol{R}}_{\alpha\alpha}, k')$ that effects the transfer of the origin of the $\boldsymbol{H}$-multipole fields from O''_α to O'_α and $\bar{\boldsymbol{R}}_{\alpha\alpha} = \boldsymbol{R}'_\alpha - \boldsymbol{R}''_\alpha$. In turn, F_α is the reflection matrix for $\boldsymbol{H}$-multipole fields with origin at O'_α. It is to be calculated as in Sect. 5.3.

The last expansion that we need to consider is that of the field within the αth sphere, namely

$$\boldsymbol{E}_{\mathrm{T}\eta\alpha} = E_{0\eta} \sum_{plm} \boldsymbol{J}^{(p)}_{lm}(\boldsymbol{r}'_\alpha, k_\alpha) \mathcal{C}^{(p)}_{\alpha\eta lm} .$$

Now we reexpand $\boldsymbol{E}_{\mathrm{S}\eta}$ and $\boldsymbol{E}_{\mathrm{SR}\eta}$ in terms of multipole fields with origin at O'_α through the use of the addition theorem of Sect. 1.8 [14]. The result for the scattered field $\boldsymbol{E}_{\mathrm{S}\eta}$ is

$$\boldsymbol{E}_{\mathrm{S}\eta} = E_{0\eta} \sum_{plm} \Bigg[\boldsymbol{H}^{(p)}_{lm}(\boldsymbol{r}'_\alpha, k') \mathcal{A}^{(p)}_{\eta\alpha lm} + \boldsymbol{J}^{(p)}_{lm}(\boldsymbol{r}'_\alpha, k')$$
$$\times \sum_{p'l'm'} \sum_{\alpha' \neq \alpha} \mathcal{H}^{(pp')}_{lml'm'}(\boldsymbol{R}'_{\alpha'\alpha}, k') \mathcal{A}^{(p')}_{\eta\alpha'l'm'} \Bigg] , \qquad (5.45)$$

where $\boldsymbol{R}'_{\alpha'\alpha} = \boldsymbol{R}'_\alpha - \boldsymbol{R}'_{\alpha'}$. For the reflected scattered field $\boldsymbol{E}_{\mathrm{SR}}$ we obtain

$$\boldsymbol{E}_{\mathrm{SR}\eta} = E_{0\eta} \sum_{plm} \boldsymbol{J}^{(p)}_{lm}(\boldsymbol{r}'_\alpha, k') \sum_{p'l'm'} \Bigg[\mathcal{H}^{(pp')}_{lml'm'}(\bar{\boldsymbol{R}}_{\alpha\alpha}, k') \mathcal{A}^{(p')}_{\mathrm{R}\eta\alpha l'm'}$$
$$+ \sum_{\alpha' \neq \alpha} \mathcal{H}^{(pp')}_{lml'm'}(\bar{\boldsymbol{R}}_{\alpha'\alpha}, k') \mathcal{A}^{(p')}_{\mathrm{R}\eta\alpha'l'm'} \Bigg],$$

where $\bar{R}_{\alpha'\alpha} = R'_\alpha - R''_{\alpha'}$. Then, with the help of (5.43) and (5.44), and taking into account that $\mathcal{H}^{(pp')}_{lml'm'}(\bar{R}_{\alpha\alpha}, k') = \mathcal{H}^{(pp')}_{ll';m}(\bar{R}_{\alpha\alpha}, k')\delta_{mm'}$, we get the required expression

$$E_{\text{SR}\eta} = E_{0\eta}\sum_{plm}J^{(p)}_{lm}(r'_\alpha, k')\sum_{p'l'}\left[\mathcal{F}^{(pp')}_{\alpha ll';m}\mathcal{A}^{(p')}_{\eta\alpha l'm}\right.$$
$$\left.+\sum_{\alpha'\neq\alpha}\sum_{p''l''m'}\mathcal{Q}^{(pp'')}_{lml''m'}(\bar{R}_{\alpha'\alpha}, k')\mathcal{F}^{(p''p')}_{\alpha'l''l';m'}\mathcal{A}^{(p')}_{\eta\alpha'l'm'}\right], \quad (5.46)$$

where

$$\mathcal{Q}^{(pp'')}_{lml''m''}(\bar{R}_{\alpha'\alpha}, k') = \sum_{p'l'}\mathcal{H}^{(pp')}_{lml'm'}(\bar{R}_{\alpha'\alpha}, k')\left[H^{-1}_{\alpha'}\right]^{(p'p'')}_{l'l'';m'}.$$

A few words of comment on the physical content of (5.45) and (5.46) are in order. Equation (5.45) shows that the field scattered by the whole aggregate includes the effect of all the multiple scattering processes that occur among the spheres of the aggregate. In turn, according to (5.46), E_{SR} collects the contribution from all the multiple scattering processes occurring between each of the spheres of the aggregate and the substrate.

The boundary conditions across the surface of each of the scattering spheres can now be imposed and we get, for each α, p, l, and m, four equations among which the amplitudes of the internal field C can be easily eliminated. As a result, we get for the amplitudes $\mathcal{A}^{(p)}_{\eta\alpha lm}$ the system of linear nonhomogeneous equations

$$\sum_{\alpha'}\sum_{p'l'm'}\mathcal{M}^{(pp')}_{\alpha lm\alpha'l'm'}\mathcal{A}^{(p')}_{\eta\alpha'l'm'} = -\mathcal{W}^{(p)}_{\eta\alpha lm}, \quad (5.47)$$

where

$$\mathcal{M}^{(pp')}_{\alpha lm\alpha'l'm'} = \left[R^{(p)}_{\alpha l}\right]^{-1}\delta_{\alpha\alpha'}\delta_{pp'}\delta_{ll'}\delta_{mm'} + \mathcal{H}^{(pp')}_{lml'm'}(R'_{\alpha'\alpha}, k')$$
$$+ \mathcal{F}^{(pp')}_{\alpha ll';m}\delta_{\alpha\alpha'}\delta_{mm'} + \sum_{p''l''}\mathcal{Q}^{(pp'')}_{lml''m'}(\bar{R}_{\alpha'\alpha}, k')\mathcal{F}^{(p''p')}_{\alpha'l''l';m'}.$$

Again, the quantities $R^{(1)}_{\alpha l}$ and $R^{(2)}_{\alpha l}$ coincide with the Mie coefficients b_l and a_l, respectively, for a homogeneous sphere of refractive index n_α embedded into a homogeneous medium of refractive index n'. In turn,

$$\mathcal{W}^{(p)}_{\eta\alpha lm} = \sum_{p'l'm'}\mathcal{J}^{(pp')}_{lml'm'}(R'_\alpha, k')W^{(p')}_{E\eta l'm'},$$

where the multipole amplitudes of the exciting field $W^{(p)}_{E\eta lm}$ are still defined by (5.6).

5.5.2 Transition Matrix for an Aggregate in the Presence of a Surface

Once the amplitudes $\mathcal{A}^{(p)}_{\eta\alpha lm}$ of $\boldsymbol{E}_{S\eta}$ have been calculated by solving (5.47), the reflected-scattered field, $\boldsymbol{E}_{SR\eta}$, is also determined by (5.42)–(5.44). Then, we resort again to the addition theorem of Sect. 1.8 to refer $\boldsymbol{E}_{S\eta}$ and $\boldsymbol{E}_{SR\eta}$ to a common origin that we choose to be the point O that is defined in Fig. 5.5. The observed field then becomes

$$\boldsymbol{E}_{\text{Obs}\,\eta} = \sum_{plm} \boldsymbol{H}^{(p)}_{lm}(\boldsymbol{r}, k') A^{(p)}_{\eta lm} ,$$

where

$$A^{(p)}_{\eta lm} = \sum_{\alpha} \sum_{p'l'm'} \left[\mathcal{J}^{(pp')}_{lml'm'}(-\boldsymbol{R}'_{\alpha}, k') \mathcal{A}^{(p')}_{\eta\alpha l'm'} + \mathcal{J}^{(pp')}_{lml'm'}(-\boldsymbol{R}''_{\alpha}, k') \mathcal{A}^{(p')}_{R\eta\alpha l'm'} \right] . \tag{5.48}$$

The amplitudes $A^{(p)}_{\eta lm}$ can be related to the amplitudes of the exciting field, $W^{(p)}_{E\eta lm}$, through the equation

$$A^{(p)}_{\eta lm} = \sum_{p'l'm'} \mathcal{S}^{(pp')}_{lml'm'} W^{(p')}_{E\eta l'm'} ,$$

where the quantities $\mathcal{S}^{(pp')}_{lml'm'}$ can be interpreted as the elements of the transition matrix for the aggregate in the presence of the plane substrate. The transition matrix S can be defined by recalling that the formal solution to (5.47) is

$$\mathcal{A}^{(p)}_{\eta\alpha lm} = -\sum_{\alpha'} \sum_{p'l'm'} [\mathrm{M}^{-1}]^{(pp')}_{\alpha lm\alpha'l'm'} W^{(p')}_{\eta\alpha'l'm'} .$$

When this formal solution is introduced into (5.48), we find

$$\mathcal{S}^{(pp')}_{lml'm'} = -\sum_{\alpha\alpha'} \sum_{qLM} \sum_{q'L'M'} \left[\mathcal{J}^{(pq)}_{lmLM}(-\boldsymbol{R}'_{\alpha}, k') [\mathrm{M}^{-1}]^{(qq')}_{\alpha LM\alpha'L'M'} \right.$$
$$\left. + \mathcal{J}^{(pq)}_{lmLM}(-\boldsymbol{R}''_{\alpha}, k') \sum_{q''L''} a^{(qq'')}_{\alpha LL'';M} [\mathrm{M}^{-1}]^{(q''q')}_{\alpha L''M\alpha'L'M'} \right]$$
$$\times \mathcal{J}^{(q'p')}_{L'M'l'm'}(\boldsymbol{R}'_{\alpha'}, k') . \tag{5.49}$$

Then, substituting (5.49) into (5.40) we get the components of the scattering amplitude of the aggregate in the presence of the interface.

At this stage, the orientational averages over an assembly of particles deposited on the interface can be performed along the lines of Sect. 5.2.1, as the procedure that we expounded there applies independently of the reflecting power of the interface. Here we stress that the considerations on the identity of the effective scatterers (actual particles plus images) hold true also in the present case. In fact, (5.49) that defines the elements of the transition matrix depends on the distance between the spheres and their respective images. Therefore, two identical clusters with a different canting angle with respect to the interface should be considered as different particles and the respective transition matrix elements need to be calculated anew. However, if orientational averages are not required, we can resort to the appropriate phase factors to effect the translation of the multipole fields from and towards origin.

A few words are still necessary on the determination of l_I both for the single sphere and for the aggregate. Of course, the considerations that we already made in Chap. 4 for the case of clusters in free space still hold. First, the presence of an interface may increase the value of l_I. For instance, in case of a perfectly reflecting surface we should deal with a cluster of $2N$ spheres. Second, the calculation either of H^{-1} or of H_α^{-1} becomes rather critical when l_I increases. As a result, the convergence and stability problems that are met in calculating the scattering properties of particles in the presence of a dielectric interface are the most complex among those considered thus far.

5.6 Perfectly Reflecting vs. Dielectric Surface: Similarities and Differences

We conclude this chapter by pointing out the similarities and the differences that one meets when calculating the scattering pattern from particles in the vicinity of a plane surface. We start by remarking that the expression of the scattering amplitude is given by (5.8) and holds true for both perfectly reflecting and dielectric surface. This is a formal result, however. In fact, the transition matrix to be used does depend on the dielectric properties of the surface. We already remarked that, when the surface is perfectly reflecting, the transition matrix is that of the effective scatterer that is a $2N$-spheres cluster when the actual scatterer is a N-spheres cluster. Nevertheless, once the transition matrix has been calculated, it would be possible to perform any kind of averages over orientations of the effective scatterer.

The transition matrix of a particle in the presence of a dielectric surface never coincides with that of the effective scatterer neither when we chose for the Fresnel coefficients the limiting value $F_\eta = (-)^{\eta-1}$ so that the surface becomes perfectly reflecting. Nevertheless, in this limiting case there is coincidence of the scattering amplitude $f_{\eta\eta'}$ and of the intensity of the scattered field as well as of the average of the intensity around an axis orthogonal to the surface. This can be explained by remarking that, for a perfectly reflecting surface, the transition matrix of the effective scatterer yields a description of the scattered field that holds true (in the far zone) in the whole space. On

120   5. Particles on a Plane Surface

the contrary, for a dielectric surface the description yielded by the transition matrix is meaningful in the accessible half-space only.

Another related problem regards the validity of the optical theorem when a surface is present. In fact, (3.9) has been obtained on the assumption that the scattering particles are in free space and that the incident field is polarized along $\hat{u}_{I\eta}$ and $\hat{k}_S = \hat{k}_I$. When a perfectly reflecting surface is present, however, the forward-scattering direction coincides with the direction of reflection. In fact, an optical instrument aimed at the particle along $-\hat{k}_R$ detects, besides the scattered field, also the reflected wave

$$\boldsymbol{E}_{R\eta} = E_0(-)^{\eta-1}\hat{\boldsymbol{u}}_{R\eta}\exp(\mathrm{i}\boldsymbol{k}_R \cdot \boldsymbol{r}) \ .$$

Therefore, by applying to $\boldsymbol{E}_{R\eta}$ the same reasoning that in Sect. 2.5 started from the incident field (2.17), we can write the optical theorem in the modified form

$$\sigma_{T\eta} = \frac{4\pi}{k}\mathrm{Im}\bigl[(-)^{\eta-1}f_{\eta\eta}(\hat{\boldsymbol{k}}_R,\hat{\boldsymbol{k}}_I)\bigr] \ . \tag{5.50}$$

6. Applications: Aggregated Spheres, Layered Spheres, and Spheres Containing Inclusions

The theory developed so far is suitable to investigate the optical properties both of single particles and of dispersions of particles that are of interest in several fields of physics. With the proviso that careful tests are necessary to verify the adequacy of the model in the sense explained in Sect. 1.1, aggregates of spheres may be suitable to model the nonspherical particles of atmospheric aerosols and the ice crystals that occur in the high atmosphere, as well as the cosmic dust grains that are of interest in astrophysics. In turn, spheres containing one or more spherical inclusions may be used to approximate the scattering properties of biological cells or of water droplets containing pollutants, but also to approximate the properties of the porous and fluffy particles that occur in the cosmic dust. Therefore we felt it convenient to devote the present and the next chapter to discussing applications of the theory developed so far with the purpose of highlighting the scattering features of the model particles, either in free space or in the presence of a plane surface. While the applications we present in this chapter can be easily specialized to atmospheric physics or to astrophysics, Chaps. 8 and 9 are specifically addressed to the study of the atmospheric ice crystals and of the cosmic dust, respectively.

Modeling actual particles as clusters of spherical scatterers or as spheres containing inclusions is made possible by the flexibility of the theory that does not imply any limitation of principle on the radii, refractive indices, and geometrical arrangement of the component spheres. This kind of modeling gives only an approximation to the actual shape of the particles, however, but the applications that we are going to describe support our opinion that in several cases the optical response of particles depends more on their general symmetry and on the quantity of refractive material than on the details of their shape.

Taking into account that the scatterers we will deal with either are spherical or are aggregates of spherical scatterers, it is convenient to consider the quantities

$$P'_\eta = \frac{4\pi}{kS} \operatorname{Re}\bigl[f_{\eta\eta}(\hat{\boldsymbol{k}}_\mathrm{S}, \hat{\boldsymbol{k}}_\mathrm{I})\bigr], \quad Q'_\eta = \frac{4\pi}{kS} \operatorname{Im}\bigl[f_{\eta\eta}(\hat{\boldsymbol{k}}_\mathrm{S}, \hat{\boldsymbol{k}}_\mathrm{I})\bigr], \tag{6.1}$$

where S is either the geometrical cross section of the spherical scatterer or the sum of the geometrical cross sections of the spheres that form the aggregate.

According to (2.7), $Q' = Q_T$, i.e., Q' coincides with the extinction efficiency, when $\hat{k}_S = \hat{k}_I$ and $S = G$. Of course, the latter condition is automatically fulfilled for a scatterer with a spherical shape.

6.1 General Features of Scattering from Aggregated Spheres

We start by applying the theory in Chaps. 3 and 4 to a few simple clusters whose morphology is chosen so as to highlight the main features of scattering from these nonspherical model particles. All the results discussed in this section were obtained by assuming that the incident light is circularly polarized with $\eta - 1$. Almost all of them, however, are independent of η, either because of the orientation of clusters or because of the averaging over orientations. Anyway, index η is omitted throughout. Since a complex refractive index is not essential for our purpose, we assume that the component spheres have a real (nonabsorptive) index of refraction. As a consequence, according to Sect. 2.2, $\sigma_S = \sigma_T$. As a matter of fact, (4.24) and (4.25), which give the multipole amplitudes of the scattered field, show that the difference between an assembly of independent spheres and a true aggregate of spheres stems from the matrix elements $\mathcal{H}^{(pp')}_{\alpha l m \alpha' l' m'}$ that describe the multiple scattering processes among the components of the aggregate. This suggests that the effects due to clustering are well described by the quantity $\Gamma = \sigma_S/\sigma_0$, where σ_S is the scattering cross section of the cluster and σ_0 is the sum of the scattering cross sections of the constituent spheres considered as independent scatterers. σ_0 is calculated through the Mie theory. As σ_S depends on the orientation of the cluster with respect to the incident field, whereas σ_0 does not, Γ gives full information on the effects due to the lack of sphericity and was, therefore preferred to the scattering efficiency Q_S (2.7).

As a first example [79], we consider a cluster composed of two identical spheres of radius ρ and (frequency-independent) refractive index $n_0 = 1.3$. The surrounding medium is assumed to be the vacuum ($n = 1$). The scattering features of the constituent spheres are known to depend on the size parameter $x = k\rho$, but, in view of the multiple scattering processes that occur between them, it is to be expected that the features of the cluster as a whole depend on the parameter kd, where d is the distance between the centers of the spheres. Accordingly, in Fig. 6.1 we report Γ as a function of ϑ_I, the angle between the incident wavevector and the axis of the cluster, for several values of kd. The incident wave is left-circularly polarized and the size parameter of the spheres is held fixed at $x = 0.1$. Since this choice of the size parameter coincides with that of Levine and Olaofe [80], who gave an analytic solution for the dependent scattering from two small spheres, the results that we are going to discuss are directly comparable with theirs. Actually, spheres with so small a size parameter can be dealt with in the RSA (see Sects. 3.5.1

6.1 General Features of Scattering from Aggregated Spheres

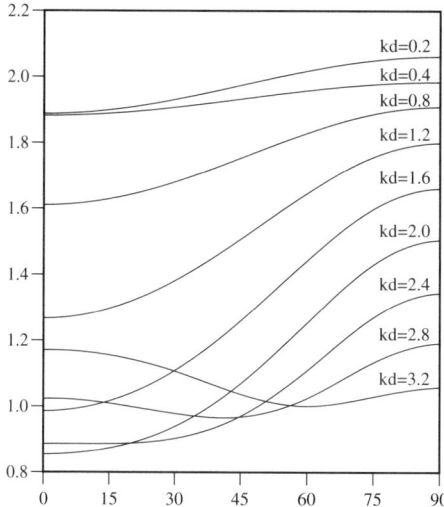

Fig. 6.1. Γ for the binary cluster as a function of ϑ_I for several values of kd

and 4.1), and, because $kd = 0.2$ when the spheres are in mutual contact, one could argue that the whole cluster also can be dealt with $l_\mathrm{M} = 1$. The results that we present in Fig. 6.1 show that it is not so. Actually, comparison of our results with those of Levine and Olaofe [80] shows that Γ decreases with increasing kd. Nevertheless, the rate of diminution of Γ is noticeably smaller than that predicted by Levine and Olaofe. Indeed, the approximation that they use amounts just to truncating the multipole expansions to $l = 1$, whereas we had to use (4.29) and (4.31), and up to $l_\mathrm{I} = 3$ (see Sect. 4.4), to get fully convergent results. We are thus led to conclude that the RSA cannot be extended to the cluster as a whole, even though it would certainly apply to each of the constituent spheres if they were not in the presence of one another. This conclusion is in substantial agreement with the considerations of Barber and Wang [81].

A further consideration that highlights the role of the multiple scattering processes stems from the remark that, for the highest values of kd reported in Fig. 6.1, the ratio Γ may be smaller that unity. This somewhat surprising result is a manifestation of the so-called shadow effect. In fact, for $\vartheta_\mathrm{I} = 0°$ the incidence is head-on, so that the incident beam *sees* one sphere only. The spheres are seen as two distinct objects when the angle of incidence increases to a sufficient value. Ultimately, one should bear in mind that the details of the geometry of a scattering particle become well distinguishable only for sufficiently small wavelength.

As a second example [79], we present Γ for a cluster of three identical spheres at the vertices of an equilateral triangle. This cluster, that will be referred to as "ozone" because its geometry recalls the stick-and-ball model of

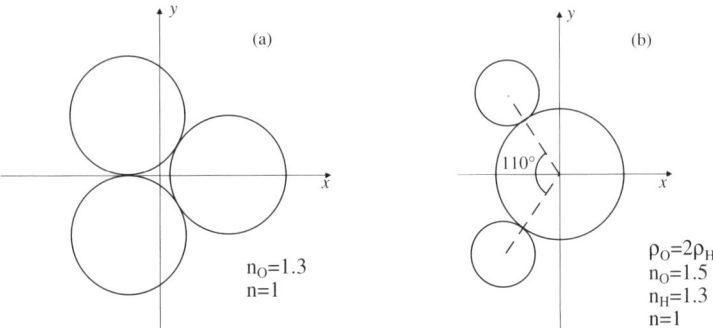

Fig. 6.2a,b. Geometry for **(a)** "ozone" and **(b)** "water"

this molecule, is depicted in Fig. 6.2 (a). In this case the component spheres are held in mutual contact and the calculations are performed for several directions of incidence as a function of $x = k\rho$. Therefore, we had to use $l_M = 5$ to achieve full convergence. The results are presented in Fig. 6.3. We first notice that for all values of x there is an evident dependence of Γ on ϑ_I that becomes more striking with increasing value of x, whereas we found that the dependence on φ_I is rather weak because of the high symmetry around the z axis. For this reason, results are reported for $\varphi_I = 0°$ only. The most important feature of Fig. 6.3 is the intersection of all the curves at about the same value of x. The dotted line is the average of Γ for random distribution of the orientations and passes almost exactly through the crossing point. We will see below that this feature does not pertain to "ozone" only.

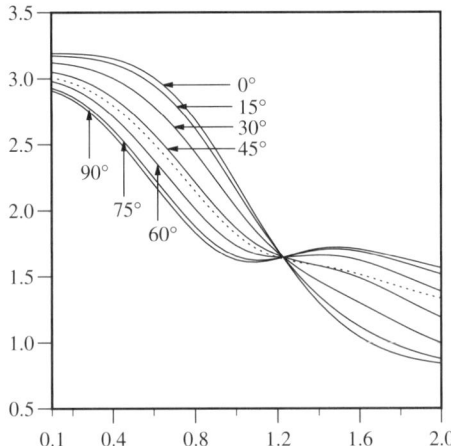

Fig. 6.3. Γ for "ozone" as a function of x for several values of ϑ_I and $\varphi_I = 0°$. The *dotted curve* is the average for random distribution of the orientations

6.1 General Features of Scattering from Aggregated Spheres 125

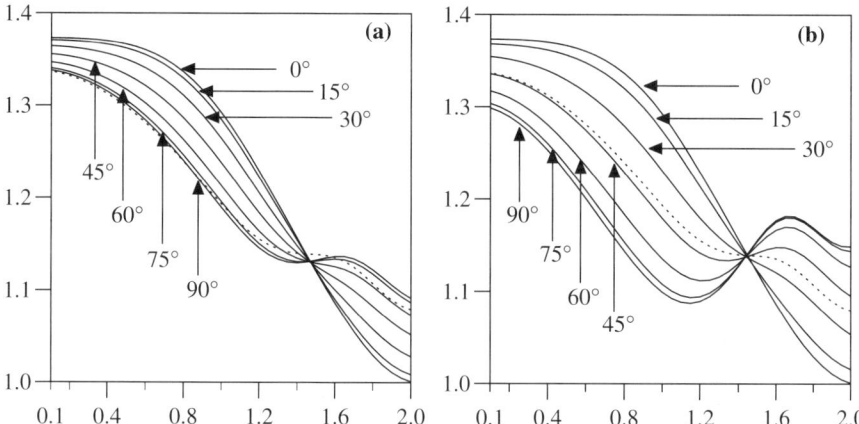

Fig. 6.4a,b. Γ for "water" as a function of $x = k\rho_\mathrm{O}$ for several values of ϑ_I. In (a) $\varphi_\mathrm{I} = 0°$ and in (b) $\varphi_\mathrm{I} = 90°$. The *dotted curve* is the average for random distribution of the orientations

The third cluster that we consider has a geometry depicted in Fig. 6.2 (b) and will be referred to as "water" [24, 79]. Full convergence required us to use $l_\mathrm{M} = 5$. The curves of Γ as a function of $x = k\rho_\mathrm{O}$ for several values of ϑ_I are reported in Fig. 6.4 (a) for $\varphi_\mathrm{I} = 0°$ and in Fig. 6.4 (b) for $\varphi_\mathrm{I} = 90°$. The curves in Fig. 6.4 (b), with the obvious exceptions of the dotted curve and of the 0°-curve, depend on η. Such a dependence is very slight, however, so that we felt unnecessary to report also the results for $\eta = 2$. Beyond the evident dependence of Γ on φ_I, we see again that all the curves intersect at about the same value of $x \approx 0.725$. The dotted curve, that also passes through, or almost through, the crossing point is the average of Γ for a randomly-oriented dispersion of the clusters. This result, together with analogous result in Fig. 6.3, suggests that there exist wavelengths at which nonspherical particles behave as if they were spherical and explains the success of the Mie theory in some cases in which the scattering particles are nonspherical. Of course, actual extinction spectra may be complicated by the possible frequency dependence of the refractive index.

6.1.1 Comparison with Experimental Data

Before we discuss further applications, it is convenient to assess the reliability of our results by comparison with experimental data. In recent times this task has been successfully performed by Mishchenko and Mackowski [82] in the optical range in spite of the experimental difficulties connected with an accurate determination of the geometrical and physical parameters of the scatterers [83]. We prefer, instead, to compare our results with the experimental data of Wang and Schuerman who performed a large set of experiments aimed at the measurement of the forward-scattering amplitude of *macroscopic* clusters of up to 8 spherical scatterers in the microwave range [84]. The specific purpose of the authors was the assessment of the contribution of the multiple scattering processes that occur among the component spheres to the scattering amplitude of the clusters as a whole [85]. The interested reader may find in the original technical report all the experimental details on the determination of the refractive index of the spheres and of their exact diameter. These are, indeed, critical input factors for any calculation meant to compare experiments with theoretical predictions. Note that the principle of electromagnetic scaling [43] that we already mentioned in Sects. 4.1 and 4.4 makes the results in the microwave range applicable to small clusters of spheres of the kind that are often found in atmospheric or industrial aerosols. This fact was borne in mind by the authors of the measurements. They point out that the refractive indexes of the clusters considered are near to $n_0 = 1.365$ that is appropriate for water and ice crystals in the optical range. Moreover, all their results are reported in terms of the adimensional parameters x_α and $x_{\alpha\beta}$ that we defined in Sect. 4.4.

In order to prove the reliability of our predictions, we compare the results of our calculations [86] with the experimental measurements for clusters of two identical spheres that Wang, Greenberg, and Schuerman report in their paper [85]. More precisely, we report P'_η vs. Q'_η for $\hat{k}_S = \hat{k}_I$ as a function of ϑ_I, the angle between the incident wavevector and the axis of the aggregate. The plane of scattering is defined by k_I and the axis of the aggregate and P'_η and Q'_η were calculated both for parallel and perpendicular polarization. In this respect we remark that Wang, Greenberg, and Schuerman do not report data for parallel polarization because of some technical problems that are fully explained in the original technical report [84]. We notice that the quantities P'_η and Q'_η were expressed in the original report in terms of the quantity $S_1(0)$ defined by van de Hulst, and the incident plane wave was assumed to depend on time through the factor $\exp(i\omega t)$. This implies the mutual exchange of the real and imaginary parts of the scattering amplitude and a redefinition of the normalization factors. The sign of the imaginary part of the refractive index must also be changed [16, 53]. Anyway, the vector drawn from the origin of the axes to any point on the curves gives the forward-scattering amplitude of the cluster. The length of the vector is the magnitude of the scattering amplitude whereas the angle that the vector forms with the Q'-axis gives the phase. In all the cases that we consider below the estimated experimental error, according to Wang and Schuerman, is 12° in phase and 10% in magnitude.

In Fig. 6.5 we compare our predictions for two binary clusters whose component spheres are mutually contacting. Figure 6.5 (a) refers to clusters of acrylic with complex refractive index. The clusters in Fig. 6.5 (b) are composed of expanded polystyrene and have a real refractive index. In both cases, full convergence required $l_M = 12$. Note that although in both figures our curves were calculated anew for each direction of incidence, their behavior

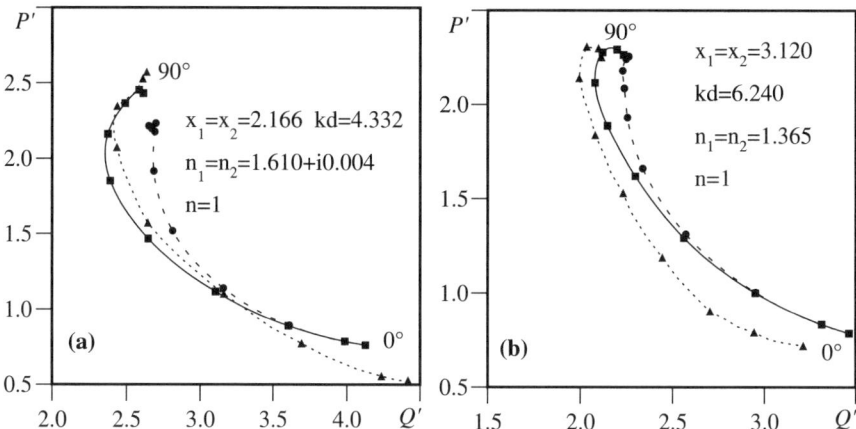

Fig. 6.5a,b. Comparison of the calculated and experimental dependence on ϑ_I of the forward-scattering amplitude of a binary cluster of acrylic (**a**) and of expanded polystyrene (**b**). The *dotted curve* shows the experimental results for polarization orthogonal to the plane of scattering. The *solid curve* is the corresponding result of our calculations. We also report for the sake of completeness our results for parallel polarization. The *symbols* on all the curves mark the value of ϑ_I at 10° intervals. For the component spheres $Q' = 2.7031$, $P' = 2.4154$ in (**a**) and $Q' = 2.2777$, $P' = 2.2956$ in (**b**)

turns out to be consistent with the transformation properties of the transition matrix. The dots both on the theoretical and on the experimental curves mark the value of ϑ_I at 10° intervals. For completeness, we also report the curves for parallel polarization that, as expected, do not differ much from those for perpendicular polarization. In particular the calculated curves for parallel and perpendicular polarization stick together for $\vartheta_\mathrm{I} = 0°$, as they must for head-on incidence. The general agreement of the theoretical and experimental curves appears to be rather good, the differences being well within the experimental error.

A further comparison is effected in Fig. 6.6 with the measurements for head-on incidence on a binary cluster whose component spheres were allowed to increase their separation. In this respect, we recall that the rate of convergence depends on kd, i.e., on the overall size of the aggregate. According to Sect. 4.4, to get well convergent results at the highest values of kd, calculations would require a large l_E. However, performing the calculation by use only of (4.22)–(4.24) and (4.26) with $l_\mathrm{M} = l_\mathrm{E}$ would require an undue effort. Since this kind of calculation does not imply orientational averages, we had the opportunity to perform the calculations with the minimum effort by using (4.29) and (4.31) and comparatively small values for l_I. In fact, l_I decreases from 13 for contacting spheres to 7 for well-separated spheres.

Actually, our curves show, in even greater detail, the wiggles that were observed in the experiment when the component spheres are well separated. In

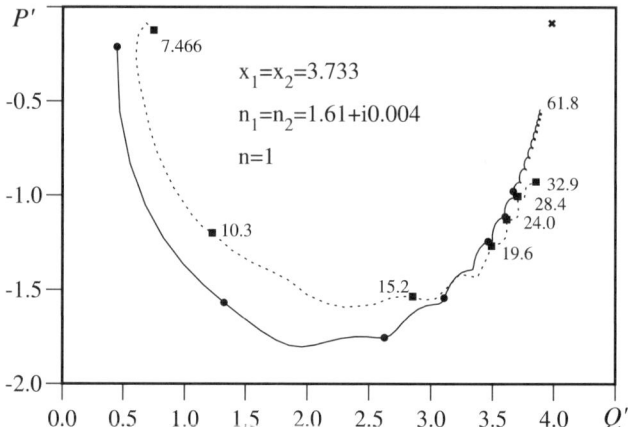

Fig. 6.6. Comparison of calculated and experimental dependence of the forward-scattering amplitude of a binary cluster of acrylic for head-on incidence on the mutual separation of the component spheres. The *dotted curve* reports the experimental data, whereas our predictions are reported by the *solid curve*. The *symbols* on all the curves report the value of kd. The *cross* at $Q' = 3.9807$, $P' = -8.4684 \cdot 10^{-2}$ gives the forward-scattering amplitude of the cluster for infinite separation of the component spheres

Fig. 6.6 the value of the scattering amplitude for infinite separation is marked by a small cross. Of course, neither the experiment nor our calculations could be stretched to this limit, but we were able to extend our calculations to a larger value of kd than allowed by the experimental setup. On account of the experimental error, the comparison effected in Fig. 6.6 shows that even in this case the reliability of the theory and of our codes is rather good.

6.1.2 Effect of the Structural Changes

The results in Figs. 6.1, 6.3, and 6.4 show that the extinction cross section of a cluster has a noticeable dependence on the geometry. It is therefore to be expected that if the clusters of a dispersion are allowed to recombine to form clusters of different geometry, the process should produce detectable effects on the extinction. Of course, studying these effects is not of academic interest only because such recombination with consequent changes of morphology are known to occur among the aerosol particles.

Let us recall that the refractive index of a tenuous dispersion of particles is given by (2.26) and that, when more than one kind of particle is present, the refractive index of the whole dispersion is the linear combination of the indexes of the partial dispersions [87]. In particular, the absorption coefficient for a dispersion of N particles, turns out to be

$$\gamma_\eta = 2k \operatorname{Im}[\mathcal{N}_{\eta\eta}] = N \frac{4\pi}{k} \sum_s c_s \operatorname{Im}[\langle f_{s,\eta\eta}(0)\rangle] ,$$

6.1 General Features of Scattering from Aggregated Spheres

where Nc_s is the number density of the particles of the sth species and the average of the forward-scattering amplitude is given by the formulas in Sect. 3.4. The extent to which the optical properties of a dispersion of clusters is influenced by the rearrangement will be studied by comparing γ_η before and after the rearrangement has taken place. To this end, let

$$\gamma_\eta^{(i)} = \sum_s c_s \gamma_{s,\eta}$$

be the initial absorption coefficient and

$$\gamma_\eta^{(f)} = \sum_{s'} c_{s'} \gamma_{s',\eta}$$

the absorption coefficient after the rearrangement, where s' denotes the species of the particles after the rearrangement. Note that the particles may be quite different both in structure and in concentration. From the preceding equations, one easily sees that a useful quantity for the detection of the changes is

$$\Gamma_\eta = \frac{p\gamma_\eta^{(f)} + (1-p)\gamma_\eta^{(i)}}{\gamma_\eta^{(i)}},$$

where $0 \leq p \leq 1$ is the completion index of the process. The effect of the clustering for the single species is given by

$$g_{s,\eta} = \gamma_{s,\eta}/\gamma_{0s},$$

where γ_{0s} is the sum of the extinction cross sections, calculated by the Mie theory, of the spheres that form the cluster of the sth species.

The remainder of this section is devoted to the discussion of the results that we obtained for a few processes that are listed in Table 6.1. In our calculations we assumed circular polarization of the incident wave and random distribution of the orientations for all the clusters. The surrounding medium is assumed to be the vacuum ($n = 1$). Note that, since the clusters recall the popular stick-and-ball model of molecules, we indicated for each process the actual chemical reaction by which both the clusters and their rearrangements were inspired. The reader is warned, however, that the spheres and the clusters we deal with are macroscopic objects and in no case they should be understood as actual atoms and molecules. The clusters are individuated by their "chemical formulas" in terms of the "elements" A, B, and C and their structures are built so as to match those of the actual chemical compounds listed under the heading Example in Table 6.1. The processes are partitioned into three groups: the group 1–4 implies the spheres A and B only; the group 5 and 6 also implies the spheres C; the group 7–10 includes processes that from the same "reagents" produce different structures.

130   6. General Applications

Table 6.1. List of the processes. The refractive indexes and the radii of the elemental spheres A, B, and C are $n_A = 1.30$, $n_B = 1.50$, $n_C = 1.40$, and $\rho_A = 1.0\,\mathrm{u_l}$, $\rho_B = 0.50\,\mathrm{u_l}$, $\rho_C = 0.75\,\mathrm{u_l}$, where $\mathrm{u_l}$ is an arbitrary unit of length

	Process	Example
1	$3B_2 \to 2B_3$	$3O_2 \to 2O_3$
2	$2AB + B_2 \to 2AB_2$	$2CO + O_2 \to 2CO_2$
3	$2AB_2 + B_2 \to 2AB_3$	$2SO_2 + O_2 \to 2SO_3$
4	$A_2 + 3B_2 \to 2AB_3$	$N_2 + 3H_2 \to 2NH_3$
5	$AC + 3B_2 \to AB_4 + B_2C$	$CO + 3H_2 \to CH_4 + H_2O$
6	$2AB_3 + C_2 \to 2ACB_3$	$2PH_3 + O_2 \to 2POH_3$
7	$C_2 + AB_4 \to AB_3C + BC$	$Cl_2 + CH_4 \to CH_3Cl + HCl$
8	$2C_2 + AB_4 \to AB_2C_2 + 2BC$	$2Cl_2 + CH_4 \to CH_2Cl_2 + 2HCl$
9	$3C_2 + AB_4 \to ABC_3 + 3BC$	$3Cl_2 + CH_4 \to CHCl_3 + 3HCl$
10	$4C_2 + AB_4 \to AC_4 + 4BC$	$4Cl_2 + CH_4 \to CCl_4 + 4HCl$

In Table 6.2 we list the clusters as well as the processes in which they are involved according to Table 6.1. We also report the coordinates of the centers of the spheres in each cluster. In this respect, we stress that neighboring spheres are always in mutual contact. The quantities Γ and g_s were calculated for k in the range from $0.002\,\mathrm{u_l^{-1}}$ to $2.0\,\mathrm{u_l^{-1}}$ but in the following figures k goes from $0.02\,\mathrm{u_l^{-1}}$ to $2.0\,\mathrm{u_l^{-1}}$ because in this range the results show the most striking k-dependence. We had to use at most $l_M = 10$ to get full convergence.

In Figs. 6.7, 6.9, and 6.11 we report Γ for the processes listed in Table 6.1. The values of g_s for the clusters concerned are reported in Figs. 6.8, 6.10, and 6.12. Note that, due to the different normalization, the corresponding curves of Γ and g_s are not directly comparable. Nevertheless the chosen normalization makes Γ independent of N and g_s independent of $N_s = Nc_s$. It should be borne in mind, however, that the results we are going to discuss hold true as long as we consider a tenuous dispersion. We start remarking that, from a general point of view, Figs. 6.8, 6.10, and 6.12 show that the multiple scattering processes that occur among the spheres in a cluster cannot be neglected without dramatically affecting the correctness of the results. Therefore it is an easy prediction that the rearrangement of the clusters should cause detectable changes in the absorption coefficient of the dispersion. This is actually shown by the curves in Figs. 6.7, 6.9, and 6.11, where Γ is significantly different from unity in the whole range of k concerned, although with increasing k the effect of the rearrangement becomes smaller and smaller.

Of course, the more the process changes the structure of the clusters the more Γ remains different from unity. A similar consideration also applies to the curves of g_s. Figures 6.8, 6.10, and 6.12, indeed, show that the curves differ

6.1 General Features of Scattering from Aggregated Spheres   131

Table 6.2. Coordinates of the centers of the spheres constituting the clusters. The coordinates refer to each sphere in the same order in which they appear in the name of the cluster. The processes in which the clusters are involved are also listed

Cluster	Coordinates	Processes
B_2	(-0.5, 0, 0) (0.5, 0, 0)	1, 2, 3, 4, 5
A_2	(-1.0, 0, 0) (1.0, 0, 0)	4
C_2	(-0.75, 0, 0) (0.75, 0, 0)	6, 7, 8, 9, 10
AB	(-1.0, 0, 0) (0.5, 0, 0)	2
AC	(-1.0, 0, 0) (0.75, 0, 0)	5
BC	(-0.5, 0, 0) (0.75, 0, 0)	7, 8, 9, 10
B_3	(0, 0.577, 0) (0.5, -0.289, 0) (-0.5, -0.289, 0)	1
AB_2	(0, 0, 0) (-1.5, 0, 0) (1.5, 0, 0)	2
AB_2	(0, 0, 0) (1.229, -0.860, 0) (-1.229, -0.860, 0)	3
B_2C	(1.024, -0.717, 0) (-1.024, -0.717, 0) (0, 0, 0)	5
AB_3	(0, 0, 0) (0, 1.5, 0) (1.299, -0.75, 0) (-1.299, -0.75, 0)	3
AB_3	(0, 0, 1.384) (0.5, -0.289, 0) (-0.5, -0.288, 0) (0, 0.577, 0)	4, 6
AB_4	(0, 0, 0) (-0.866, -0.866, 0.866) (-0.866, 0.866, -0.866) (0.866, -0.866, -0.866) (0.866, 0.866, 0.866)	5, 7, 8, 9, 10
ACB_3	(0, 0, 1.384) (0, 0, 3.134) (0, 0.577, 0) (0.5, -0.289, 0) (-0.5, 0.289, 0)	6
AB_3C	(0, 0, 1.384) (0, 0.577, 0) (0.5, -0.289, 0) (-0.5, -0.289, 0) (0, 0, 3.134)	7
AB_2C_2	(0, -0.375, 1.452) (0, 0, 0) (0, -0.375, 2.952) (0.75, -1.0, 0) (-0.75, -1.0, 0)	8
ABC_3	(0, 0, 3.021) (0, 0, 1.521) (0, 0.866, 0) (0.75, -0.433, 0) (-0.75, -0.433, 0)	9
AC_4	(0, 0, 0) (-1.010, -1.010, 1.010) (-1.010, 1.010, -1.010) (1.010, -1.010, -1.010) (1.010, 1.010, 1.010)	10

from each other the more the structures of the clusters are different, so that similar structures have virtually the same behavior. This is confirmed, e.g., by the curves for CO_2 and SO_2 that are indistinguishable from each other. The spheres that compose the two clusters AB_2 are identical, and the differences of structure are too small to give appreciably different spectra, especially on account of the randomness of the orientations. We also notice the identity of the spectra of NH_3 in Fig. 6.8 and PH_3 in Fig. 6.10. According to Table 6.2, both clusters have AB_3 composition and the same pyramidal structure. The difference from the spectrum of SO_3 in Fig. 6.8 is easily explained on account that the latter cluster is also AB_3 but has a planar structure.

The considerations above led us to conclude that the change of structure of clusters, whatever their cause, have well detectable effects on the absorption coefficient of a dispersion even when the orientations of the particles are randomly distributed.

132   6. General Applications

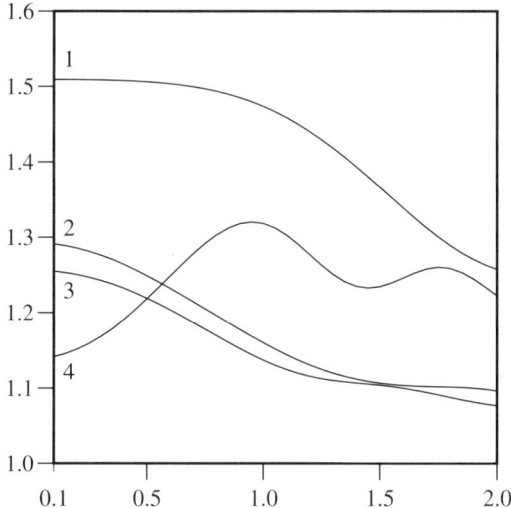

Fig. 6.7. Γ vs. k (in arbitrary units) for $p = 1$, for the first group of processes in Table 6.1

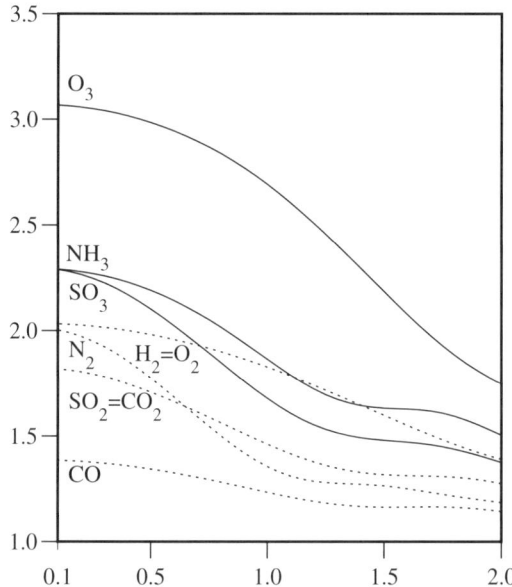

Fig. 6.8. g_s vs. k (in arbitrary units) for the first group of processes in Table 6.1. The *solid lines* refer to the "reaction products," whereas the *dotted lines* refer to the "reagents"

6.1 General Features of Scattering from Aggregated Spheres   133

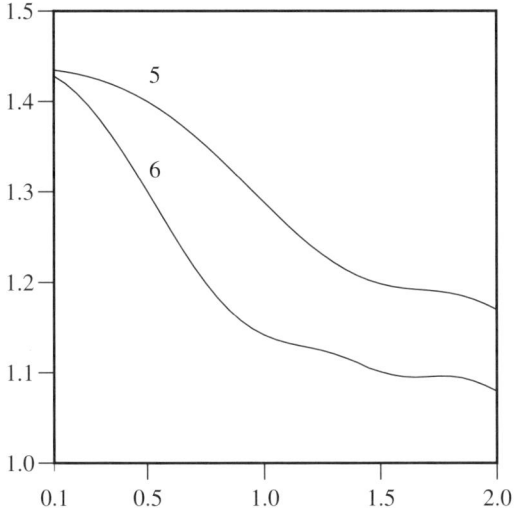

Fig. 6.9. Same as Fig. 6.7 but for the second group of processes in Table 6.1

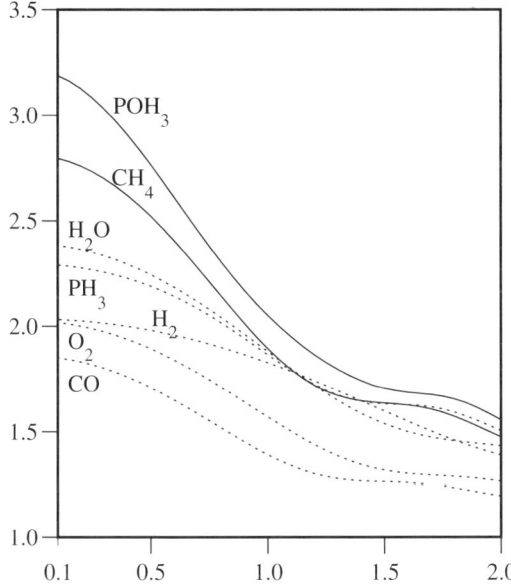

Fig. 6.10. Same as Fig. 6.8 but for the second group of processes in Table 6.1

134   6. General Applications

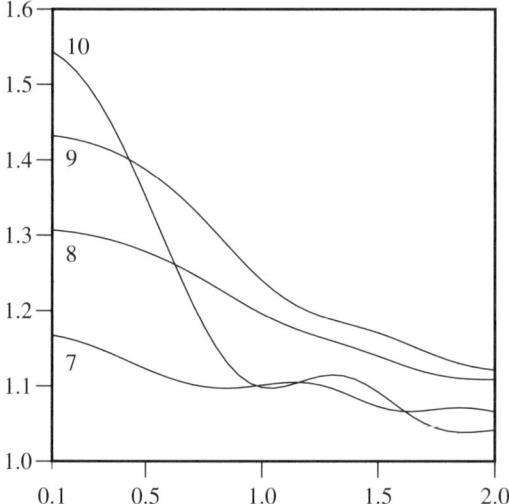

Fig. 6.11. Same as Fig. 6.7 but for the third group of processes in Table 6.1

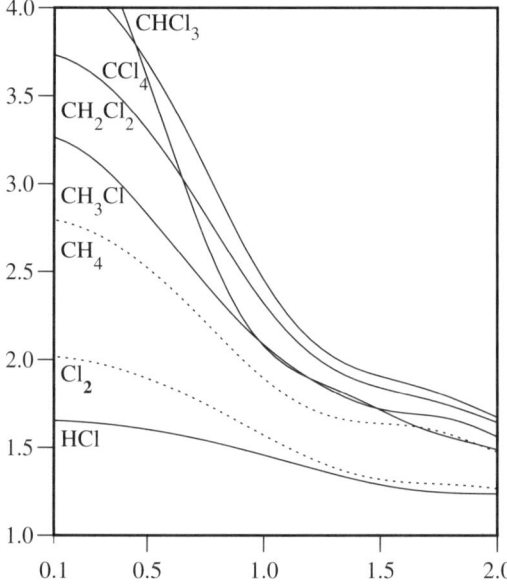

Fig. 6.12. Same as Fig. 6.8 but for the third group of processes in Table 6.1

6.2 Clusters in an Electrostatic Field

In Sect. 3.5 we stressed that a possible cause of nonrandom orientational distribution could be the presence of an external electrostatic field. The theory developed so far makes us able to investigate the extent to which the optical properties of a dispersion of clusters are influenced by the presence of an external electrostatic field [34]. As a representative example we chose aggregates of four identical spheres each of radius $\rho = 0.1\,\mu$m. Using these building blocks two kinds of aggregates were formed: the first kind, which should simulate a prolate particle, was built by arranging the four spheres in a linear chain; the second kind, which was shaped by arranging the spheres at the vertices of a square, should model an oblate particle. In both kinds of cluster the neighboring spheres are in mutual contact. The wavevector of the incoming plane wave was chosen to be either parallel or orthogonal to the electrostatic field and, in the latter case, the polarization was chosen either parallel or perpendicular to the field itself.

The dielectric function of the spheres was assumed in the form [88]

$$\frac{\epsilon - \epsilon_\infty}{\epsilon_0 - \epsilon_\infty} = \frac{\omega_T^2}{\omega_T^2 - \omega(\omega + i\Gamma\omega_T)}$$

that is appropriate for a dielectric material, where ω_T is the (circular) frequency of the transverse phonons. In our calculations, the parameters of MgO were assumed according to [89], i.e., $\omega_T = 7.5 \cdot 10^{13}\,\text{s}^{-1}$, $\epsilon_0 = 9.8$, $\epsilon_\infty = 2.95$, and $\Gamma = 0.27$. As a result, using (3.46)–(3.48), we obtained $\alpha_\perp = 0.259 \cdot 10^{-14}\,\text{cm}^3$, $\alpha_\parallel = 0.435 \cdot 10^{-14}\,\text{cm}^3$ for the prolate particles and $\alpha_\perp = 0.336 \cdot 10^{-14}\,\text{cm}^3$, $\alpha_\parallel = 0.245 \cdot 10^{-14}\,\text{cm}^3$ for the oblate particles. In this respect, let us recall that, according to (3.47), the calculation of the polarizability tensor requires the elements of the transition matrix for $l = l' = 1$ only, whereas the calculation of the optical properties requires well converged values of the whole transition matrix. The results that we are going to discuss required at most $l_M = 11$.

In view of the cylindrical symmetry of the linear chains, their transition matrix can be made diagonal in m by choosing the local reference frame with the z axis parallel to the cylindrical axis. For the square clusters we notice that, by choosing the local reference frame with the z axis along the quaternary axis, the elements with $l, l' = 1$ of their transition matrix, i.e., the elements that are relevant for the calculation of the polarizability, are nonzero for $m' = m$ only. This is not a numerical accident. In fact, group theoretical considerations show that it is a necessary consequence of the D_{4h} symmetry of the square clusters [90, 79]. It is for this reason that (3.47) that we established in Sect. 3.5.1 for axially symmetric scatterers apply to both kinds of clusters and the orientational distribution function depends on the Eulerian angle β only for both kinds of clusters. The resulting function $v(\beta)$ is reported in Fig. 6.13 at $T = 300°$K for some values of the strength

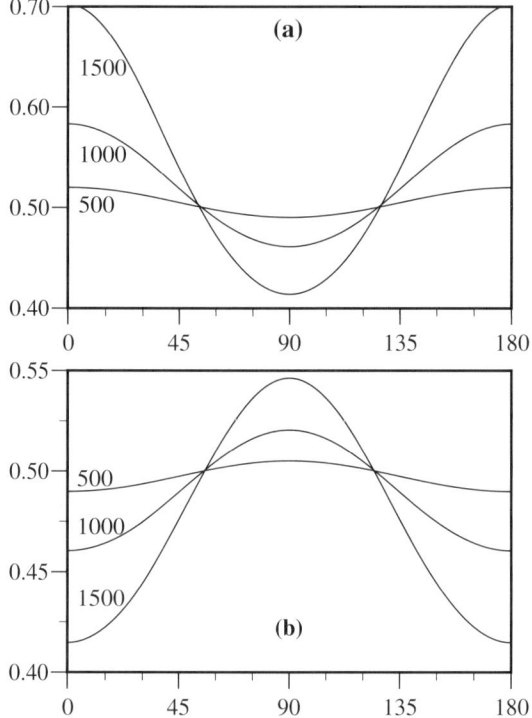

Fig. 6.13a,b. Weight function $v(\beta)$ for the linear chains (**a**) and for the square clusters (**b**). The curves are labelled by the strength of the external electrostatic field (in V/cm)

of the static field, that, as usual, is given in Volt/cm whereas electrostatic units were used in (3.37). Figure 6.13 shows that the difference $\alpha_\| - \alpha_\perp$ is paramount in determining the distribution of the orientations. It is easily seen that the prolate particles tend to orient their symmetry axis along the electrostatic field whereas the quaternary axis of the oblate particles tends to become perpendicular to the field. Thus, in the limit of infinite strength of the field, all the prolate particles orient with their cylindrical axes parallel to the field itself, whereas all we can state on the orientation of the oblate particles is that their quaternary axes lie in a plane that is perpendicular to the field but that the particles are otherwise randomly oriented.

In view of the results reported in Fig. 6.13, it is convenient to see preliminarily to what extent the orientational distribution influences the optical properties. For a dispersion with number density N_0 we define the quantity

$$\Delta G_\eta = \frac{1}{N_0} \left[\mathrm{Im}(\mathcal{N}_{\eta\eta}^\infty) - \mathrm{Im}(\mathcal{N}^0) \right],$$

where the factor $1/N_0$ has been introduced to normalize ΔG_η to unit number density and $\mathcal{N}_{\eta\eta}^\infty$ and $\mathcal{N}^0$ were calculated by using in (3.24) $N = N_0 P(\Theta)$, where $P(\Theta)$ is the weight function appropriate for infinite and for zero strength of the electrostatic field, respectively. Since for $E = \infty$ the linear chains are all likely oriented, one has $\langle f_{\eta\eta} \rangle = f_{\eta\eta}$. For $E = 0$ they are randomly oriented so that the relevant $\langle f_{\eta\eta} \rangle$ is given by (3.27a). We also notice that $\langle f_{\eta\eta} \rangle_{\text{random}}$ is actually independent of the polarization. Consequently, the index η has been dropped from $\mathcal{N}^0$. It is also convenient to define the quantity

$$g_\eta = \frac{1}{N_0} \text{Im}(\mathcal{N}_{\eta\eta}) ,$$

where $\mathcal{N}_{\eta\eta}$ is calculated at the actual strength of the applied electrostatic field.

We define ΔG_η even for the square clusters, but in this case the appropriate weight function is that reported in Sect. 3.4, Case 5. With the help of (3.30a) and (3.31) we get

$$\langle f_{\eta\eta'} \rangle_{\beta = 90°} = -\frac{i}{4\pi k} \sum_L \sum_{pp'} \sum_{ll'} \sum_{m\mu} (-)^{m-\mu} C(l,l',L;m,-m)$$

$$\times C(l,l',L;\mu,-\mu) W_{S\eta lm}^{(p)*} \bar{S}_{l\mu l'\mu}^{(pp')} W_{l\eta' l'm}^{(p')} P_L(0) .$$

The curves in Fig. 6.14, where ΔG_η is plotted as a function of the size parameter of the component spheres, are very useful for the choice of the value of x at which the effect of the electrostatic field is more evident. Indeed, for both kinds of clusters, ΔG_η vanishes for some value of x in agreement with the results in Sect. 6.1 [24] that there exist values of the size parameter at which the optical behavior of a dispersion of clusters is identical to that of a dispersion of spherical particles. Accordingly, in Fig. 6.16 we report g_η for a dispersion of linear chains of spheres as a function of the strength of the external field. Both figures show that the effect of the external field on the extinction coefficient of the dispersion is fairly visible even at relatively small values of E, i.e., when the number of the chains with their axis along the field is still relatively small. The corresponding results for the square clusters are reported in Fig. 6.17

Figures 6.16 and 6.17 show that the saturation of the effect for the linear chains is achieved for values of the field larger than the value that is needed for the dispersion of prolate particles. Anyway, the effect of the electrostatic field is quite noticeable even when its strength is so low that only a relatively small number of square clusters is oriented with the quaternary axis orthogonal to the field itself. Although all the curves in Fig. 6.16 and 6.17 go steadily towards the saturation, this should not be taken as a general rule. In fact, for the linear chains, we found that there are some ranges of x at which the

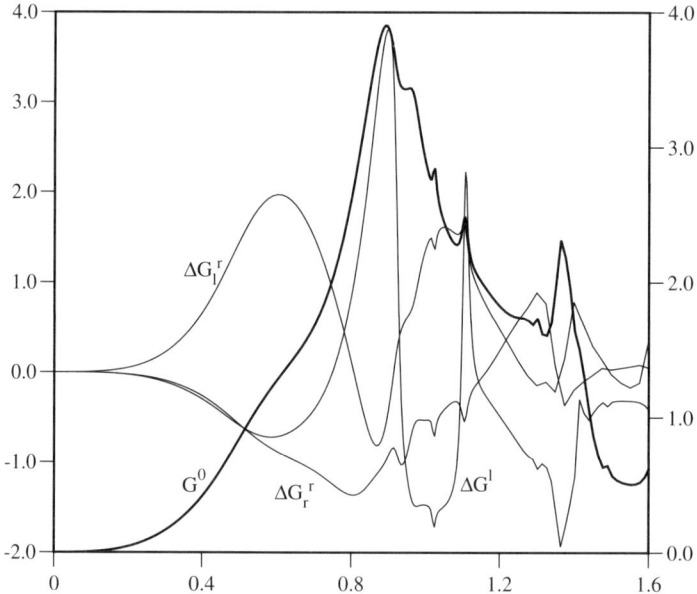

Fig. 6.14. $\Delta G_\eta \cdot 10^{14}$ for the linear chains as a function of the size parameter of the component spheres. The upper indices r and l refer to incidence perpendicular or parallel to the static field, respectively. The scale to the right refers to $G^0 = (1/N_0)\text{Im}[\mathcal{N}^0] \cdot 10^{14}$ (*thick line*) that is reported for the sake of completeness

curves of g_η as a function of the static field strength are nonmonotonic; for instance, this behavior occurs for $x > 1.2$, as is shown in Fig. 6.18(a) for $x = 1.35$ and in Fig. 6.18(b) for $x = 1.5$. In particular, Fig. 6.18(a) shows that g_η for perpendicular incidence is almost independent of the polarization up to ≈ 200 V/cm. On the contrary, in Fig. 6.18(b) the nonmonotonicity of g_l^r and g_r^r results in the crossing of the curves between 200 and 500 V/cm. This kind of behavior never occurs for the square clusters.

The examples and the results that were presented above show that our four-sphere clusters are suitable both to achieve an effective description of the optical phenomena that are produced by the electrostatic field and to contain the computational effort. We stated earlier in this section that the convergence of the calculations required at most $l_M = 11$. Therefore, the matrix to be inverted was of order 1144. The computational situation is rather different if only the polarizability need to be calculated. As the polarizability is a static quantity, indeed, one must take the limit for $k \to 0$ of the calculated transition matrix, according to (3.45). In this limit only the elements with $l = l' = 1$ need to be included in (4.22). The matrix to be inverted to calculate the transition matrix is thus of order $d = 3N = 12$ on account of the vanishing of the magnetic terms. As a result, the calculation of the polarizability tensor can be performed even for clusters that include a large number of spheres without undue computational effort.

6.2 Clusters in an Electrostatic Field 139

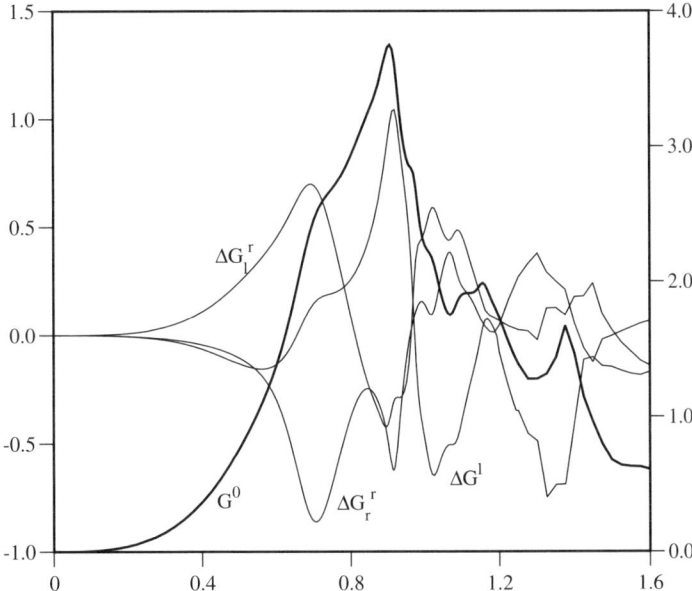

Fig. 6.15. $\Delta G_\eta \cdot 10^{14}$ for the square clusters as a function of the size parameter of the component spheres. The upper indices r and l refer to incidence perpendicular or parallel to the static field, respectively. The scale to the right refers to $G^0 = (1/N)\mathrm{Im}[\mathcal{N}^0] \cdot 10^{14}$ (*thick line*) that is reported for the sake of completeness

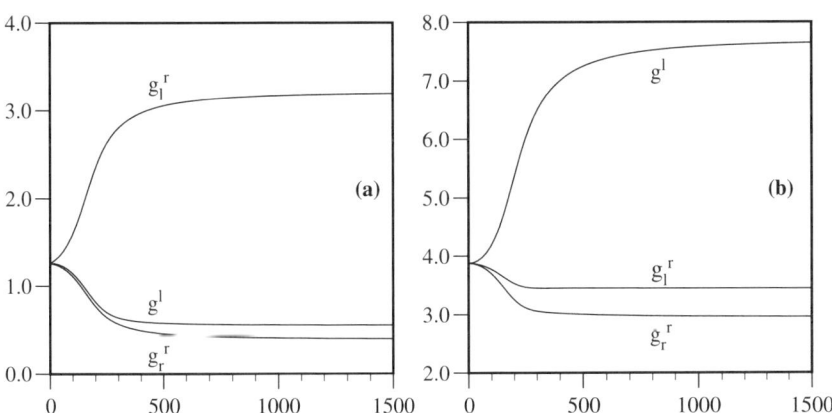

Fig. 6.16a,b. g_η in $\mathrm{cm}^3 \cdot 10^{-14}$ for the linear chains of spheres with size parameter $x = 0.6$ (**a**) and $x = 0.9$ (**b**) as a function of the electrostatic field strength (in V/cm)

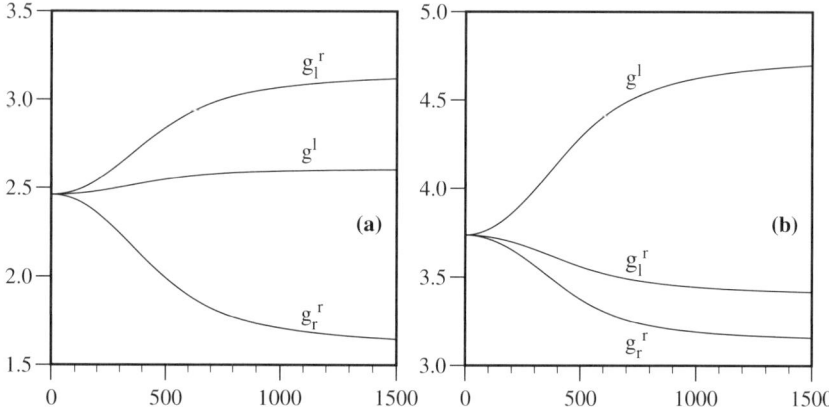

Fig. 6.17a,b. g_η in cm$^3 \cdot 10^{-14}$ for the square clusters with size parameter of the component spheres $x = 0.7$ **(a)** and $x = 0.91$ **(b)** as a function of the strength of the electrostatic field (in V/cm)

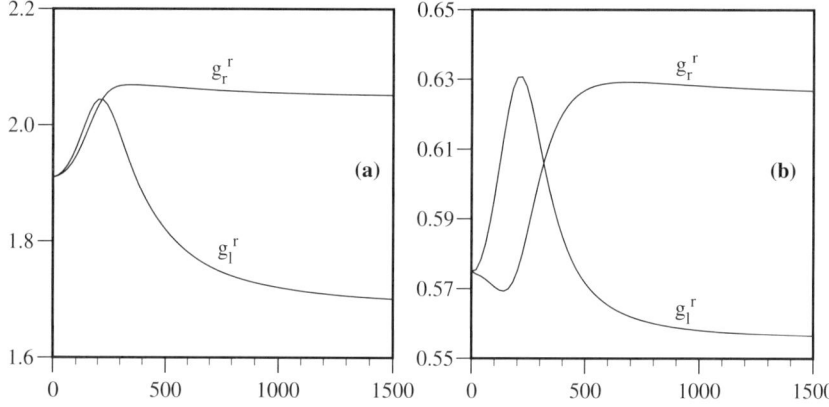

Fig. 6.18a,b. g_η^r in cm$^3 \cdot 10^{-14}$ for the linear chains of spheres with size parameter $x = 1.35$ **(a)** and $x = 1.5$ **(b)** as a function of the strength of the electrostatic field (in V/cm)

On account of the structure of the polarizability tensor, the modifications of the macroscopic optical properties that are induced by an electrostatic field cannot be used to gain information on the shape of the particles. In fact, it is not possible, on general grounds, to distinguish, e.g., a cone from a cylinder because the axial symmetry of both figures implies that only two components of their polarizability tensor are different. Moreover, one is neither able to distinguish a disk from a square because the polarizability tensor of both figures has an identical axially symmetric structure. The electrostatic field, however, may help in discriminating the optical properties of oblate particles from those of prolate particles. This is suggested by the crossing of the g_r^r

and of the g_l^r curves for the linear chains when $x > 1.2$. As stressed above, such crossings never occur for the square clusters. Actually, it is to be borne in mind that, in general, the collective optical behavior of the prolate and of the oblate particles looks rather similar. This is due to the fact that the different orientational response of the two kinds of particles to the electrostatic field tend to compensate for the different orientational dependence of their forward-scattering amplitudes.

6.3 Extinction from Single and Aggregated Layered Spheres

This section is devoted to discussing the extinction properties of single and aggregated concentrically layered spheres composed of materials with a frequency-dependent dielectric function. We also consider single and aggregated spheres in which the change from the refractive index of the bulk material to that of the surrounding medium is not sharp but occurs with continuity through a transition layer of suitable thickness according to (4.16) in Sect. 4.2. We say that these spheres have a soft surface. A previous study of the properties of all these kinds of radially nonhomogeneous spheres and of their binary aggregates had the specific purpose of assessing the effect of the aggregation on the resonances in their extinction spectrum [46]. Here we also focus on the effect of the imaginary part of the refractive index on the convergence of the calculations. In this respect, we recall that both the real and the imaginary part of the refractive index of metal particles may be rather large.

Hereafter, we consider spheres of two different radii, i.e., $\rho_1 = 10$ nm and $\rho_2 = 50$ nm, inclusive of the thickness of the covering layer or of the soft surface when present [46]. The dielectric function of the metallic material was assumed to be of the free electron Drude form [91, 92]

$$\epsilon = 1 - \frac{\omega_p^2}{\omega(\omega + i\Gamma\omega_p)}, \qquad (6.2)$$

where ω_p is the plasma frequency of the metal concerned. In turn, for the dielectric material we chose the damped-oscillator form [92] that we already used in Sect. 6.2. The relevant parameters were taken from Kittel [89] and are listed in Table 6.3. The calculations for the layered spheres where performed through the technique that we expounded in Sect. 4.2, i.e., by interposing between each pair of layers a thin transition layer within which the dielectric function varies continuously as a function of r according to (4.16). In all cases, the thickness of the transition layer was chosen to be 0.1 nm. We report, for $\hat{k}_S = \hat{k}_I$, Q'_{sp} for the single spheres and Q' averaged over randomly oriented aggregates as a function of $\nu = \omega/\omega_p$. When more than one metal is present,

Table 6.3. Parameters used in the calculations in this section

Material	ω_p (s^{-1})	ω_T (s^{-1})	Γ	ϵ_0	ϵ_∞
K	6.500×10^{15}		0.02		
Mg	1.610×10^{16}		0.01		
Al	2.400×10^{16}		0.01		
Ag	5.743×10^{15}		0.01		
MgO		7.5×10^{13}	0.27	9.8	2.95

ω_p is the highest of the implied plasma frequencies. All the results that we are going to discuss were obtained using $l_\mathrm{M} = 12$ in the whole range $0.1 \leq \nu \leq 1.1$, the corresponding wavelength ranging, in all cases, from the far infrared to the ultraviolet.

In the range of frequency we considered, all the scatterers present one or more resonance peaks. Since a resonance peak may be rather sharp, locating its maximum requires a fine scanning on the ν axis lest the peak be missed at all. Often, the preliminary location of the resonances is done by resorting to the so called Güttler formula [3] that estimates the polarizability of a small sphere coated by a thin concentric layer through effective medium considerations. For a coated sphere in vacuo Güttler formula reads

$$\alpha = \frac{(\epsilon_\mathrm{s} - 1)C_+ + (2\epsilon_\mathrm{s} + 1)C_-}{(\epsilon_\mathrm{s} + 2)C_+ + (2\epsilon_\mathrm{s} - 2)C_-}\rho^3,$$

with

$$C_+ = \epsilon_\mathrm{N} + 2\epsilon_\mathrm{s}, \quad C_- = (\epsilon_\mathrm{N} - \epsilon_\mathrm{s})(\rho_\mathrm{N}/\rho)^3 ,$$

where ϵ_N and ρ_N are the dielectric constant and the radius of the core, respectively, and ϵ_s is the dielectric constant of the coating. Then, with the help of the Lorentz–Lorenz formula [3]

$$\alpha = \frac{\epsilon - 1}{\epsilon + 2}\rho^3,$$

the effective dielectric constant of a coated sphere can be defined as

$$\epsilon_\mathrm{eff} = \frac{C_+ + 2C_-}{C_+ - C_-}\epsilon_\mathrm{s}$$

and the condition for the occurrence of a resonance reads

$$\epsilon_\mathrm{eff} + 2 = 0 . \tag{6.3}$$

Of course, (6.3) can safely be applied to small spheres only. In fact, the Güttler formula, from which it stems, can be applied only to spheres that can be dealt

6.3 Extinction from Single and Aggregated Layered Spheres

with in the small particle approximation, viz., when only dipole terms need to be considered. Moreover, it should be borne in mind that, besides the latter approximation, (6.3) encompass an effective medium approximation also [93]. Therefore, if the particle is not sufficiently small and the coating layer is not sufficiently thin, neither can the resonance peaks be accurately located, nor can ϵ_{eff} be introduced into the Mie theory to calculate the extinction spectrum. Equation (6.3) was successfully used in [46] to estimate the frequency of the resonances of 5 nm spheres. Nevertheless, our calculations show that (6.3) becomes unreliable when applied to the 10 nm spheres.

6.3.1 Metal Spheres with a Soft Surface

Considering spheres with a soft surface may be a useful tool to describe scatterers whose radius is not perfectly known, for instance because of smooth irregularities of the surface. This kind of sphere is often invoked in astrophysics to explain the optical behavior of dust particles originated by accretion by interstellar gas atoms.

As far as we know, spheres with a soft surface were first dealt with by Ruppin [47] who calculated the extinction from 5 nm spheres with a soft surface of 0.4 nm thickness within which the transition from the bulk material to the surrounding medium occurs according to a linear law. On the contrary, we assume a law of transition given by (4.16) and in Fig. 6.19 report Q'_{sp} for spheres of potassium with a soft surface whose thickness is 1 nm for the 10 nm spheres and 5 nm for the 50 nm spheres, and Q' for their binary aggregates. We first remark that the results for the 10 nm spheres are practically identical to those for the 5 nm spheres with a soft surface of 0.4 nm thickness that are reported in [46] and that were in fair agreement with the results of Ruppin [47]. It is therefore not surprising that the curves in Fig. 6.19 (a) and (b) are also in good agreement with those of Ruppin, both in the position of the peaks and in their width, in spite of our choice of a different law of variation of the refractive index within the transition layer. This suggests that the effect of the diffuseness is qualitatively almost independent of the details of the radial dependence of ϵ and, to a certain extent, also of the radius of the spheres. The interested reader can find in [45] a thorough analysis of the effects that are due to the shape of the radial dependence. The results in Fig. 6.19 show that, as a general rule, the presence of a diffuse surface attenuates the extinction peaks and makes noticeably smoother the spectra both of the single spheres and of their binary aggregates. We do not present here specific results but we can state that, according to our calculations, this behavior is accentuated by the increase of the thickness of the transition layer. In turn, the curves of g for the binary aggregates of spheres with a sharp surface show the duplication of the peaks that is expected as far as only dipole terms ($l_{\text{M}} = 1$) are considered. Nevertheless, we also see the appearance of further peaks whose maxima are rather small, however. When we go to consider the binary aggregates of spheres with a soft surface, we

144   6. General Applications

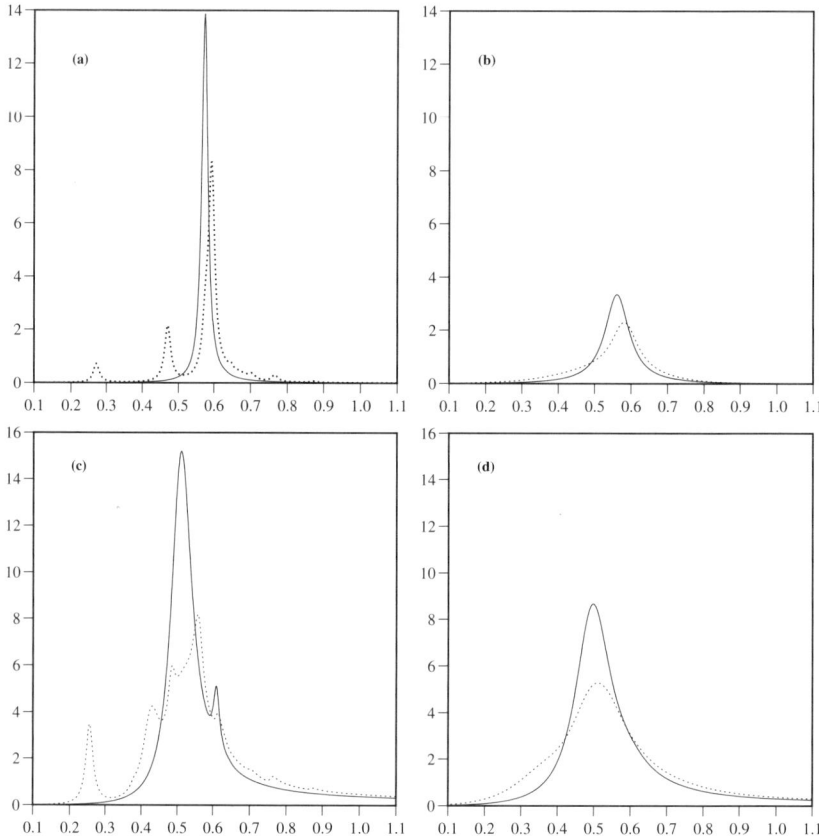

Fig. 6.19a–d. $Q'_{\rm sp}$ (*solid curves*) and Q' (*dotted curves*) as a function of ν for spheres of potassium and for their binary aggregates in random orientation. (**a**) and (**c**) refer to 10 nm and to 50 nm spheres with a sharp surface. (**b**) and (**d**) refer to 10 nm and to 50 nm spheres with a soft surface whose thickness is 1 nm and 5 nm, respectively

see that the duplication of the peaks occurs in the form of smooth shoulders, especially for the 50 nm spheres.

6.3.2 Metal Spheres with a Dielectric Coating

In Fig. 6.20 we report the extinction spectrum for spheres of Mg coated by MgO. The radius of the spheres is 10 nm and 50 nm inclusive of the respective thickness of a coating of 2 nm and 5 nm, when present.

For the 10 nm spheres, the presence of the coating produces two main effects on the resonance peaks: a shift toward lower frequencies and an attenuation that does not depend on the normalization. In fact, the attenuation remains even when $\sigma_{\rm T}$ is normalized to the geometrical cross section of the metal core rather than to S. The spectrum of the 50 nm spheres shows sharper

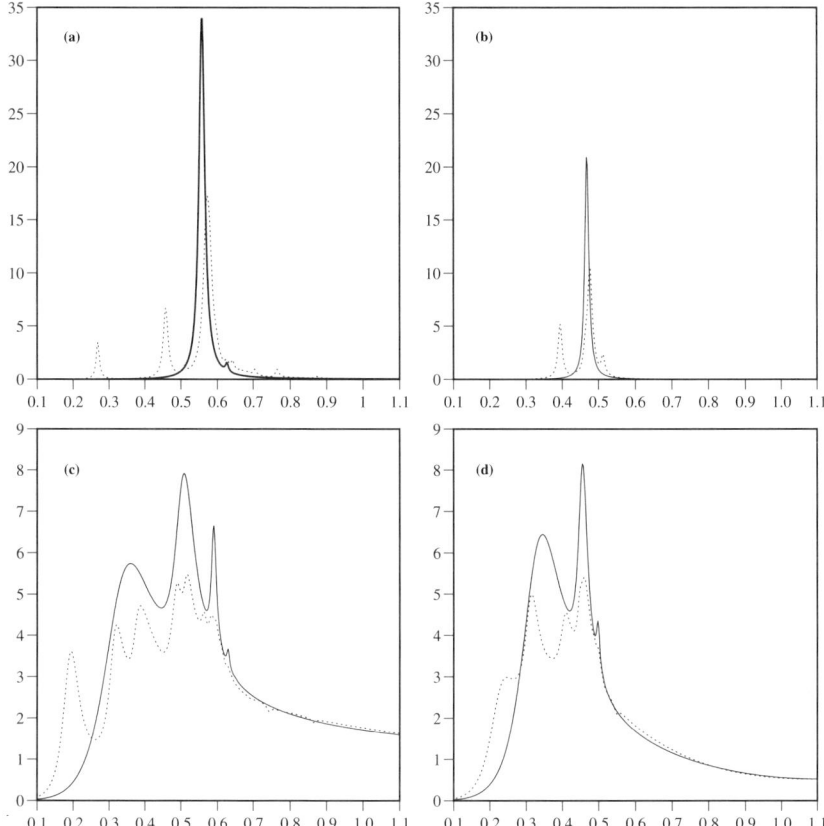

Fig. 6.20a–d. Q'_{sp} (*solid curves*) and Q' (*dotted curves*) as a function of ν for spheres of radius 10 nm and 50 nm of solid Mg in **(a)** and **(c)** and of Mg coated by a layer of MgO in **(b)** and **(d)**. The radius of the spheres is inclusive of the thickness of the coating of 2 nm in (b) and 5 nm in (d)

peaks for the coated spheres, i.e., the coating affects more the width than the heigth of the resonances. However, the qualitative features are on the whole unaffected.

A look to Fig. 6.20 shows that the aggregation produces analogous effects both for the metal-only spheres and for the coated ones. The duplication of the peaks that is due to the aggregation occurs in complete analogy with the case of spheres with a soft surface. Note that, although we do not report the specific results, the change of the thickness of the coating does not produce any further effect.

For the sake of argument, it may be of interest to remark that (6.3) predicts for the 10 nm spheres with coating, the occurrence of three peaks, only one of them occurring in the range of interest. Unfortunately, the location of this peak is inaccurately estimated by (6.3).

6.3.3 Dielectric Spheres with a Metal Coating

The spectrum for 10 nm spheres of MgO with a coating of Al are reported in Fig. 6.21, together with the spectrum of solid Al spheres of the same radius. The thickness of the coating varies form 1.6 nm to 6.4 nm and, for each choice of the thickness there occur two resonance peaks whose frequency is lower and higher than the resonance peak at $0.57\,\omega_p$ that is characteristic of a sphere of solid Al. This pair of peaks, when the thickness increases, tends to converge towards the peak of the solid Al sphere. Of course, the overall radius of the spheres influences the behavior of the peaks. In fact, we also calculated the extinction spectrum of 50 nm spheres, but the features that we described above disappear, especially when the thickness of the coating is a nonnegligible fraction of the radius. This confirms that the presence of the resonance peaks depends critically on the radius of the spheres in analogy with the well known fact that resonances of homogeneous spheres depend on the size parameter and thus on the combination of radius and frequency, not to speak of the refractive index.

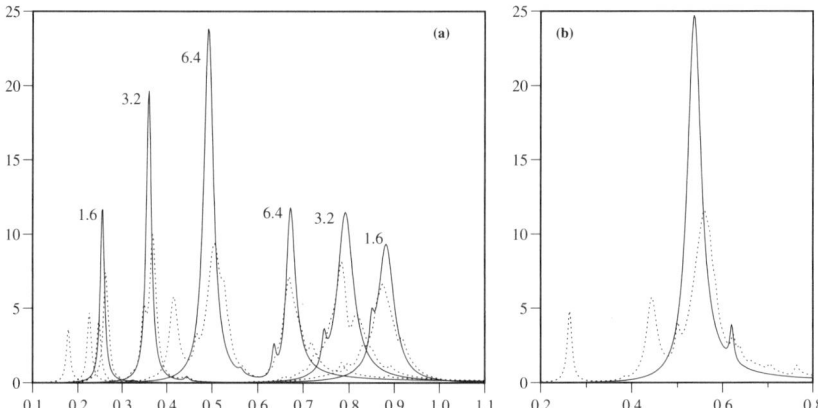

Fig. 6.21a,b. (a) Q'_{sp} (*solid curves*) and Q' (*dotted curves*) as a function of ν for spheres of MgO of radius 10 nm inclusive of a coating of Al of thickness 1.6, 3.2, and 6.4 nm. In (b) we report Q'_{sp} (*solid curves*) and Q' (*dotted curves*) as a function of ν for spheres of solid Al with 10 nm radius

The results that we presented above suggest that, with a wise choice of the thickness of a coating layer, we may produce strong extinction in any chosen region of the spectrum. This conclusion is supported by the results of analogous calculations that we performed by substituting the Al coating with one of Ag. As a consequence, we can safely state that, as far as (6.2) is valid, the extinction spectrum is only slightly affected by the choice of ω_p and Γ.

6.3.4 Metal Spheres with a Metallic Coating

As a last example, we consider spheres of Al with a coating of Ag and spheres of Ag with a coating of Al. In both cases, the radius is 10 nm inclusive of a coating of 2 nm. The resulting extinction is shown in Fig. 6.22, where we also include the extinction for Ag-only spheres. The choice of these two metals has been suggested by the remark that $\omega_p(\text{Ag})/\omega_p(\text{Al}) = 0.239$, i.e., their plasma frequencies are rather different. Comparison of Figs. 6.22 (a) and (b) shows that the most striking effect of this difference is the different intensity of the peaks when the role of Al and Ag is exchanged. Actually, we made a detailed study of the position and intensity of the peaks as a function of the thickness of the coating. We do no report the specific results of our study that can,

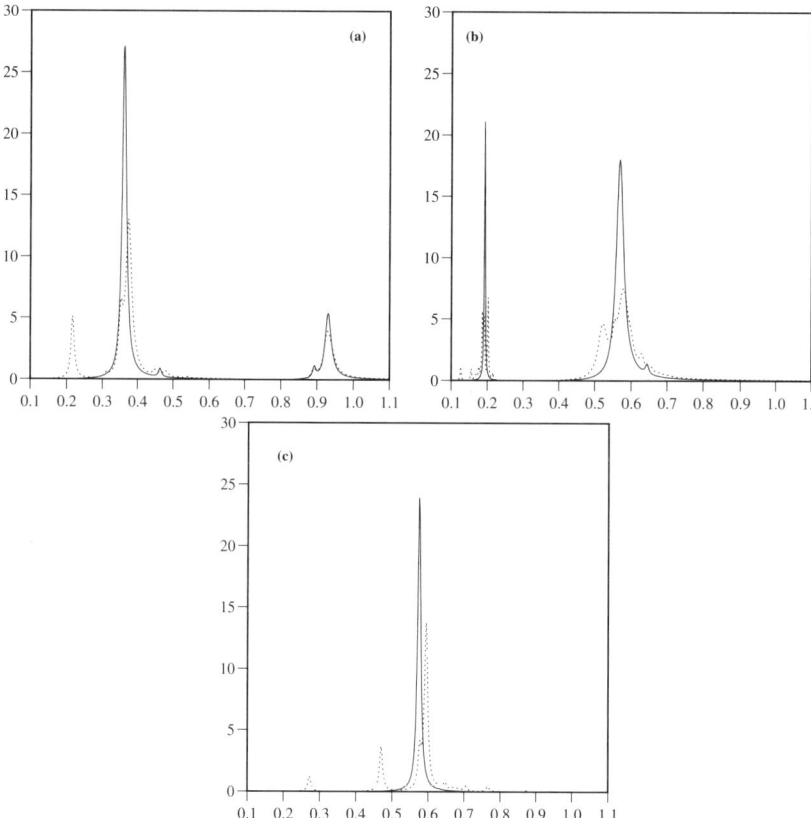

Fig. 6.22a–c. Q'_{sp} (*solid curves*) and Q' (*dotted curves*) as a function of ν for spheres of radius 10 nm of Ag coated by a layer of Al **(a)** and of Al coated by a layer of Ag **(b)**. The radius of the spheres in (a) and (b) is inclusive of the thickness of the coating of 2 nm. In **(c)** we report Q'_{sp} and Q' as a function of ν for spheres of solid Ag with 10 nm radius

however, be summarized as follows. When the metal with the higher $\omega_{\rm p}$ coats a core of the metal with the lower $\omega_{\rm p}$, the peak at $0.365\,\omega_{\rm p}$ in Fig. 6.22 (a), with increasing thickness, moves from $0.574\,\omega_{\rm p}({\rm Ag}) = 0.137\,\omega_{\rm p}({\rm Al})$ (characteristic of Ag-only spheres) to $0.536\,\omega_{\rm p}({\rm Al})$ (Al only spheres). This shift is quite significant, especially in units of $\omega_{\rm p}({\rm Ag})$. At the same time, the less intense peak becomes smaller and smaller with increasing thickness of the coating up to complete disappearance, as expected, when the sphere becomes of Al only.

When the roles of two metals are exchanged and the higher-$\omega_{\rm p}$ metal forms the core (coating of Ag), neither of the two peaks is the more important within a large range of the thickness of the coating. The mobility of the peak is smaller than in the preceding case, although in units of $\omega_{\rm p}({\rm Ag})$ it is still considerable. We are thus lead to the conclusion that the optical behavior of spheres composed of two metals is governed by the ratio of the plasma frequencies. As the mobility of the peaks as a function of the thickness of the coating can be enhanced by a convenient choice of the ratio of the respective $\omega_{\rm p}$, the considerations made for the case of dielectric spheres with a metal coating about designing particles intended to produce extinction in selected regions of the spectrum also apply.

6.3.5 Considerations on Convergence

We conclude this section with some specific remarks regarding the convergence of our calculations. In fact, the analogous calculations in [46] were performed at most with $l_{\rm M} = 4$, which was sufficient to ensure the convergence of the single spheres. The results that we presented above were obtained with $l_{\rm M} = 12$ in order to check the degree of convergence also for the aggregates in case they contain some metallic material. In fact, the large imaginary part of the refractive index may affect the rate of convergence at resonance. We already remarked that in the case of the binary aggregates of metal spheres with a dielectric coating the increase of $l_{\rm M}$ yields further peaks of small intensity in the lower part of the spectrum. A similar situation occurs for the aggregates of metal spheres with a metallic coating. Since this is the worst case we dealt with, further investigation was felt to be in order. Therefore we report in Fig. 6.23 the results of our study for the aggregates of spheres of Ag with a coating of Al. First, we notice that the increase of $l_{\rm M}$ from 12 to 16 has no influence on the high-frequency part of the spectrum. Actually, Fig. 6.23 shows that the peak at $0.93\,\omega_{\rm p}$ remains unaffected. On the contrary, at low frequency the increase of $l_{\rm M}$ yields small but still noticeable shifts of the peaks toward lower frequencies; no new peaks appear in the region of interest, however. Therefore, one may wonder what value of $l_{\rm M}$ would ensure the true convergence. In practice, since with increasing $l_{\rm M}$ the shift of the low-frequency peaks is comparatively small on a wavelength scale, we are led to conclude that the choice

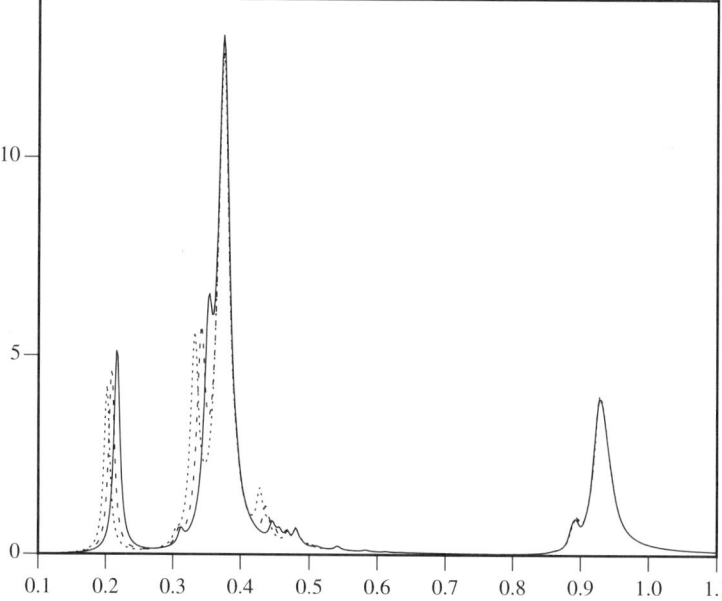

Fig. 6.23. Q' as a function of ν for spheres of Al of radius 10 nm inclusive of a coating of Ag of 2 nm thickness, for $l_\mathrm{M} = 12$ (*solid curve*), $l_\mathrm{M} = 14$ (*dashed curve*), and $l_\mathrm{M} = 16$ (*dotted curve*)

of l_M depends on the experimental capabilities. In other words, one must consider that the shifts may be smaller than the experimental error and that, when further small peaks are invoked by the calculation, their intensity may be hard to be discriminated against the unavoidable background noise.

6.4 Spheres Containing Inclusions

Spheres containing one or more spherical inclusions may be a good model for several interesting objects, such as biological cells, water droplets containing pollutants, and complex interstellar dust grains, e.g. fluffy grains.

The calculations we are going to discuss are based on the theory of Sect. 4.5 that is subject to no limitation of principle as regards the size of the host sphere and the number and chemical composition of the inclusions [53, 54]. Nevertheless, in the present section, we consider spheres containing at most two inclusions that may be either metallic, dielectric, or also empty. In fact, it would be an impossible task to extract from the results for spheres containing many inclusions the general features that depend on

150   6. General Applications

the configuration of the inclusions present. The case of many inclusions will be dealt with as a specific application in Chap. 9.

Our calculations are performed with reference to the following geometry. The host sphere is centered at the origin of a system of Cartesian axes and the center of one of the inclusions always lies on the z axis. The incident field is a plane wave whose wavevector is individuated by the polar angles ϑ_I and φ_I, whereas ϑ_S and φ_S are the polar angles of the direction of observation. Our calculations were performed with $\vartheta_\mathrm{I} = 0$, $\pi/4$, $\pi/2$, and $\varphi_\mathrm{I} = 0$, and, indicating by Φ ($0 \leq \Phi \leq \pi$) the angle of scattering,

$$\vartheta_\mathrm{S} = \vartheta_\mathrm{I} + \Phi \text{ and } \varphi_\mathrm{S} = 0 \quad \text{for } \vartheta_\mathrm{I} + \Phi < \pi,$$
$$\vartheta_\mathrm{S} = 2\pi - (\vartheta_\mathrm{I} + \Phi) \text{ and } \varphi_\mathrm{S} = \pi \quad \text{for } \vartheta_\mathrm{I} + \Phi > \pi.$$

Therefore the plane of scattering coincides with the zx plane to which the polarization vector both of the incident and of the scattered field is always parallel ($\hat{e} = \hat{u}_1$) or orthogonal ($\hat{e} = \hat{u}_2$).

The scattering properties of a sphere containing spherical inclusions are described through its normalized scattering amplitude. More precisely, we report the quantities [85] P' and Q' that are defined in (6.1). We also report the averages $\bar{P}$ and $\bar{Q}$ for random orientational distribution. Of course, the latter are meaningful for forward scattering only ($\Phi = 0$) (see Sect. 3.4). The quantities $\bar{P}$ and $\bar{Q}$ give information on the macroscopic optical properties of a tenuous dispersion of the scatterers we deal with.

Since the objects we deal with have a size parameter $x = k\rho_0 > 1$, the value of l_M that is necessary to get fully converged results turns out to be rather large. As a criterion of full convergence, we required that, with respect to the any increase of l_M, our results be stable at least to 4 significant digits. This accuracy is by far higher than required for any graphical display. It may be interesting to notice that, in agreement with the remarks by other workers dealing with dependent scattering from aggregated spheres, we met the slowest convergence when the surfaces of one of the inclusions and of the host sphere touch each other [94].

In Sects. 6.4.1–6.4.3, we present the results of our calculations for a dielectric host sphere with refractive index $n_\mathrm{D} = 1.61 + \mathrm{i}0.004$, which describes satisfactorily the optical properties of spheres of acrylic in a large frequency range. In all cases, the external medium is assumed to be the vacuum ($n = 1$). The dielectric properties of the metallic inclusion are assumed to be well described by the free-electron Drude function with $\Gamma = 0.01$ and $\omega = 0.1\omega_\mathrm{p}$. The latter for most metals corresponds to a frequency in the visible or in the infrared range [89], and the resulting refractive index of the inclusion is $n_\mathrm{M} = \sqrt{\epsilon_\mathrm{M}} = 0.4994 + \mathrm{i}9.9126$. Note that no resonance occur for this value of the refractive index. The resonances of a dielectric sphere containing a metallic inclusion will be discussed in Sect. 6.4.4.

6.4.1 Metallic Inclusion in a Dielectric Sphere

The size parameter of the host dielectric sphere and of the inclusion are $x_D = 3$ and $x_M = 1$, respectively, so that the ratio of their radii is $\rho_D/\rho_M = 3$. The *eccentricity* of the inclusion is accounted for by the parameter $x_E = k_v z$, where z is the coordinate of the center of the inclusion. Therefore $|x_E|$ can range from $|x_E| = 0$, when the inclusion is concentric, to $|x_E| = 2$, when it is tangent to the surface of the host sphere [53]. In our calculations for $\hat{k}_S = \hat{k}_I$, i.e., when forward scattering is considered, z is allowed to assume positive values only, on account of the symmetry properties of the scattering amplitude that are discussed, e.g., by van de Hulst [3]. The results we are going to illustrate required at most $l_M = 10$ to reach full convergence.

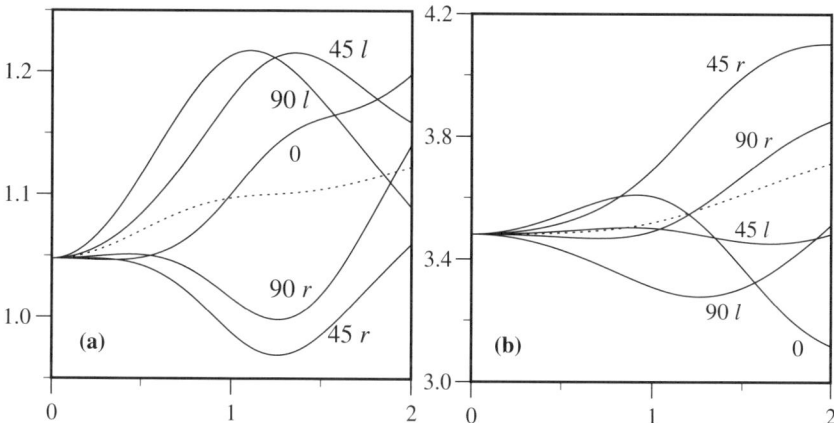

Fig. 6.24a,b. P' (a) and Q' (b) for $\hat{k}_S = \hat{k}_I$ as a function of the eccentricity x_E for a dielectric sphere containing a metallic spherical inclusion (*solid curves*). These curves are labeled by the value of ϑ_I and by l or r to denote polarization parallel or perpendicular to the plane of scattering, respectively. The *dotted curves* report $\bar{P}$ in (a) and $\bar{Q}$ in (b)

The quantities P' and Q' as a function of x_E for the three directions of incidence referred to above and for polarization both parallel and orthogonal to the plane of scattering are reported in Fig. 6.24. In the same figure we also report the averages $\bar{P}$ and $\bar{Q}$ that, for random orientational distribution, are independent both of the direction and of the polarization of the incident wave. Of course, P' and Q' are also independent of the polarization when the direction of incident wavevector lies along the symmetry axis ($\vartheta_I = 0$), but, when $\vartheta_I = \pi/4$ and $\vartheta_I = \pi/2$, they become strongly dependent on the polarization. A further interesting remark is at hand in the results of Figs. 6.24. If we consider only one state of polarization, either parallel or orthogonal, the curves for $\bar{P}$ and $\bar{Q}$ do not always lie within the curves for the various incidences, contrary to what would be expected of averaged quantities. A similar effect that also occurred for clusters in free space (see Sect. 6.1) is due to the contributions to $\bar{P}$ and $\bar{Q}$ coming from scatterers so oriented that their symmetry axis do not lie in the plane of scattering. These contributions are never explicitly computed but are automatically accounted for by the averaging procedure. Indeed, the response of a scatterer with its symmetry axis not lying in the scattering plane is easily recognized as identical to the response of a scatterer with its axis in that plane when this latter object is excited by a wave with an appropriate state of polarization that, in general, is neither parallel nor orthogonal. Therefore, the above mentioned response must be a linear combination of the responses for parallel and orthogonal polarization and, since the scattering properties of the objects we are dealing with here show a noticeable sensitivity to the state of polarization, we get the seemingly anomalous behavior of $\bar{P}$ and $\bar{Q}$ described above.

As a reference we also calculated P' and Q' for a homogeneous dielectric sphere either with the same radius of the host sphere ($P'_\mathrm{h} = 1.3173$ and $Q'_\mathrm{h} = 3.9117$) or containing the same quantity of dielectric material as the sphere with the inclusion ($P'_\mathrm{e} = 1.3151$ and $Q'_\mathrm{e} = 3.9044$). Indeed, a comparison of these values with those of the curves in Fig. 6.24 will give a better insight into the effect of the very presence of the inclusion and of its eccentricity as well. Let us remark, first of all, that P'_h and P'_e as well as Q'_h and Q'_e differ very little from each other because the ratio of the volumes of the homogeneous spheres defined above is 27/26 on account of our choice $\rho_\mathrm{D}/\rho_\mathrm{M} = 3$. Moreover, the values of P' and Q' for a sphere with a centered inclusion ($x_\mathrm{E} = 0$) are remarkably different from the corresponding values for the homogeneous dielectric spheres of both sizes considered above. When x_E increases the difference of $\bar{P}$ and P'_h and, especially, that of $\bar{Q}$ and Q'_h, tend to decrease while the behavior of the difference of P' and P'_h as well as that of Q' and Q'_h depends on the direction of incidence and, in general, also on the polarization. In fact, the spread of the curves of P' and Q' is rather large and reaches its maximum in the range $x_\mathrm{E} \simeq 1.15$–1.35 [Fig. 6.24 (a)] and for $x_\mathrm{E} = 2$ [Fig. 6.24 (b)], respectively.

In Fig. 6.25 we show $P'(\varPhi)$ vs. $Q'(\varPhi)$ for parallel polarization and for $\vartheta_\mathrm{I} = 0$ (a), $\vartheta_\mathrm{I} = \pi/4$ (b), and $\vartheta_\mathrm{I} = \pi/2$ (c). In each figure, we report the curves for the extreme values of the eccentricity ($x_\mathrm{E} = 2$ and $x_\mathrm{E} = -2$) as well as the curve for the centered inclusion ($x_\mathrm{E} = 0$). As a reference, the curve of $P'_\mathrm{h}(\varPhi)$ vs. $Q'_\mathrm{h}(\varPhi)$ is also reported. Of course, both the curves for the centered inclusion and for the homogeneous sphere do not depend on the incidence and are thus the same in all these figures. It is easily seen that the curves for the spheres with an eccentric inclusion are rather different from each other and differ also from those both for the sphere with a centered inclusion and for the homogeneous sphere. For forward scattering ($\varPhi = 0$), the curves for the two eccentric positions stick together as expected on account of the symmetry properties of the scattering amplitude. At all incidences, the position of the

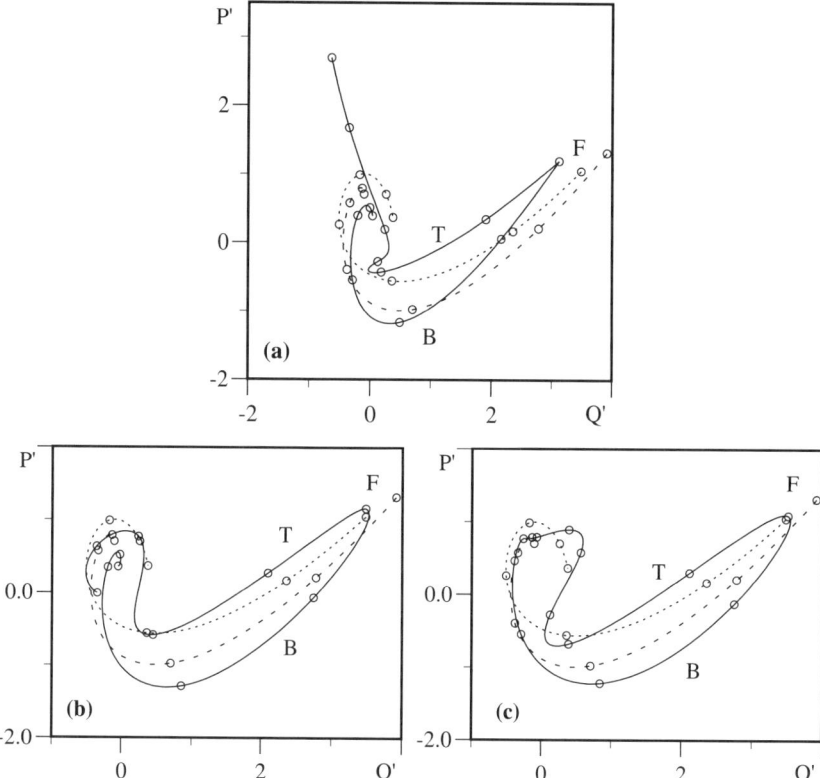

Fig. 6.25a–c. $P'(\Phi)$ vs. $Q'(\Phi)$ for $\vartheta_I = 0$ **(a)**, $\pi/4$ **(b)** and $\pi/2$ **(c)** for the dielectric sphere containing the metallic inclusion. The polarization is parallel to the plane of scattering. The *solid curves* labeled T and B refer to the inclusion at eccentricity $x_E = 2$ and $x_E = -2$, respectively. The *dotted curve* refers to a centered inclusion and the *dashed curve* to $P'_h(\Phi)$ vs. $Q'_h(\Phi)$. The *circles* on all the curves mark the value of the scattering angle Φ at $30°$ intervals. The forward direction $\Phi = 0°$ is labeled F

inclusion within the external sphere has an evident effect on the shape of the Q'-P' curves, although the general behavior remains unchanged. When the incidence is orthogonal to the symmetry axis ($\vartheta_I = \pi/2$), the curves for $x_E = 2$ and for $x_E = -2$ stick together, not only for forward scattering ($\Phi = 0$) but also at backscattering ($\Phi = \pi$) as required by an obvious symmetry property of the matrix of the scattering amplitude [Fig. 6.25 (c)].

We do not report the curves for orthogonal polarization because, although the numerical values are rather different from those for parallel polarization, the general shape and properties of the curves are identical to those reported in Figs. 6.26 and do not deserve, in our opinion, a separate comment.

6.4.2 Empty Cavity in a Dielectric Sphere

The size parameter of the dielectric external sphere and of the empty cavity ($n_C = 1$) are $x_D = 4.3410$ and $x_C = 2.1705$, respectively, and the ratio of their radii is $\rho_D/\rho_C = 2$. As a consequence, the eccentricity can range from $x_E = 0$ to $x_E = 2.1705$ [53]. Our choice of $x_D = 4.341$ is due to the fact that the experimental scattering properties of a solid dielectric sphere with these features are known [84], so that we often used such an object as a reference scatterer. In the present case the convergence of the results required at most $l_M = 8$.

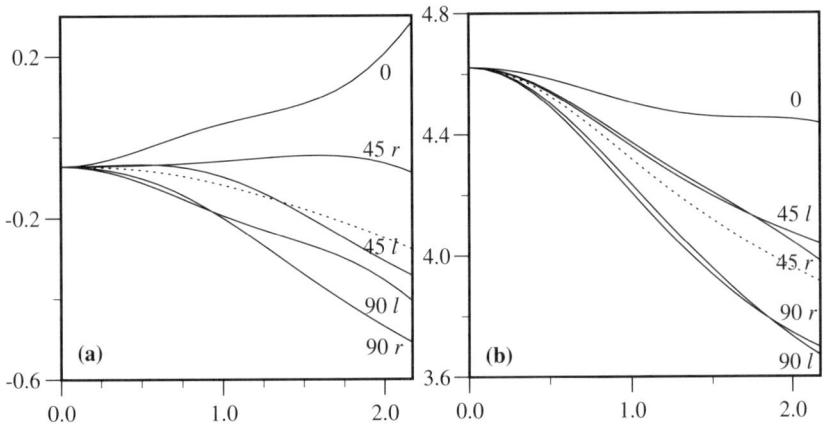

Fig. 6.26a,b. P' (a) and Q' (b) as a function of x_E for dielectric spheres containing the empty spherical cavity. The curves are labelled by ϑ_I and by l or r to denote polarization parallel or perpendicular to the plane of scattering, respectively. The *dotted curve* refers to the averages for random orientational distribution

The main body of our results is displayed in Figs. 6.26 (a) and (b) that are analogous to Figs. 6.24 (a) and (b), respectively. As compared with the results for a metallic inclusion, the present results show a less strong dependence on the polarization and, in particular, this dependence is rather weak for Q' [Fig. 6.26 (b)]. As an immediate consequence, we notice that the curve of $\bar{Q}$ always lies within the curves of Q' for a single polarization although they refer only to orientations with the symmetry axis in the scattering plane. The results both for the solid dielectric sphere with radius equal to that of the external sphere (the reference sphere mentioned above) and of the sphere containing the same quantity of dielectric material as the sphere with the inclusion are $P'_h = -0.9238$, $Q'_h = 3.8373$, and $P'_e = -0.7796$, $Q'_e = 4.1082$, respectively. In the present case these values are noticeably different from each other since the ratio of the radii $\rho_D/\rho_C = 2$ implies that the ratio of the volumes is 8/7. Furthermore, the relative positions of these values show that P' and Q' for a homogeneous sphere are, in this range, decreasing functions of the size parameter. Figures 6.26 (a) and (b) show that even in the present case, $\bar{P}$ and especially $\bar{Q}$, as the eccentricity increases, tend to reduce their difference both from P'_h and Q'_h and from P'_e and Q'_e. The spread of the values for the different incidences we considered reaches a rather large maximum at the highest value of the eccentricity. Although we also performed calculations analogous to those displayed in Fig. 6.25, we resolved not to report the results because they do not show any new significant feature worthy of a separate comment.

6.4.3 Spheres Containing Two Metallic Inclusions

We now go to consider a homogeneous sphere containing two identical spherical inclusions. In fact, the subdivision of the included material into two identical inclusions proved to be quite sufficient to appreciably change the scattering properties in comparison to the case of a single inclusion with the same volume.

The size parameter of the host sphere is $x_D = 3$. Both the inclusions have size parameter $x_I = k_v \rho_I = 0.7937$, so that their total volume equals that of a single inclusion with unit size parameter [54]. The centers of the two inclusions lie either both on the z axis or one on the z axis and the other outside the z axis in the plane $\varphi = 0$. The latter choice prevents, of course, the existence of a cylindrical symmetry. In the following figures, we report, for each of the three directions of incidence mentioned above, the efficiencies Q'_η for $\hat{k}_S = \hat{k}_I$ and $\bar{Q}$ as a function of the position of the inclusions within the external sphere. The geometrical arrangements that we consider in this section are such that the position of the two inclusions can be defined by the parameters $x_{E_1} = k_v z_1$ and $x_{E_2} = k_v z_2$, where z_1 and z_2 are the z coordinates of their centers. As x_{E_1} and x_{E_2} turn out to be mutually dependent, Q'_η and $\bar{Q}$ are reported as a function of the eccentricity x_{E_2} only. Q'_η and $\bar{Q}$ are also reported for the sphere containing a single spherical inclusion, with its center on the z axis and volume equal to the sum of the volumes of the two inclusions above, as a function of its eccentricity $x_E = k_v z$. This kind of inclusion, that we dealt with in Sect. 6.4.1, hereafter will be referred to as the equivalent inclusion. It has unit size parameter so that its eccentricity always range between 0 and 2. The comparison of the curves for the case of two inclusions and the case of the equivalent inclusion will give a clear view of the effects of the subdivision of the included material.

In Figs. 6.27 (a), (b), and (c) we report, for $\vartheta_I = 0$, $\pi/4$ and $\pi/2$, respectively, the results for spheres containing two inclusions with the centers on the z axis and lying symmetrically with respect to the center of the host sphere. Therefore $x_{E_1} = -x_{E_2}$ and the smallest value of $x_{E_2} = 0.7937$ occurs when the inclusions contact each other at the center of the external sphere; the maximum value of $x_{E_2} = 2.2063$ occurs when the inclusions touch the surface of the host sphere.

156   6. General Applications

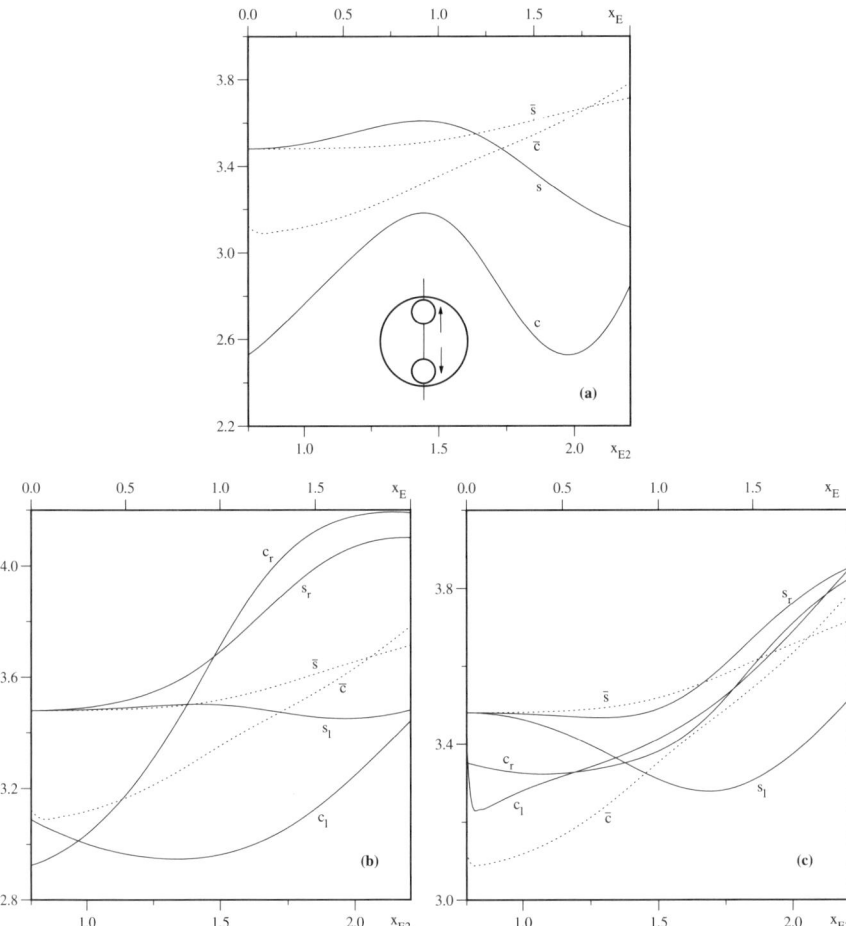

Fig. 6.27a–c. Q'_η (*curves* c) and $\bar{Q}$ (*curves* $\bar{\text{c}}$) for spheres containing two identical inclusions arranged as in the inset in (**a**), plotted versus x_{E_2} in the range from 0.7937 to 2.2063. For the sake of comparison, Q'_η (*curves* s) and $\bar{Q}$ (*curves* $\bar{\text{s}}$) for spheres containing the equivalent inclusion are also plotted versus x_E. The angles of incidence are $\varphi_I = 0$ in all cases and $\vartheta_I = 0$, $\pi/4$ and $\pi/2$ in (**a**), (**b**), and (**c**) respectively. The subscripts l and r refer to polarization parallel and perpendicular the plane of scattering, respectively

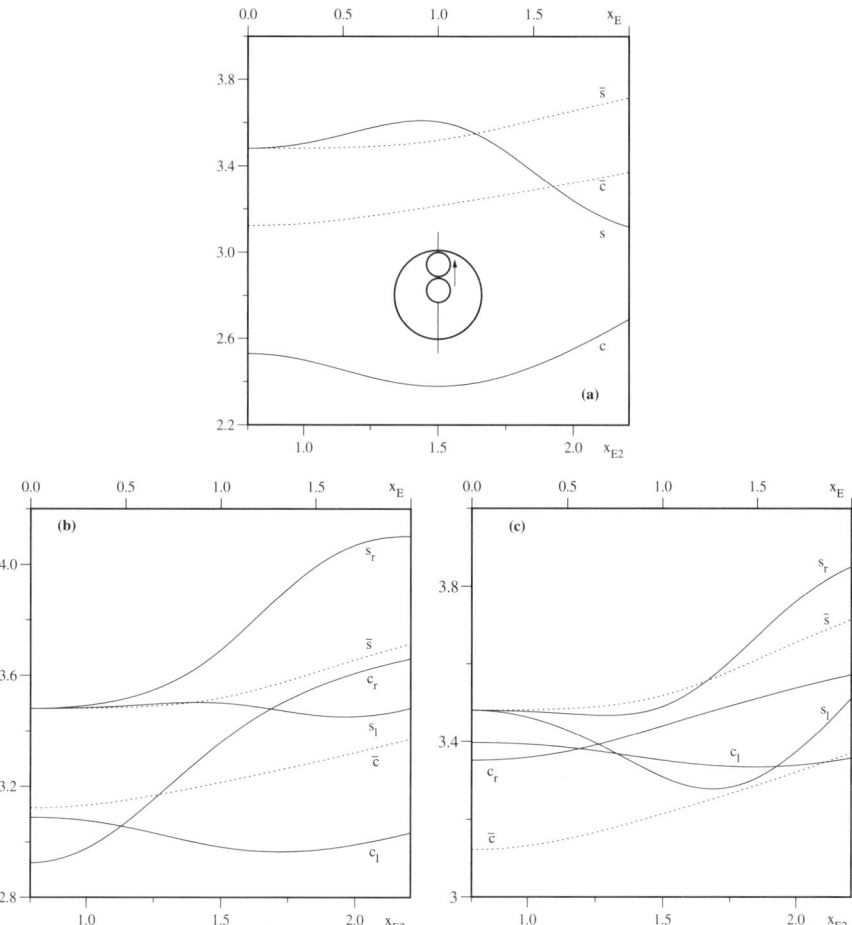

Fig. 6.28a–c. Q'_η and $\bar{Q}$ same as in Fig. 6.27 but for the geometric arrangement of the inclusions illustrated in the inset in (**a**)

In Fig. 6.28 the centers are still on the z axis but the two inclusions are always in contact and are translated along the z axis. In this case, $x_{E_1} = x_{E_2} - 2x_I$ and $x_{E_2} = 0.7937$, when the point of contact coincides with the center of the external sphere, and $x_{E_2} = 2.2063$, when inclusion 2 touches the surface of the external sphere. The true minimum of x_{E_2} actually occurs at the negative value $x_{E_2} = 3x_I - x_D = -0.6189$, i.e., when inclusion 1 touches the surface of the external sphere. Nevertheless, this choice would add no new information on account of the symmetry properties of the scattering amplitude [3]. In Fig. 6.29 we consider a configuration in which, unlike the preceding cases, no cylindrical symmetry is present. In fact, we put the two inclusions in contact and the center of inclusion 2 always lies on the z axis

158   6. General Applications

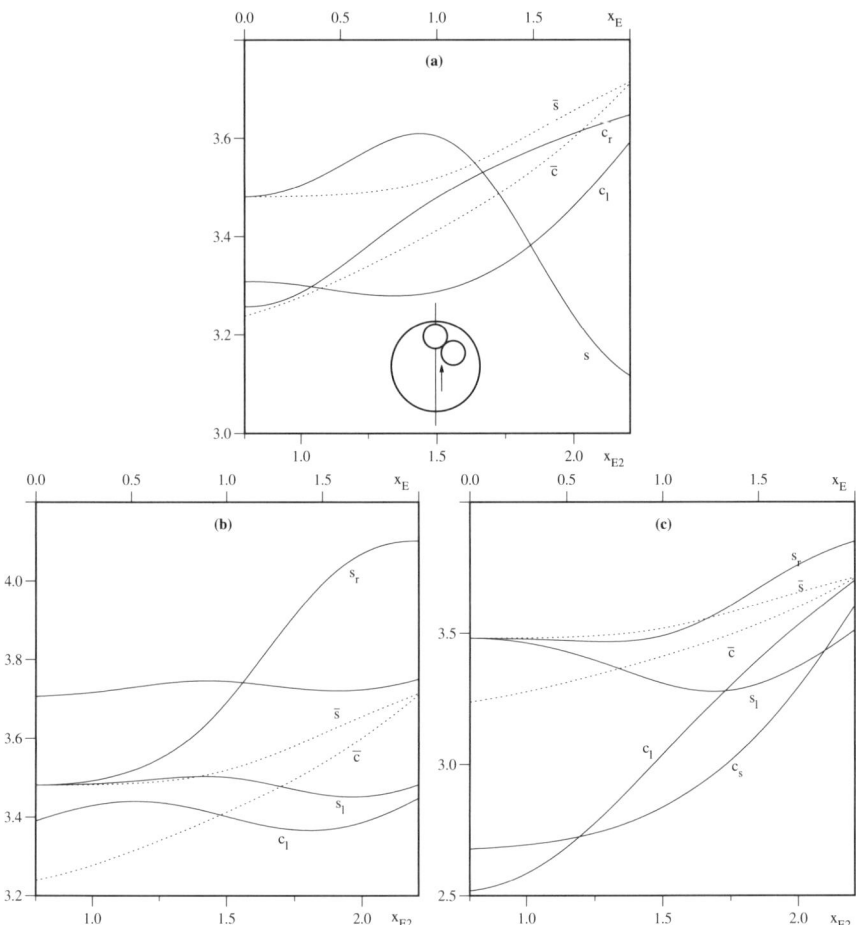

Fig. 6.29a–c. Q'_η and $\bar{Q}$ same as in Fig. 6.27 but for the geometric arrangement of the inclusions illustrated in the inset in **(a)**.

while the center of inclusion 1 is in the plane $\varphi = 0$ with $z_1 = z_2 - \rho_\mathrm{I}$ so that its distance from the z axis is $d = \rho_\mathrm{I}\sqrt{3}$; x_{E_2} ranges from $x_{\mathrm{E}_2} = 0.7937$ to $x_{\mathrm{E}_2} = 2.2063$. In this case, our choice of the minimum value of x_{E_2} privileges the uniformity of the presentation of our results even at the cost of excluding some new information that, in our opinion, is not particularly significant, however. Note that $\bar{s}$ at $x_\mathrm{E} = 2$ and $\bar{c}$ at $x_{\mathrm{E}_2} = 2.2063$ seem to coincide. This is a mere accident, however, that is due to the scale of the figure.

As regards the value of l_M that is necessary to get fairly convergent results, we found that $l_\mathrm{M} = 10$ is quite sufficient to get the required accuracy.

When considering the results reported in Figs. 6.27–6.29 it should be borne in mind that a rather complex mechanism displays its effects. In fact, when the inclusions become more and more distant, the effects of the multiple scattering processes occurring between them may become faint. When the inclusions are going to touch the surface of the host sphere it is easy to foresee a strong interaction. Both the decreasing symmetry of the scatterers we dealt with and the direction of $\hat{k}_I$ concur with the above mentioned multiple scattering processes in determining the calculated dependence on the polarization.

In Fig. 6.30 we summarize all the preceding results for $\bar{Q}$ as a function of x_{E_2} so as to give a general view of the optical properties of the model media composed of a tenuous dispersion of the scatterers considered so far. In particular, since Fig. 6.30 also includes the curve of $\bar{Q}$ as a function of x_E for a dispersion of spheres containing the equivalent inclusion (curve a), one can easily see the effect on the extinction of the subdivision of the included material. At least for the cases we considered in this section the subdivision slightly reduces the extinction efficiency; $\bar{Q}$, however, recovers the loss when the separation between the inclusions increases (curve b). Anyway, even a careful examination of the results does not reveal a trend that allows us to understand whether the included material is subdivided or not. Of course, the optical properties of the random dispersions do not show any dependence on the polarization as it is compensated for by the average over the orientations. On the contrary, the dependence on the polarization is present in the spectra

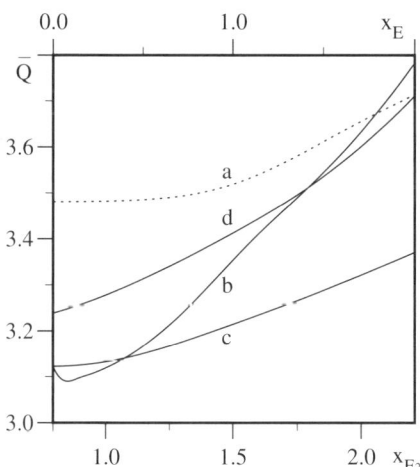

Fig. 6.30. Summary of the curves for $\bar{Q}$ versus x_{E_2}. For the sake of comparison we also include the curve for $\bar{Q}$ (*curve* a) versus x_E for the sphere containing the equivalent inclusion. The labels b, c, and d refer to the curves coming from Figs. 6.27, 6.28 and 6.29, respectively

of the single spheres and is, as expected, dominated by the symmetry of the distribution of the included material. We can summarize our findings by remarking that the optical properties of a tenuous dispersion are determined by the strength of the multiple scattering processes that occur within each sphere.

We do not compare our findings with specific experimental data that, as far as we know, are not available, but we are confident that our results have a high degree of reliability. In fact, our results were reproduced by Ioannidou and Chrissoulidis [57] as a test of their reformulation of the theory of scattering by spheres containing several inclusions. Moreover, even in the present case, all the tests that we described in our previous work [53] were successfully performed. In particular, by putting $n_0 = 1$, i.e., by assuming that the interstitial medium is the vacuum, we got the same results as for bare clusters.

The results that we presented above do not pretend to be exhaustive, as their purpose is to give relevant examples of the effects of the subdivision of the included material according to known geometries. In fact, even a cursory examination of the results shows that the subdivision has rather visible effects on the properties both of single objects and of their dispersions. It is also apparent that the optical behavior is so dependent on the geometry that understanding how the included material is subdivided seems to be excluded, at least by a simple examination of the extinction spectrum.

We considered spheres containing at most two inclusions, whereas, in actual cases, the included material is often more minutely subdivided. Such a choice is not dictated by any limitation of the formalism but rather by the consideration that a rigorous treatment of spheres containing a very large number of small inclusions may be a nonsense. In fact, when the number of the inclusions is very large, the spheres may present a spongy appearance and could be more efficiently described, for instance, by simulating their properties through an effective dielectric function and, possibly, by assuming the presence of a soft surface [46]. Nevertheless, some experimental results show that sometimes, when the number of the inclusions is large but not too large, there occur effects that cannot be explained in terms of an effective dielectric constant. We refer to the correlation-spectroscopy experiments of Bronk, Smith, and Arnold on $30\,\mu$m droplets containing more than 100 small inclusions [95] and on the suppression of the lasing effect from ethanol droplets by the presence of several submicrometer latex inclusions as reported by Armstrong et al. [96]. Furthermore a recent perturbation technique has been devised by Ng, Leung, and Lee [97] just to deal, with not too heavy computational effort, with the case of spheres containing many inclusions. In view of these experimental findings as well as on account of the results in this section, we are led to conclude that a lot of both experimental and theoretical work is still necessary in order to gain understanding of the effects of included material within an otherwise homogeneous particle.

6.4.4 Resonances of a Sphere Containing a Spherical Inclusion

In this section we resort to the potentialities of the transition matrix approach to study the resonances of spheres containing a spherical eccentric inclusion. This kind of scatterer, in spite of its seemingly spherical appearance, is intrinsically nonspherical so that it may be useful to highlight the differences of its resonance spectrum from that of truly spherical particles. According to (4.35) that gives, in compact form, the transition matrix of a homogeneous sphere containing a single spherical inclusion,

$$\left(R_0^{-1} - R_W J_{0\leftarrow 1} R_1 J_{1\leftarrow 0}\right)^{-1} \tag{6.4}$$

is responsible for resonances, just as R_0 does in the case of homogeneous spheres. Nevertheless, (4.35) contains the translation matrices $J_{0\leftarrow 1}$ and $J_{1\leftarrow 0}$ that account for the eccentricity of the inclusion. As a result, we should expect an evolution of the resonance spectrum when a centered inclusion becomes more and more eccentric.

The particle that we choose for our study is of interest in astrophysics. In fact, we consider the extinction spectrum of spheres of MgO with a radius $\rho_0 = 25$ nm containing a spherical inclusion of Mg with a radius of $\rho_1 = 16.6$ nm. In other words, we are considering a metal sphere coated by its oxide [98]. ρ_0 and ρ_1 were chosen so as to lie within the estimated ranges for the particles in the interstellar medium [99]. Furthermore, our choice of the radii implies that the particles, although small, can have resonances also for $l > 1$. As in Sect. 6.3.2, the dielectric function of Mg was assumed to be of the free-electron Drude form [89], that of MgO was assumed of the damped oscillator form [88], and the related parameters are reported in Table 6.3. The geometry that we assumed is identical to what we used in Sects. 6.4.1–2. Accordingly, the center of the inclusion always lies on the z axis and the plane of scattering was assumed to be the zx plane.

The results are presented by reporting the extinction efficiency Q'_η as a function of the ratio $\nu = \omega/\omega_p$. Of course, Q'_η is independent of the polarization both when the incidence is along the z axis and when the inclusion takes a centered position so that the scatterer as a whole becomes spherically symmetric. In these cases, the polarization index η will be dropped. The results are better understood by recalling (3.12) for the elements of the scattering amplitude in terms of the transition matrix elements and of the amplitudes both of the incident and of the scattered field. It is an easy matter to show that, on account of the assumed geometry,

$$W^{(p)}_{I\eta lm} = (-)^{\eta+p+m-1} W^{(p)}_{I\eta l, -m}$$

and that the transition matrix elements have the property

$$S^{(pp')}_{ll';m} = (-)^{p-p'} S^{(pp')}_{ll'; -m}\,.$$

162   6. General Applications

An immediate consequence of the equations above is that the resonances for given η, p, and l that are associated to $\pm m$ are degenerate, i.e., they occur at the same frequency. These considerations are useful when we separate the contributions from each l to the scattering amplitude. To this end, we carry to convergence the calculation of the transition matrix and write the corresponding scattering amplitude as

$$f_{\eta\eta'} = \sum_{l} B_{\eta\eta'l} , \qquad (6.5)$$

where

$$B_{\eta\eta'l} = -\frac{i}{4\pi k} \sum_{pm} \sum_{p'l'} W^{(p)*}_{S\eta lm} S^{(pp')}_{ll';m} W^{(p')}_{I\eta'l'm} .$$

We then define the partial efficiencies

$$q_{\eta l} = \frac{4}{k\rho_0^2} \text{Im}\left[B_{\eta\eta l}\right] ,$$

which give the contribution to the extinction from multipoles of order l.

In Fig. 6.31 we report Q' for a sphere of solid Mg with radius ρ_1. The reader may find useful the comparison with Fig. 6.20 (a). In order to check the convergence of our calculations and to associate the computed resonances to the appropriate value of l, the extinction efficiency is reported for $l_M = 1$ and $l_M = 2$, l_M being the maximum value of l included into (6.5). We notice that taking $l_M > 2$ neither yield any higher order resonance nor produces visible changes on the scale of the figure. Therefore, although all the calculations

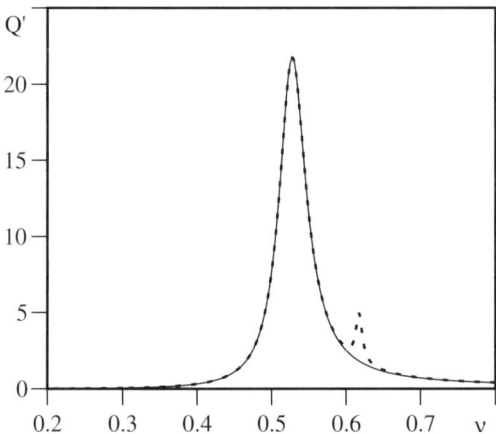

Fig. 6.31. Extinction efficiency for a sphere of solid Mg with radius $\rho_1 = 16.6$ nm. The *continuous curve* was obtained with $l_M = 1$ and the *dashed curve* with $l_M = 2$. The incidence is along the z axis

were performed with $l_M = 10$, we can state that using $l_M = 5$ is sufficient to achieve a numerical accuracy to four significant digits. The sphere of Mg presents two resonances: the lowest-frequency peak is associated with $l = 1$ whereas the second peak is associated with $l = 2$. An analysis of the elements of the transition matrix suggests that both peaks are electric resonances, and, due to the independence of m of the elements of the transition matrix for a sphere, the value of m need not be specified. We do not report Q' for the sphere of MgO of radius ρ_0 because this scatterer does not present any resonance in the range of interest.

As one should expect, we found that the effects that are due to the eccentricity achieve the maximum of evidence when the inclusion becomes tangent to the surface of the external sphere so that in the following figures only the results for this extreme case are reported. Accordingly, Fig. 6.32 shows, for incidence along the z axis, Q'_η for the scatterer with the inclusion at maximum eccentricity together with the partial efficiencies $q_{\eta l}$ with $l = 1, 2$. Of course, due to the cylindrical symmetry of the scatterer, both the total and the partial extinction efficiencies are independent of η. Figure 6.32 also reports Q' for the centered inclusion. In the latter case there occurs a single electric resonance that is associated to $l = 1$, whereas the resonance that for a sphere of solid Mg is associated to $l = 2$ turns out to be so damped as to reduce to a small shoulder. This behavior is easily understood in the framework of the theory of Johnson [48] that we summarized in Sect. 4.3. In fact, the presence of a coating may modify the radial shape of the refractive index $n_0(r)$, and thus of the effective potential (4.18), to such an extent that the potential well for the appropriate value of l (in the present case for $l = 2$) may not occur at all, thus preventing the occurrence of the corresponding resonance, or may become so shallow as to produce the damping that is observed in Fig. 6.32. Considerations of this kind do not apply when the inclusion is off center because of the lack of spherical symmetry. Indeed, Fig. 6.32 shows four resonances. The one at the lowest frequency is associated to $l = 1$ and is the evolution of the main peak that occurs when the inclusion is centered. The three further resonances can be attributed to the multiple scattering processes between the inclusion and the external sphere [53]. Due to the off-center position of the inclusion, these processes are asymmetric and thus yield the observed multiplication of peaks. The behavior of the quantities q_l clearly shows that all these peaks arise from the superposition of contributions for $l = 1$ and $l = 2$. The occurrence of this superposition calls for a word of caution about the classification of the resonances into electric and magnetic as well as about the order of the resonating multipoles. In [100] we noticed that this classification depends on the origin to which the multipole fields are referred. For a sphere with a centered inclusion there exists a natural origin, namely the center of the external sphere that coincides with the center of symmetry of the whole scatterer. Of course, when the inclusion is eccentric such a center of symmetry does not exist. Now, we remarked in Sect. 1.8.3

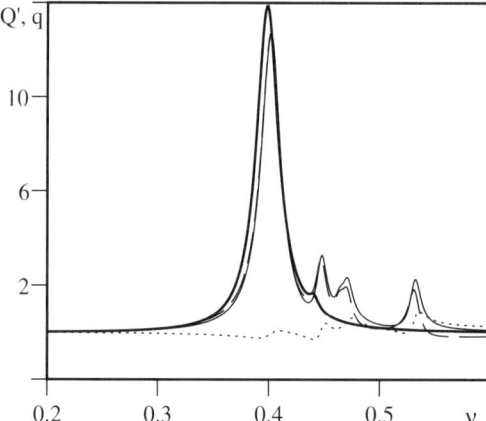

Fig. 6.32. Extinction efficiency for the scatterer with the inclusion centered (*thick solid line*) and with the inclusion at maximum eccentricity (*solid line*). The *dashed curve* and the *dotted curve* report the partial efficiencies for $l = 1$ and $l = 2$, respectively, for the case of maximum eccentricity. The incidence is along the z axis

that a multipole field that is purely electric or magnetic when referred to a given origin becomes a superposition of both electric and magnetic multipole fields, in general of all the multiplicities, when referred to a different origin. On the other hand, in the framework of our present approach, the addition theorem of Sect. 1.8, gives rise to the nondiagonal matrices $J_{0 \leftarrow 1}$ and $J_{1 \leftarrow 0}$ in the expression (4.35) of the transition matrix. As these matrices account for the multiple scattering processes between the components of the scatterer, we feel authorized to state that the high frequency peaks are due to multiple scattering processes and that their electric or magnetic nature cannot be stated in an absolute sense but only with reference to the chosen origin that, in the present case, coincides with the center of the host sphere. With this exception in mind, we can safely state, however, that all the resonances in Fig. 6.32 belong to $|m| = 1$ because, for any η, p, and l, the multipole amplitudes $W_{\eta l m}^{(p)}$ of a plane wave that propagates along the z axis do not vanish for $m = \pm 1$ only. Furthermore, the cylindrical symmetry of the scatterer implies that the resonances associated to m and to $-m$ must occur at the same frequency. We finally notice that the same spectrum is obtained by changing the direction of incidence from $\vartheta_I = 0°$ to $\vartheta_I = 180°$. This is an expected result that stems from the symmetry properties of $f_{\eta \eta}$ for forward scattering [3].

The results that we have presented so far show the expected independence of the polarization either because the scatterer is spherically symmetric, Fig. 6.31, or because the incidence is along the cylindrical axis, Fig. 6.32. In this respect let us recall that the dependence of the scattering amplitude on the polarization as well as on $\hat{\boldsymbol{k}}_I$ and on $\hat{\boldsymbol{k}}_S$ is entirely due to the amplitudes

$W^{(p)}_{I\eta lm}$ and $W^{(p)}_{S\eta lm}$, which are independent of frequency, however. On the contrary, the ability to produce resonances is contained in the frequency-dependence of the elements of the transition matrix, which do not depend either on η or on $\hat{\boldsymbol{k}}_I$ or on $\hat{\boldsymbol{k}}_S$. With this in mind, let us go on to discuss the results of Fig. 6.33 where the dependence of the spectrum on the choice of the polarization shows to its full extent. In Fig. 6.33, indeed, we superpose to the spectrum for a sphere with the centered inclusion (that we already reported in Fig. 6.32) the extinction efficiency for the case of the inclusion at maximum eccentricity for polarization both parallel and perpendicular to the plane of scattering. The direction of incidence is individuated by the polar angles $\vartheta_I = 52°$, $\varphi_I = 0°$. This choice of the incidence does not bear any particular significance thus ensuring the true generality of our results. In the same figure we also report the result of the average over the particles with the eccentric inclusion on the assumption that they are randomly distributed in orientation. Of course, the latter spectrum, which gives information on the extinction from a dispersion of identical scatterers, is independent of the polarization. We first notice that, except for the leftmost peak, all resonances for the case of an eccentric inclusion arise from a superposition of contributions from $l = 1$ and for $l = 2$. This is shown by the values of the partial efficiencies $q_{\eta 1}$ and $q_{\eta 2}$, which are not reported, however. Therefore our previous considerations both on the origin and on the classification of the resonances apply. Comparison of the spectra for the individual particles shows

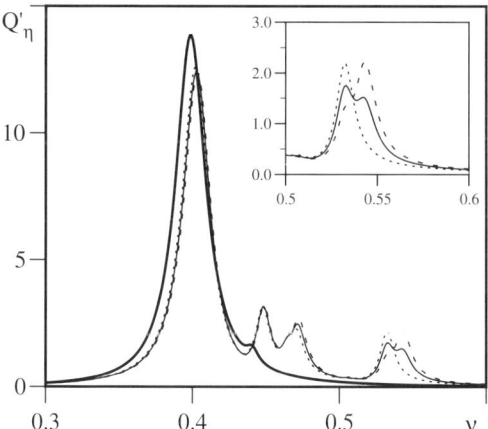

Fig. 6.33. Extinction efficiency for the actual scatterer with the inclusion at maximum eccentricity for polarization parallel (*dashed curve*) and perpendicular (*dotted curve*) to the plane of scattering. The incidence is individuated by the polar angles $\vartheta_I = 52°$, $\varphi_I = 0°$. The *thin continuous curve* reports the extinction efficiency of a dispersion of scatterers with random distribution of the orientations. A detail of the rightmost part of the spectrum is shown in the inset. For comparison, the efficiency for the case of the centered inclusion (*thick continuous curve*) is also reported

that the resonance at $\nu_1 = 0.535$ occurs only in perpendicular polarization, whereas the resonance at $\nu_2 = 0.545$ occurs only in parallel polarization. The separation $\nu_2 - \nu_1$, although small on the scale of the figure, is quite noticeable in wavelength. The peak at ν_1 is the same that appears in Fig. 6.32 for $\vartheta_I = 0°$. In fact, this resonance mainly stems from the elements $\mathcal{S}^{(2\,2)}_{1\,1;\pm 1}$ that, for $\vartheta_I = 0°$, are excited for both polarizations as the relevant amplitudes $W^{(2)}_{\eta 1,\pm 1}$ are independent of the choice of η. On the contrary, for general direction of incidence $W^{(2)}_{\eta 1,\pm 1}$ are unchanged for $\eta = 2$, whereas they are smaller for $\eta = 1$. The peak at ν_2 originates instead from the leading element $\mathcal{S}^{(2\,2)}_{1\,1;0}$ that is excited for parallel polarization as the exciting amplitude $W^{(2)}_{\eta 10}$, provided that the incidence is not along the z axis, does not vanish for $\eta = 1$ only.

When the particles are randomly oriented, the averaged spectrum shows that both peaks at ν_1 and ν_2 are present. This is an expected result in connection to what we already noticed in Sect. 6.4.1 about the averaging of the optical properties of spheres containing an eccentric inclusion [53]. Since the extinction efficiency of the individual scatterers displays a peak whose frequency changes with the polarization (ν_2 for $\eta = 1$, ν_1 for $\eta = 2$), we get the spectrum for random orientation in Fig. 6.33. Ultimately, a dependence on the polarization in the resonance spectrum from a dispersion of particles of the very kind we dealt with here may provide an indication that a noticeable fraction of them has a preferred orientation.

A careful analysis of the extinction spectrum from spheres containing a spherical inclusion may help in establishing whether the inclusion is concentric or eccentric, as the eccentricity yields a multiplicity of resonances. In case the inclusion is eccentric, measurements performed with polarized light may give an indication on the fraction of particles that are oriented alike. Getting this kind of information can be of importance, e.g. in astrophysics. Indeed, the galactic extinction shows an evident dependence on the polarization for certain lines of sight. This fact has been interpreted as being due to the presence of nonspherical particles of the kind we dealt with in this section, which are partially oriented by the general magnetic field of the galaxy. Except for the cause of the alignment, which does not concern us here, the results discussed above support the reliability of this interpretation.

6.5 Correlation Spectroscopy

Correlation spectroscopy is a useful tool for investigating the features of the diffusive motion, both translational and rotational, that may occur within a dispersion of particles whose individual optical properties are known. The observed time-domain spectra have in general an exponential-like behavior, as required by statistical mechanics. Nevertheless, it is a hard matter to deduce from the observations the number and the amplitudes of the exponentials that give an effective contribution. Actually, by modeling the scattering particles as clusters of spherical scatterers, we are able to give a better description of their optical properties and to recover, in the limit of a sufficiently long wavelength, the same results that can be obtained in the small particle approximation (SPA). According to the SPA, the elements $\mathcal{S}^{(2\,2)}_{1\mu 1\mu'}$ are sufficient for the description of the scattering properties of a particle. Since these elements and the polarizability are related, as we discussed in Sect. 3.5.1, the SPA is related to the RSA that assumes the particles to be so small that their properties can be described through their polarizability. Actually, several experiments are interpreted on the assumption that the particles concerned are so small that the RSA or, at least, the SPA applies [101], so that the optical properties of the particles can be described through their polarizability tensor. Now, it is well known that a second rank Cartesian tensor is equivalent to the sum of three irreducible spherical tensors [5] whose rank (angular momentum) is 0, 1, and 2. As a result, using the SPA at most three exponentials can enter the description of the time-domain spectral intensity. On the contrary, we have already shown in Sect. 3.6 that, when the optical properties of the individual particles are described exactly through their transition matrix, exponentials of all orders should occur. In this section, we show what kind of new information can be gained by modeling the scattering particles as clusters of spheres and comparing the results obtained by an exact description through the transition matrix method with those obtainable in the SPA. To this end, we consider axially symmetric particles built as linear chains of 2, 3, and 4 identical spheres whose radii are chosen so the total volume of each chain is independent of the number of its components [40]. We are thus able to follow the behavior of the amplitudes $R_{\eta'\eta L}$ defined in (3.51) with increasing elongation of the particles. Accordingly, the chosen radii are 90.2, 78.8, and 71.6 nm, respectively. All our calculations were performed for a wavelength range from 400 to 800 nm and the refractive index of the spheres was assumed to be $n = 1.61 + i0.004$, which is appropriate for acrylic in the visible and near infrared. Note that our choice of the structure of the particles and of the wavelength range is such that our calculations are well beyond the range of applicability of the RSA. Nevertheless, $l_\mathrm{M} = 10$ proved sufficient to ensure the convergence to four significant digits. Before we discuss our specific results we stress that, due to the axial symmetry of all the particles we deal with and to the presence of a reflection plane orthogonal to the main axis, the only terms that are nonvanishing are those with even L. A few sparse calculations on axially-symmetric particles which do not posses such reflection plane, e.g., linear chains formed of spheres with different radii or refractive indexes, show that the odd-L terms do not vanish.

In Fig. 6.34 we report R_0 calculated for the three clusters described above as a function of the angle of scattering, together with the results of the calculation in the SPA. The polarization both of the incident and of the scattered field is orthogonal to the plane of reference. As expected, the SPA gives

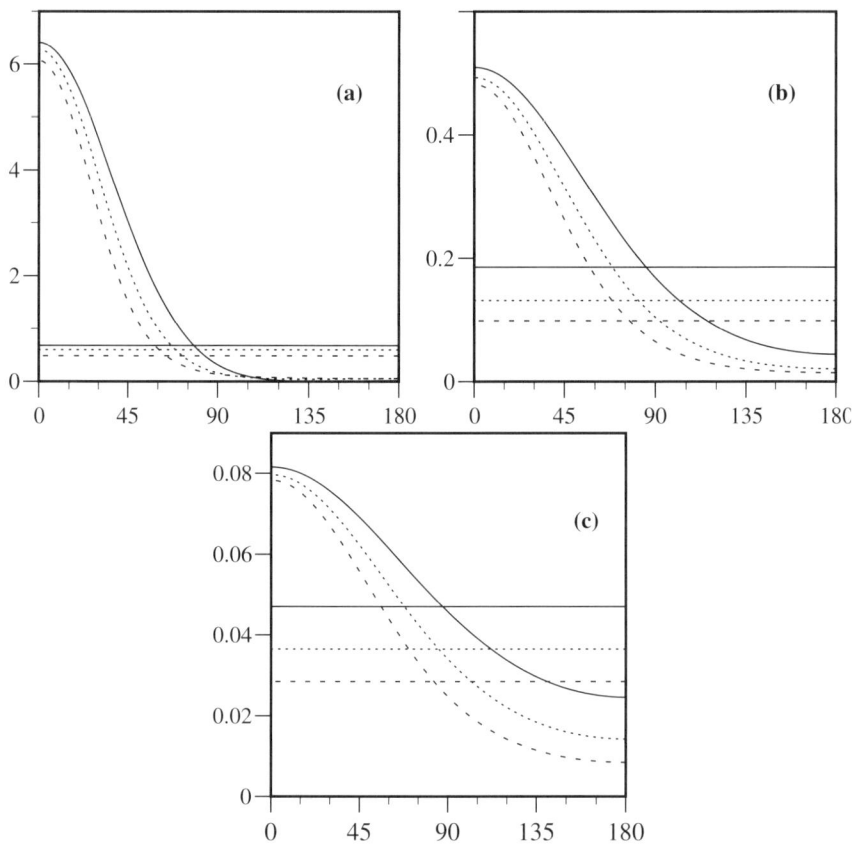

Fig. 6.34a–c. $R_0/16\pi^2$ as a function of the angle of scattering Φ for polarization both of the incident and of the scattered field orthogonal to the plane of reference. The wavelength of the incident field is 400 nm in (**a**), 600 nm in (**b**), and 800 nm in (**c**). The *solid*, *dotted*, and *dashed curves* refer to clusters of two, three, and four spheres, respectively. The *flat lines* are those obtained in the SPA

R_0 independent of the angle of scattering in contrast with the result of the exact calculations, although the dependence on Φ flattens as the wavelength increases. In this respect, it is worth stressing that at 800 nm the SPA R_0 is more or less the average of the exact results. This may explain why, at certain angles of scattering, the observations may appear to be consistent with the assumption that the particles are small. According to calculations that are not reported here, R_0 becomes practically flat at $\lambda = 1200$ nm. The trend towards smaller wavelengths shows that the dependence on Φ becomes stronger and stronger thus making the SPA quite unacceptable. The curve for the two-sphere chain always lies above those for the three- and four sphere chains. This behavior is contrary to what one would expect when taking into account that the more elongated particles should have a more striking

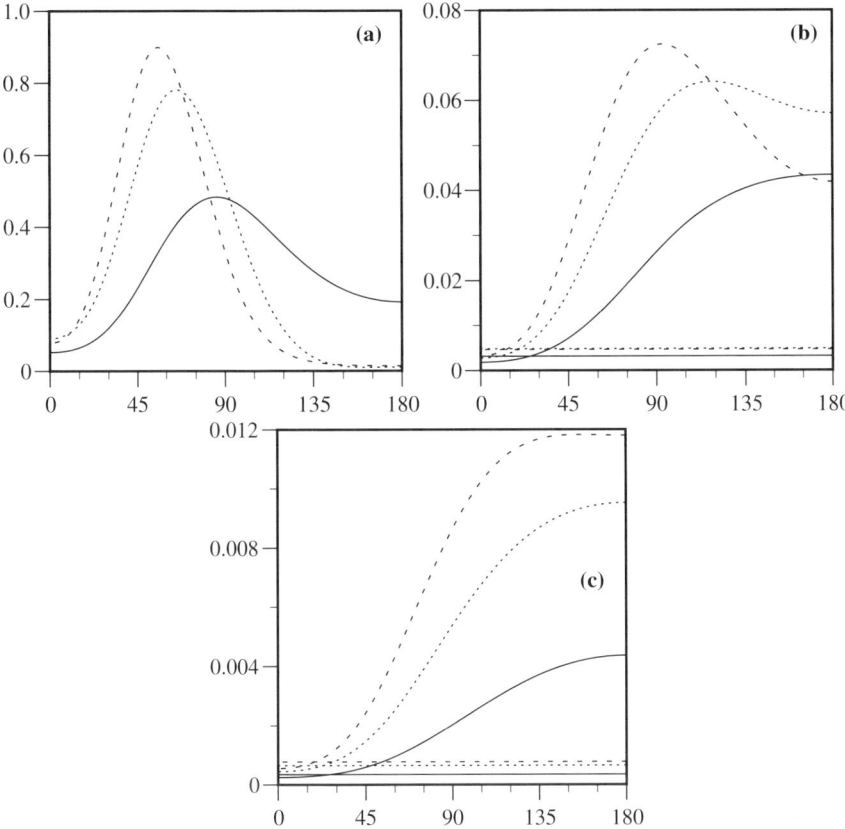

Fig. 6.35a–c. $R_2/16\pi^2$ as a function of the angle of scattering Φ for polarization both of the incident and of the scattered field orthogonal to the plane of reference. The wavelength of the incident field is 400 nm in **(a)**, 600 nm in **(b)**, and 800 nm in **(c)**. The *solid*, *dotted*, and *dashed curves* refer to clusters of two, three, and four spheres, respectively. The *flat lines* are the results yielded by the SPA, and they do not appear in (a) because R_2 is vanishingly small at $\lambda = 400$ nm

dependence on the angle of scattering. Nevertheless, this apparent contradiction is easily disentangled when one thinks that the two-sphere chains are more akin to spherical particles and that, therefore, their scattering power is almost entirely contained in the $L = 0$ term, whereas it is to be expected that the higher-L terms contain a good part of the scattering power of the more elongated three- and four-sphere chains.

That this is, indeed, the case is confirmed by Fig. 6.35, where we report R_2. The curves for the four-spheres chains are, in fact, the highest ones, except for forward scattering when $\lambda = 400$ nm or $\lambda = 600$ nm. Note that at $\lambda = 400$ nm the SPA terms are so small that they do not appear on the scale of the figure. Of course, with increasing wavelength, i.e., when the SPA

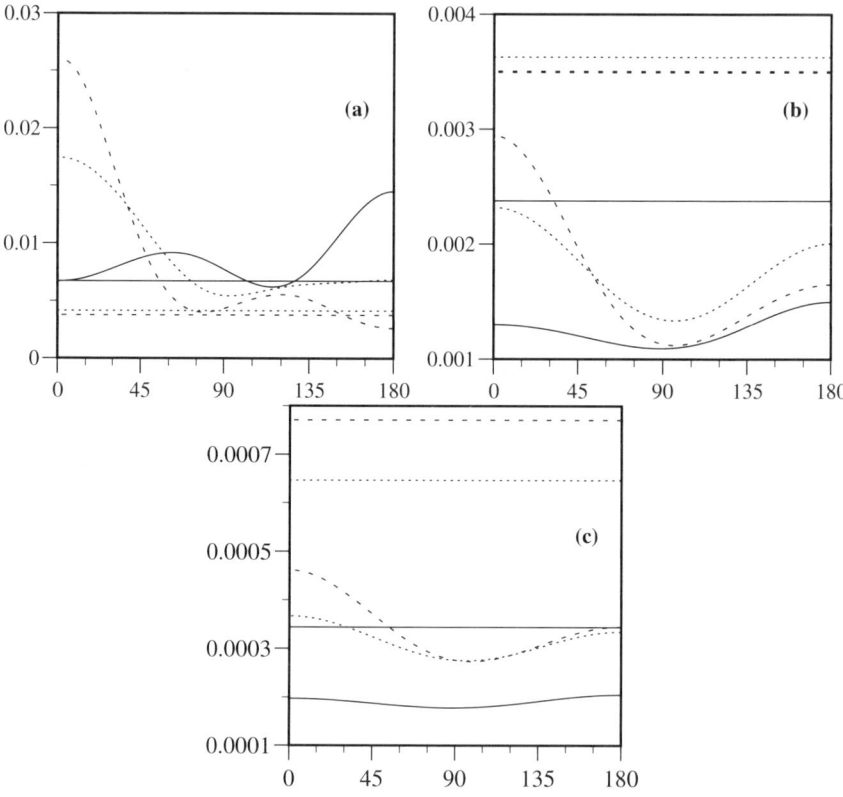

Fig. 6.36a–c. $R_2/16\pi^2$ as a function of the angle of scattering Φ when the polarization of the incident field is orthogonal, whereas that of the scattered field is parallel to the plane of reference. The wavelength of the incident field is 400 nm in (**a**), 600 nm in (**b**), and 800 nm in (**c**). The *solid*, *dotted*, and *dashed curves* refer to clusters of two, three, and four spheres, respectively. The *flat lines* are those obtained in the SPA

becomes more or less acceptable, the exact curves for R_2 assume a monotonic shape. Therefore, a first conclusion that can be drawn from the examination of Figs. 6.34 and 6.35 is that the SPA cannot, at least in principle, be applied to the nonspherical particles we deal with, neither to the less anisotropic such as the two-sphere chains. Nevertheless, the dependence of R_0 and R_2 on the angle of scattering turns out to be rather complicated, so that it is not always true that the most elongated particles yield the larger effects at a given Φ.

In Fig. 6.36 we present the cross-polarization results for R_2. We first notice that the Φ-dependence of the exact results is fully appreciable at the shortest wavelengths only. Nevertheless, as expected, the largest variation with the scattering angle occurs for the four-sphere chains, i.e., for the most elongated particles, although, in agreement with Figs. 6.34 and 6.35, it is in no way certain that these particles produce the strongest effect at a given Φ.

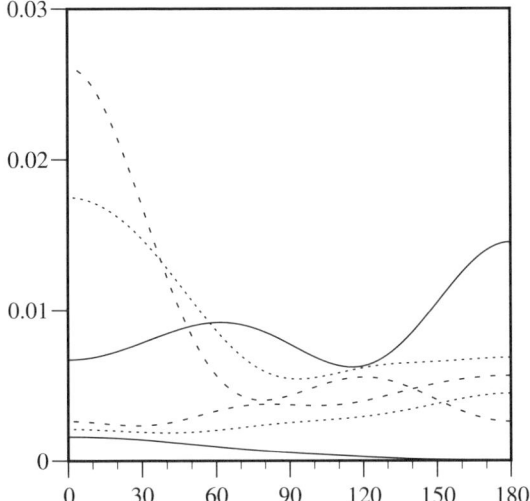

Fig. 6.37. $R_2/16\pi^2$ (*higher curves*) and $R_4/16\pi^2$ (*lower curves*) cross-polarization amplitudes yielded by the exact theory for $\lambda = 400$ nm. The *solid*, *dotted*, and *dashed curves* refer to clusters of two, three, and four spheres, respectively

Anyway, the behavior of our results is, in all circumstances, rather different from that one would expect in the SPA.

The results that we have presented so far refer to $L \leq 2$ and have their counterpart in the results that can be obtained by using the polarizability tensor of the particles. On the contrary, the results that we present in Fig. 6.37 cannot be obtained but with an exact calculation. In fact, we present the cross-polarization Φ-dependence of the R_4 terms together with the R_2 exact terms. Although on the scale of the figure the R_4 terms appear to be rather flat, specially in comparison with the R_2 terms, their range of variation is still considerable, and they appear to be in no way negligible. As the strongest Φ-dependence pertains to the four-sphere chains, we conclude that the R_4 terms should be considered when dealing with highly nonspherical particles.

7. Applications:
Single and Aggregated Spheres
and Hemispheres on a Plane Interface

The detection and characterization by nondestructive methods of particles deposited on a plane substrate is a problem that is frequently met in several fields of physics. For instance, particles are often studied after deposition on a suitable substrate. Of course, the effect of the substrate needs to be discriminated. This chapter is devoted to the study of the results that can be obtained from calculations of the scattering pattern of model particles deposited on or in the vicinity of a plane surface. Our purpose is not so much interpreting the spectra from actual particles but to inform the interested reader on the kind of information that can be gained by such calculations.

We recall that, according to Chap. 5, when an object is in the presence of a plane surface, two different approaches are in order to solve the scattering problem. The choice of the appropriate approach is dictated by the material of the surface. In the case of a metal surface it is convenient to resort to image theory (see Sect. 5.2). In the case of a dielectric surface we found it convenient to establish first the reflection rule for vector multipole fields (see Sect. 5.3). While discussing specific results yielded by the method described in Sect. 5.4, we also take the opportunity to highlight the effect of some approximations.

7.1 Aggregated Spheres and Hemispheres on a Metallic Surface

In this section we discuss the results for the pattern of the scattered intensity from particles on a metallic surface, that, due to the high conductivity, can be considered as perfectly reflecting. As a result, the calculations we are going to describe were performed by resorting to image theory as described in Sect. 5.2. We recall that image theory can deal with any choice of the incidence and polarization and does not imply any limitation of principle either on the size or on the refractive index of the particles of interest. By associating image theory and the transition matrix approach we are able to present the full pattern of the scattered intensity not only for a single sphere on the reflecting surface but also for some representative nonspherical model particles. In practice, we also present the scattering pattern for a cluster of two identical and mutually contacting spheres lying on the surface as well as for clusters of two identical mutually contacting hemispheres and for linear chains of four identical hemispheres, the neighboring hemispheres being in contact with each other, with their flat face lying on the reflecting surface.

As far as we know, the first exact calculation of this kind was performed by Johnson [102] who dealt with a single sphere in contact with the metallic surface. His calculations were restricted to the case of normal incidence, however. The computational technique devised by Johnson need not be described here because it turns out to be a particular case of the general theory of Sect. 5.2. The fact that Johnson's approach does not consider the phase factor $(-)^{\eta-1}$ that we found necessary in (5.50) to get the correct form of the scattered field and of the optical theorem is irrelevant at normal incidence.

With the exception of the case of a single sphere, almost all the patterns that we report refer to a dispersion of identical objects whose orientations are randomly distributed around an axis orthogonal to the reflecting surface. In this respect, the reader should bear in mind the remarks in Sect. 5.2.1 on the need that all the effective scatterers are identical particles. The results were obtained on the assumption that the medium that fills the accessible half-space is the vacuum ($n' = 1$) and that the wavelength of the incident radiation is $\lambda = 628.3$ nm. The refractive index of both the spheres and the hemispheres is $n_0 = 3$ while the radius both of the single sphere and of the hemispheres that form a binary cluster was chosen to be $\rho_s = 126.0$ nm. In turn, the radius both of the spheres that form the binary cluster and of the hemispheres that form the four-hemisphere chain is $\rho_h = 100.0$ nm. This choice of the radii makes the total volume of all the clusters, either of spheres or of hemispheres, equal to the volume of the single sphere. In fact, beyond studying the effect of the presence of the reflecting surface, we are also interested in the effects of the nonsphericity of the particles. Therefore, according to our previous experience [103], we found it convenient to investigate objects with a different geometry but containing the same quantity of refractive material.

The theory in Chap. 5 has been set up on the assumption that the z axis is orthogonal to the reflecting surface. This choice is hardly compulsory but, for cylindrically-symmetric particles, it yields the automatic factorization of the matrix M whose inversion is necessary to get the transition matrix. Accordingly, the polar angles of the direction of observation should be in the range $0 \leq \vartheta_S \leq 90°$ and $0 \leq \varphi_S \leq 360°$. We preferred, instead, to display our results with respect to the frame of reference that is sketched in Fig. 7.1. The reflecting surface coincides with the zx plane and the y axis is thus orthogonal to the surface and directed towards the accessible half-space [71]. Thus, the scattering pattern is reported for $0° \leq \vartheta_S \leq 180°$ and $0 \leq \varphi_S \leq 180°$. As regards the direction of incidence, once the transition matrix of the effective scatterer S is known, the generation of the scattering pattern for several values of ϑ_I and φ_I is a fast and low-cost operation that produces a large amount of data, however. Therefore, we report all the scattering patterns for a single direction of incidence, that is indicated by the arrow in Fig. 7.1. It was chosen

7.1 Aggregated Spheres and Hemispheres on a Metallic Surface

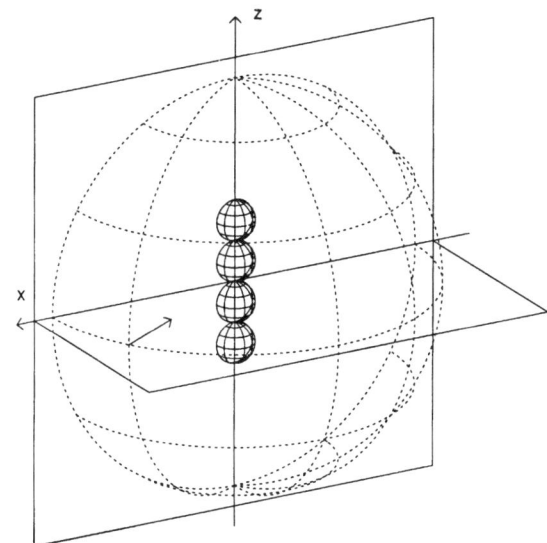

Fig. 7.1. Geometry that we adopted to display our results. The orientation of the four-hemisphere linear chain is also shown

to form an angle of 45° with the normal to the surface and has thus $\vartheta_\mathrm{I} = 90°$ and $\varphi_\mathrm{I} = 225°$. We also chose to not refer the state of polarization to the usual pair of basis vectors that are parallel and orthogonal to the plane of scattering, i.e., to the plane of $\boldsymbol{k}_\mathrm{I}$ and $\boldsymbol{k}_\mathrm{S}$. We preferred instead to project the polarization vector along the pair of unit vectors $\hat{\vartheta}$ and $\hat{\varphi}$ that are tangent to the the meridians and to the parallels, respectively, of the large spherical surface in Fig. 7.1. As a consequence, the appearance of cross-polarization effects is expected even when the scatterer is a single sphere.

The quantity that we report in Figs. 7.2–7.4 as a function of ϑ_S and φ_S is $\mathcal{I}_{\eta\eta'} = |f_{\eta\eta'}|^2 = r^2 I_{\eta\eta'}/I_0$ for the single particles and $\langle \mathcal{I}_{\eta\eta'} \rangle = \langle |f_{\eta\eta'}|^2 \rangle = r^2 \langle I_{\eta\eta'} \rangle / I_0$ for the dispersions, where r is the distance of observation, I_0 the incident intensity, $I_{\eta\eta'}$ is the scattered intensity, $\langle I_{\eta\eta'} \rangle$ is the orientationally-averaged scattered intensity (see Sect. 5.2.1). As usual, the subscripts η and η' indicate the polarization of the observed and of the incident field, respectively, and take on the symbolic values ϑ and φ to denote polarization along the meridians (ϑ-polarization) and along the parallels (φ-polarization), respectively.

We first notice that even when ϑ_S reaches its limiting values, $\vartheta_\mathrm{S} = 0°$ and $\vartheta_\mathrm{S} = 180°$, the angle φ_S is still well defined as this angle characterizes an observation with a well defined choice of the polarization, e.g. along the meridians. Thus, the limiting curves of our patterns ($\vartheta_\mathrm{S} = 0°$ and $\vartheta_\mathrm{S} = 180°$) describe the observation of the scattered beam that propagates along the reflecting surface at right angles to the plane of incidence with a polarization that depends on φ_S. So, for the ϑ-polarized component of the scattered wave, when $\varphi_S = 90°$, $\boldsymbol{E}_S$ is orthogonal to the surface, whereas when $\varphi_S = 0°$ or $\varphi_S = 180°$, $\boldsymbol{E}_S$ is parallel to the reflecting surface, and thus, as a direct consequence of the boundary conditions, the scattered intensity must vanish. For the φ-polarized component of the scattered wave, when $\varphi_S = 0°$ or $\varphi_S = 180°$, $\boldsymbol{E}_S$ is orthogonal to the surface whereas, when $\varphi_S = 90°$, $\boldsymbol{E}_S$ is parallel to the reflecting surface, and thus the scattered intensity must again vanish. Since, for any given polarization, the four extreme vertices of the pattern correspond to the same physical situation, the scattered intensity must have the same value at all these extreme points. A further consequence is that, e.g. $I_{\varphi\varphi}(\vartheta_S = 0°, \varphi_S = 0°) = I_{\vartheta\varphi}(\vartheta_S = 0°, \varphi_S = 90°)$.

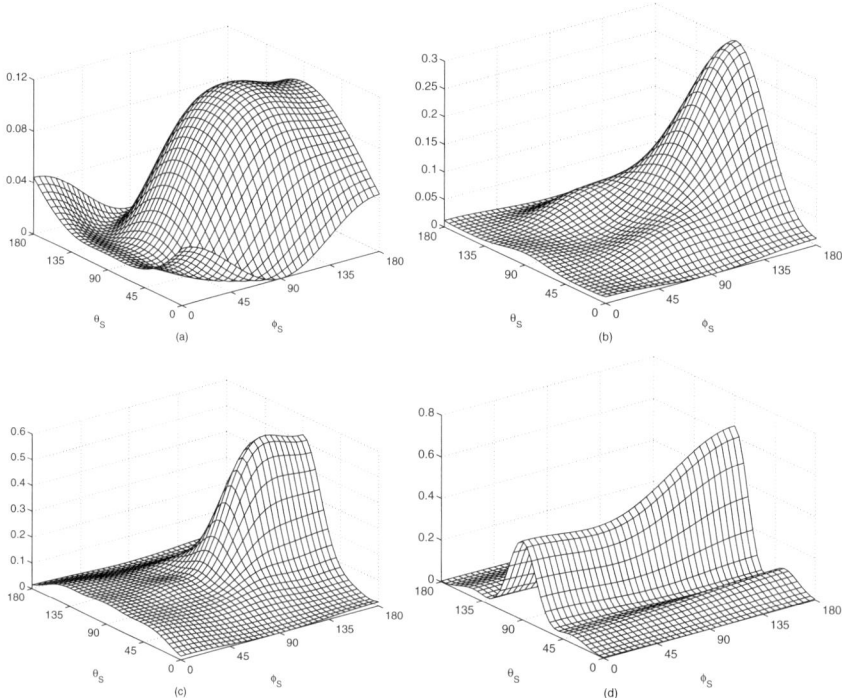

Fig. 7.2a–d. Pattern of the scattered intensity from (**a**) the single sphere on the reflecting surface, (**b**) the dispersion of randomly oriented two-sphere clusters, (**c**) the dispersion of randomly oriented four-hemisphere linear chains, and (**d**) the four-hemisphere linear chain oriented as shown in Fig. 7.1. The wavelength of the incident field is $\lambda = 628.3$ nm and the refractive index of the scatterers is $n_0 = 3$. The radius of the single sphere in (a) is $\rho_S = 126$ nm, whereas the radius of the spheres in the two-sphere clusters in (b) and of the hemispheres of the four-hemisphere linear chain in (c), and (d) is $\rho_h = 100$ nm. We actually report (in μm^2) the quantity $\mathcal{I}_{\varphi\varphi}$ in (a) and (d) and $\langle \mathcal{I}_{\varphi\varphi} \rangle$ in (b) and (c) as a function of the angles of observation ϑ_S and φ_S

7.1 Aggregated Spheres and Hemispheres on a Metallic Surface   177

All these features can be seen in Figs. 7.2–7.4 that report the results of calculations. In particular, Fig. 7.2 reports the scattering pattern (a) for a single sphere in contact with the surface, (b) for a dispersion of randomly oriented clusters of two mutually contacting spheres on the surface, (c) for a dispersion of randomly oriented linear chains of four hemispheres, and (d) for a linear chain of four hemispheres along the z axis. The incident field is φ-polarized and the φ-polarized component of the scattered wave is considered. The pattern for a single sphere in Fig. 7.2 (a) shows only a symmetry of reflection in the plane of incidence. Therefore, even for such a cylindrically symmetric object all the directions of scattering must be considered when the incidence is not normal to the reflecting plane. In all the four patterns in Fig. 7.2 we notice the presence of a strong scattered beam that propagates

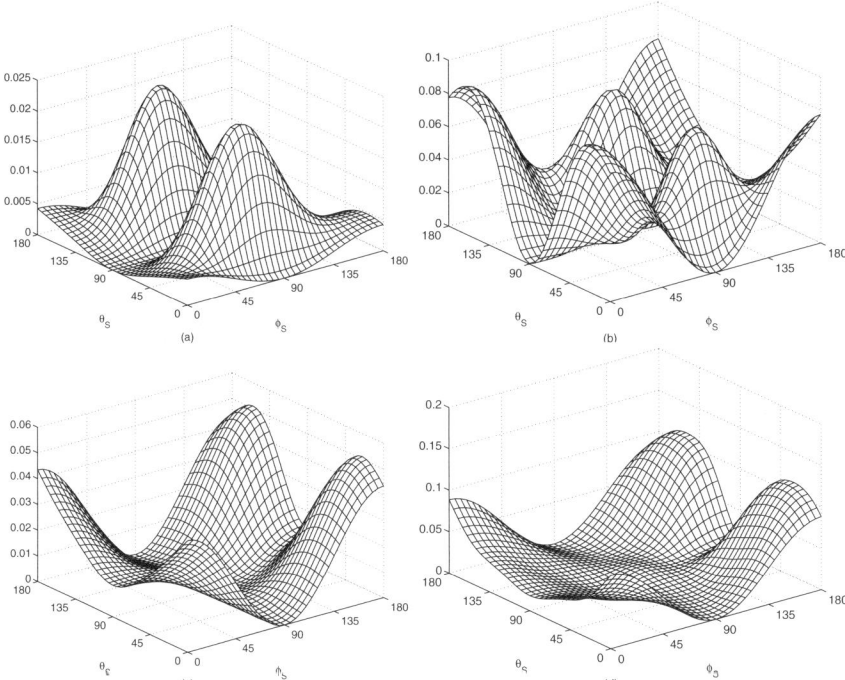

Fig. 7.3a–d. Pattern of the scattered intensity from **(a)** the single sphere on the reflecting surface, **(b)** the dispersion of randomly oriented two-sphere clusters, **(c)** the dispersion of randomly oriented two-hemisphere clusters, and **(d)** the dispersion of randomly oriented four-hemisphere linear chains. The wavelength of the incident field is $\lambda = 628.3\,\mathrm{nm}$ and the refractive index of the scatterers is $n_0 = 3$. The radius of the single sphere in (a) and of the hemispheres in (c) is $\rho_\mathrm{s} = 126\,\mathrm{nm}$, whereas the radius of the spheres in the two-sphere clusters in (b) and of the hemispheres of the four-hemisphere linear chain in (d) is $\rho_\mathrm{h} = 100\,\mathrm{nm}$. We actually report (in $\mu\mathrm{m}^2$) the quantity $\mathcal{I}_{\varphi\vartheta}$ in (a) and $\langle \mathcal{I}_{\varphi\vartheta} \rangle$ in (b), (c), and (d) as a function of the angles of observation ϑ_S and φ_S

along the reflecting surface for $\vartheta_S = 90°$ and $\varphi_S = 180°$. We also notice the two wings in the pattern from the linear chain of four hemispheres in Fig. 7.2 (d). The comparison with the orientationally averaged pattern in Fig. 7.2 (c) suggests that the wings are a result of the particular orientation of the scatterer. The limiting curves at $\vartheta_S = 0°$ and $\vartheta_S = 180°$ turn out to be almost flat for all the clusters but are in no way flat for the single sphere. This suggests that the observation of the scattered light that propagates along the reflecting surface may yield useful information on the shape of the scattering particles.

This is confirmed by Fig. 7.3 that reports the patterns for a single sphere in (a) and for the assemblies of randomly-oriented two-sphere clusters in (b), two-hemisphere clusters in (c), and four-hemisphere linear chains in (d). The incident wave is ϑ-polarized and the φ-polarized component of the scattered wave is considered. Thus, all these patterns describe cross-polarization effects. The patterns in (a) and (b), present a couple of peaks for nonlimiting values of ϑ_S and φ_S. Such peaks are not present in the patterns in (c) and (d). Furthermore, the patterns for all the clusters, either of spheres and of hemispheres reach their maximum value at the four vertices ($\vartheta_S = 0°$, $180°$ and $\varphi_S = 0°$, $180°$). This is not the case for the single sphere in (a). The differences of the patterns of the single sphere and of the clusters can be attributed on one hand to the cylindrical symmetry of the single sphere and on the other hand to the lack of sphericity of the clusters in spite of their random orientation. In other words, the process of averaging does not cancel the effects that are due to the lack of sphericity of the scatterers. The patterns in Fig. 7.3 suggest that the effects of the nonsphericity are still visible provided that the observation is made with the appropriate polarization.

In Fig. 7.4 we report the patterns for the same objects that we considered in Fig. 7.3. The only difference is the choice of the polarization, i.e., the incident wave is φ-polarized and the ϑ-polarized component of the scattered wave is considered. The patterns in (a) and (c), that refer to the single sphere and to the two-hemispheres clusters, show some similarities. Such a similarity is also shown by the patterns in (b) and (d) that refer to the two-sphere and to the four-hemisphere clusters, respectively. Thus, the general structure of the patterns seems to be related to the number of the spheres that compose the effective scatterers. In this respect, we recall that all the scatterers that we consider in this section have the same volume.

We do not report any result for the case in which both the incident and the scattered wave are ϑ-polarized because these patterns do not add any information worthy of a separate comment.

In order to achieve convergence to four significant digits, we had to extend the multipole expansions up to $l_M = 8$ for the single sphere, $l_M = 9$ for the two-hemisphere clusters, and $l_M = 10$ both for the two-sphere and the four-hemisphere clusters. The need to use so large a value of l_M even for the single sphere is due to the fact that, using image theory, we do not actually deal

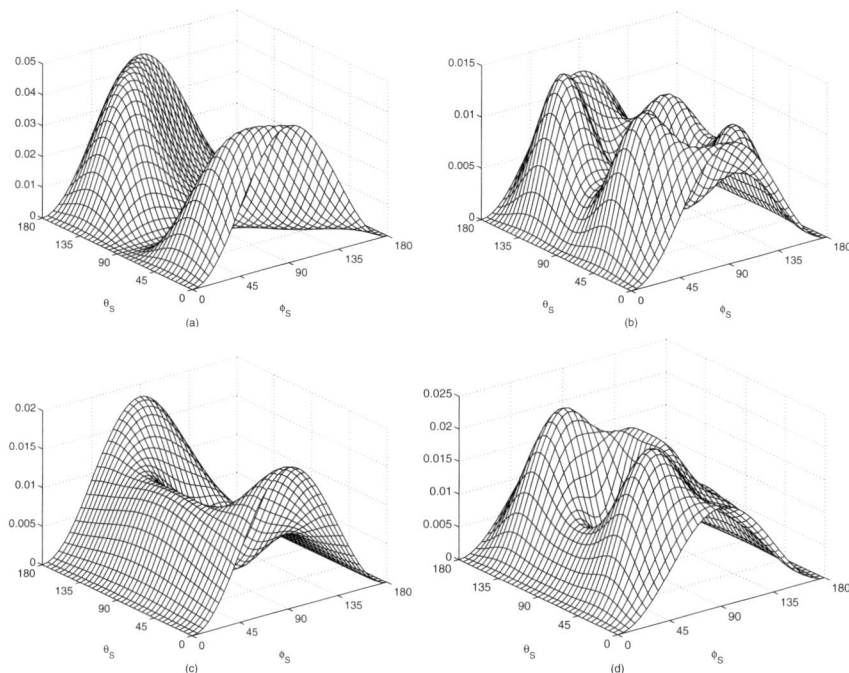

Fig. 7.4a–d. Same as Fig. 7.3, except that $\mathcal{I}_{\vartheta\varphi}$ and $\langle\mathcal{I}_{\vartheta\varphi}\rangle$ are considered

with a single sphere but rather with the two-sphere cluster that forms the effective scatterer. Now, according to Waterman [23], the size parameter x of a nonspherical object can be taken to be the size parameter of the smallest sphere that contains the object itself. Accordingly, for the single sphere we have $x = 2.52$ and even larger values for the other clusters that we consider here. If one also adds that the refractive index is $n_0 = 3$, the values of l_M that we quoted above can in no way be considered too large.

Ultimately, the results of calculations show that the scattering pattern of the model anisotropic scatterers are remarkably different from those of the spherical particles. Therefore, provided this behavior pertains also to more general anisotropic scatterers than those we considered here, a polarization analysis of the grazing scattered intensity could yield useful information on the possible nonsphericity of the particles on the surface.

7.2 Inclusion-Containing Hemispheres on a Metallic Surface

The discussion in Sect. 7.1 proves that image theory is a convenient method to calculate the optical properties of particles in the presence of a perfectly reflecting plane surface [71, 102, 104, 105]. When the particles themselves can be modeled as clusters of spheres or of hemispheres, the association of image theory with the transition matrix approach yields a safe ground for the calculations.

In this section we resort again to image theory to investigate the optical properties of a hemisphere with its flat face on a perfectly reflecting surface, and the investigation is extended to hemispheres containing either a hemispherical or a spherical inclusion. The case of an inclusion composed by a pair of mutually contacting identical spheres is also considered. The purpose is the assessment of how and to what extent the presence of inclusions of various morphologies affects the observed spectrum. Such a study, however, goes beyond the academic interest of solving a problem of electromagnetism through an exact procedure. In fact, the objects we are going to consider may be acceptable models for liquid droplets, possibly containing pollutants, deposited on a metal surface. Furthermore, independent hemispheres on a metallic surface are known to form at the early stages of the deposition of an electrolyte. The dielectric properties of such hemispheres are of interest to the electrochemists who, in general, face the problem through statistical methods [106, 107]. Our study may thus make a useful contribution to the solution of this problem as well as to the problem of assessing whether, during their growth, the hemispheres include any impurity that may be present.

We discuss the scattered intensity from homogeneous hemispheres of radius $\rho_0 = 380$ nm and refractive index $n_0 = 1.59$ on the assumption that the homogeneous medium that fills the accessible half-space is the vacuum ($n' = 1$) [108]. The geometry is depicted in Fig 7.5. The incident wavevector is taken to lie in the first quadrant of the zx plane and the angle of incidence ϑ_I, as usual in optics, is measured between the negative z axis and the positive $\hat{k}_\mathrm{I}$ direction. The angle of observation ϑ_S is always measured from the positive x axis. When inclusions are present, their refractive index is assumed to be $n_1 = 3.8$, whereas their geometrical parameters are so chosen that their total volume is $1/8$ of the volume of the host hemisphere. In all instances, the wavelength of the incident field is taken to be $\lambda_\mathrm{I} = 632.8$ nm and is thus comparable to the size of the scattering objects. Therefore, even in view of

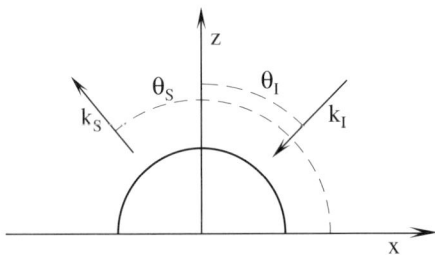

Fig. 7.5. Geometry for the hemisphere on a metallic surface

7.2 Inclusion-Containing Hemispheres on a Metallic Surface

our choice of the refractive indices n_0 and n_1, the convergence of all the multipole expansions that are involved in the calculations must be carefully checked. In this respect we found that, for any geometry that we considered for the inclusions, using $l_M = 9$ achieves convergence to three significant digits, whereas we had to use $l_M = 13$ to get a convergence to four significant digits. Of course, the correctness of our results was checked through the usual tests [53]. Thus, by letting the refractive index of the host hemisphere tend to that of the vacuum ($n_0 \to 1$), we followed the evolution of the scattered intensity towards that expected from the bare inclusions. Moreover, by letting $n_1 \to n_0$ the scattered intensity evolved to that of homogeneous hemispheres with no inclusion.

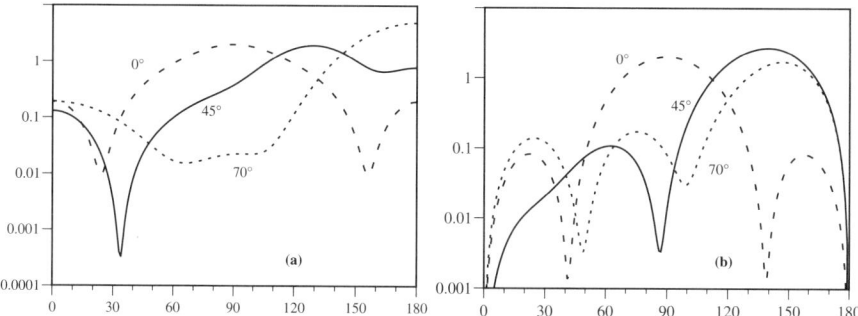

Fig. 7.6a,b. $\mathcal{I}_{\eta\eta}$ for $\eta = 1$ **(a)** and $\eta = 2$ **(b)**, in µm², for the homogeneous hemisphere as a function of the angle of observation ϑ_S. The curves are labeled by the angle of incidence ϑ_I

In all the following figures we report the quantity $\mathcal{I}_{\eta\eta}$, see Sect. 7.1, in µm² for a few selected values of ϑ_I. Accordingly, in Fig. 7.6 we report $\mathcal{I}_{\eta\eta}$ for a homogeneous hemisphere when $\vartheta_I = 0°$, $45°$ and $70°$. The polarization of the incident light is parallel ($\eta = 1$) in Fig. 7.6 (a) and perpendicular ($\eta = 2$) in Fig. 7.6 (b). As we stated in Sect. 5.2, the effective scatterer is a homogeneous sphere so that these patterns were calculated by using the Mie theory. Nevertheless, they look quite different from the patterns from a isolated homogeneous sphere because the exciting field is, in the present case, the superposition of two plane waves whose phase relation, given by (5.3), is dictated by the boundary conditions on the reflecting plane. As a result we get a strong dependence of the pattern on the polarization and in particular the vanishing of the scattered field for $\eta = 2$ (perpendicular polarization) at $\vartheta_S = 0°$ and $\vartheta_S = 180°$. The latter result stems from the boundary conditions because for polarization perpendicular to the plane of incidence the field is parallel to the metallic surface.

Figure 7.7 refers to the hemisphere containing a centered hemispherical inclusion of radius 190 nm. The angles of incidence and the polarization

182   7. Particles on an Interface

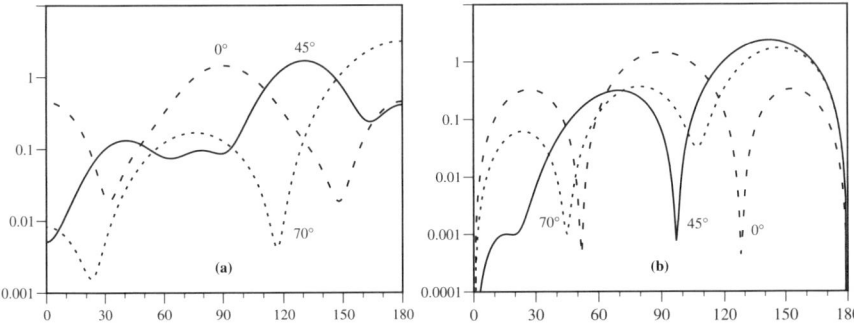

Fig. 7.7a,b. $\mathcal{I}_{\eta\eta}$ for $\eta = 1$ **(a)** and $\eta = 2$ **(b)**, in μm^2, for the hemisphere containing the centered hemispherical inclusion as a function of the angle of observation ϑ_S. The curves are labeled by the angle of incidence ϑ_I

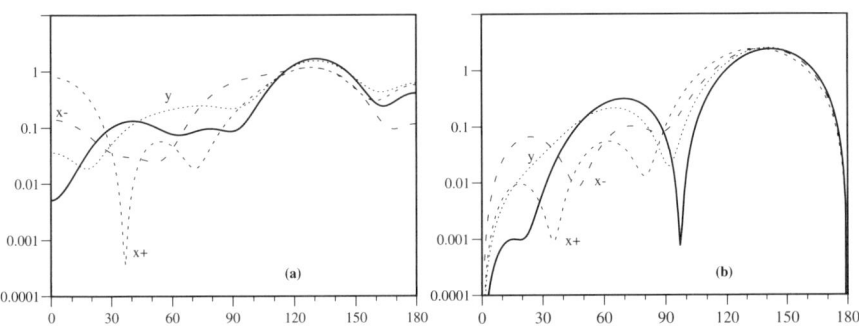

Fig. 7.8a,b. $\mathcal{I}_{\eta\eta}$ for $\eta = 1$ **(a)** and $\eta = 2$ **(b)**, in μm^2, for the hemisphere containing an eccentric hemispherical inclusion as a function of the angle of observation ϑ_S. $\vartheta_I = 45°$. The curves labeled $x\pm$ refer to the inclusion at maximum eccentricity along the positive and negative x axis, respectively. The curve labeled y refer to the inclusion at maximum eccentricity along the y axis. The solid curve is reported for comparison with Fig. 7.7

are as in Fig. 7.6 so that the respective curves can be directly compared. Actual comparison shows that the effect of the presence of the hemispherical inclusion is quite evident. In particular, we notice that the most striking changes occur in the pattern for parallel polarization ($\eta = 1$). For this choice of the polarization, indeed, the patterns for $\vartheta_I = 70°$ and $\vartheta_I = 45°$ are qualitatively different from the corresponding patterns in Fig. 7.6 (a). Less evident changes, such as shifts of the minima, are also present in the patterns for $\eta = 2$ in Fig. 7.7 (b). It may be worth noticing that the patterns in Fig. 7.7 are determined both by the properties of the multipole amplitudes of the exciting field and by the multiple scattering processes that occur between the inclusion and the internal surface of the host hemisphere. Of course, it is to be expected that when the included hemisphere is off center the multiple scattering processes are effective in accounting also for the loss of spherical symmetry of the equivalent scatterer.

To pursue this point, we calculated $\mathcal{I}_{\eta\eta}$ for $\vartheta_\mathrm{I} = 45°$ for several positions of the included hemisphere. The results are reported in Fig. 7.8 for the included hemisphere at maximum eccentricity along both the positive and the negative x axis as well as along the y axis. For the sake of comparison, we also report, from Fig. 7.7, $\mathcal{I}_{\eta\eta}$ for the case in which the included hemisphere is concentric. It is then an easy matter to see that for both choices of the polarization the eccentricity of the inclusion induces qualitative changes in the scattering pattern. In particular, for $\eta = 1$, we notice the occurrence of the deep minimum at $\vartheta_\mathrm{S} = 37°$ when the included hemisphere is at maximum eccentricity in the $+x$ direction. Such a minimum is also present in the case of the homogeneous hemisphere, however [see Fig. 7.6 (a)]. On the contrary, for $\eta = 2$, we should remark the disappearance of the deep minimum that occurs at $\vartheta_\mathrm{S} = 96°$ when the inclusion is centered, and the onset of less deep minima when the inclusion moves both in the $\pm x$ as well as in the y direction. The occurrence and the disappearance of minima as a function of the position of the inclusion are expected, on general grounds, to be due to interference effects as the field propagates back and forth between the inclusion and the internal surface of the host hemisphere.

In Fig. 7.9 we report, for $\vartheta_\mathrm{I} = 45°$, $\mathcal{I}_{\eta\eta}$ for the hemisphere containing a whole sphere of radius $\rho_1 = 150.8$ nm. The included sphere was considered touching both the reflecting surface and the internal surface of the host hemisphere along the $\pm x$ and the y directions. We also considered the included sphere with its center on the z axis either touching the reflecting surface or touching the surface of the host hemisphere. Comparison of the patterns for the corresponding configurations and polarizations in Figs. 7.8 and 7.9

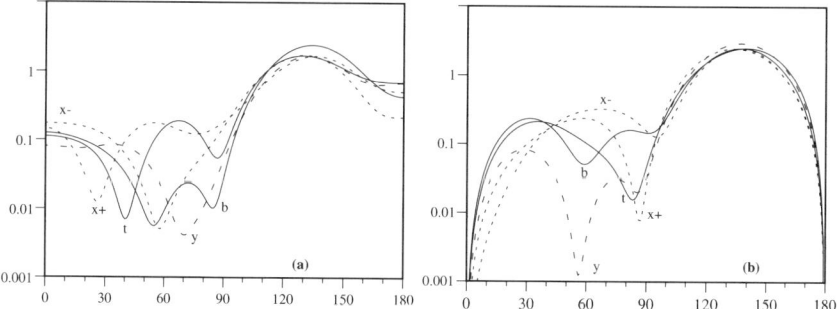

Fig. 7.9a,b. $\mathcal{I}_{\eta\eta}$ for $\eta = 1$ (**a**) and $\eta = 2$ (**b**), in µm², for the hemisphere containing a spherical inclusion as a function of the angle of observation ϑ_S. $\vartheta_\mathrm{I} = 45°$. The curve labeled b refers to the sphere on the reflecting surface with its center on the z axis. The curve labeled t refers to the sphere tangent to the top of the host hemisphere and with its center on the z axis. The curves labeled $x\pm$ and y refer to the sphere at maximum eccentricity along the $\pm x$ axis and the y axis, respectively

show that the change of the shape of the inclusion, even though with no change of its volume, produces visible qualitative changes of the scattered intensity. These changes are easily understood when one recalls that, in the framework of image theory, the equivalent scatterer of a hemisphere with a spherical inclusion is a whole sphere containing a cluster composed by two identical spheres. Accordingly, the scattering pattern of the whole object is determined not only by the multiple scattering processes that occur between the inclusion and the surface of the host sphere, but also by the processes that occur between the components of the inclusion (see also Sect. 6.4.3). For example, when the included sphere with its center on the z axis and tangent to the reflecting plane is lifted so as to become tangent to the top of the host hemisphere, the resulting configuration is equivalent to considering a pair of spheres that are somewhat separated. This nonzero separation implies a decrease of the intensity of the multiple scattering processes that occur between them.

Finally, in Figs. 7.10 and 7.11 we report, for $\vartheta_\mathrm{I} = 45°$, $\mathcal{I}_{\eta\eta}$ for the hemisphere containing two identical, mutually contacting spheres that lie on the reflecting surface. The radius of both spheres is $\rho_1 = 120\,\mathrm{nm}$. The axis of the included bisphere is considered both in the plane of incidence (the zx plane) and in the yz plane. In Fig. 7.10 the contacting point between the spheres lies on the z axis. In Fig. 7.11 one of the spheres touches the surface of the host hemisphere, so that the bisphere is as eccentric as possible. The dependence of all the curves both on the position of the cluster and on the polarization of the incident field can be traced back to the fact that the multiple scattering processes now occur between the four spheres that form the equivalent scatterer and thus give rise to patterns that may be rather complicated. It is worth noticing, however, that $\mathcal{I}_{\eta\eta}$ has a weaker

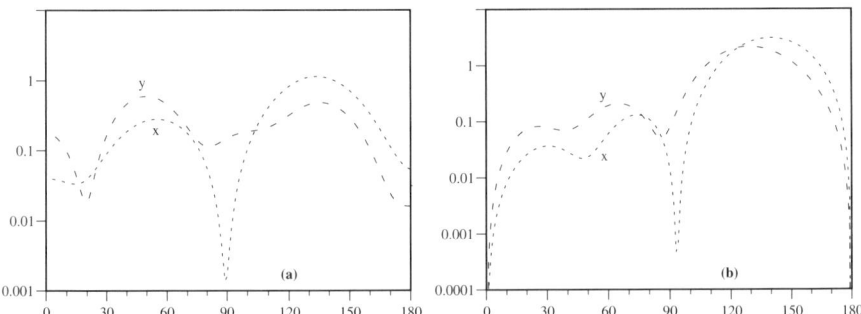

Fig. 7.10a,b. $\mathcal{I}_{\eta\eta}$ for $\eta = 1$ (**a**) and $\eta = 2$ (**b**), in $\mu\mathrm{m}^2$, for the hemisphere containing the inclusion composed by two mutually contacting identical spheres as a function of the angle of observation ϑ_S. $\vartheta_\mathrm{I} = 45°$. The point of contact between the included spheres always lies on the z axis. The curves labeled x and y refer to the bisphere with its axis in the plane of incidence (zx plane) and in the xy plane, respectively

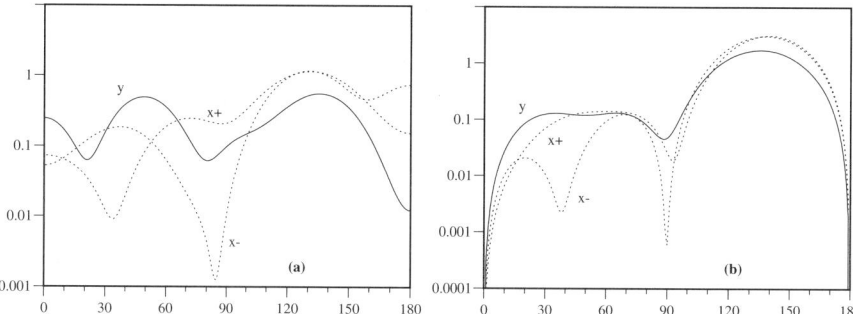

Fig. 7.11a,b. $\mathcal{I}_{\eta\eta}$ for $\eta = 1$ **(a)** and $\eta = 2$ **(b)**, in μm^2, for the hemisphere containing the inclusion composed by two mutually contacting identical spheres as a function of the angle of observation ϑ_S. $\vartheta_I = 45°$. The curves labeled $x\pm$ refer to the bisphere with its axis in the zx plane and one of its components touching the surface of the host hemisphere along the $\pm x$ direction. The curve labeled y refers to the bisphere with its axis in the yz plane and one of its components touching the host hemisphere

dependence on ϑ_S than one would expect, e.g. the number of the minima is not as large as one could expect from the calculations for bare clusters with a comparable complexity. This result is in agreement with our finding that the more minutely subdivided is the matter of the inclusions the smoother are the curves of the scattered intensity [54]. Nevertheless when we compare $\mathcal{I}_{\eta\eta}$ in Fig. 7.10 with $\mathcal{I}_{\eta\eta}$ for the eccentric cluster in Fig. 7.11, we see that the curves are quite distinguishable, in particular for $\eta = 2$, even though the maximum eccentricity is achieved with a small displacement of the contacting point between the spheres from the z axis.

The results that we reported above are aimed at informing the interested reader of the potentialities of the formalism of the transition matrix in association with image theory. Beyond the interest for the electrochemists, the formalism may also be of interest to people studying the effect of atmospheric pollution, for instance, for the interpretation of the optical observations aimed at the detection of the presence and of the nature of pollutants deposited on a metal surface. In this respect, the results that we reported above should be understood as a demonstrative sample of the patterns that can be found in actual experiments. Nevertheless, the reported results are sufficient to show that, beyond the expected dependence on the direction of incidence and on the polarization, one has also to expect a noticeable dependence on the geometry of the inclusions. Of course, these features are not sufficient to allow the experimentalists to discriminate among the possible geometries, in particular when several different inclusions are present. However, further information on the nature and on the configuration of the inclusions could stem from the comparison of patterns taken for different choices both of the angle of incidence and of the wavelength.

7.3 Resonance Suppression Mechanism

It is intuitive that the spectrum of a homogeneous hemisphere on a metallic surface can display optical resonances. According to image theory, the resonances are those of the effective scatterer, a homogeneous sphere, illuminated by the exciting field. It is not so intuitive that the hemisphere may not present some of the resonances that are present in the spectrum of the effective scatterer when illuminated by the actual incident field only. This resonance suppression mechanism can be understood when one thinks that the exciting field is the superposition of the actual incident field and of the field that comes from the image source and that the phase relation between these fields are dictated by the condition of perfect reflection on the metallic surface. As discussed in Sect. 5.2, the multipole amplitudes of the exciting field are given by (5.7). Then, it is an easy matter to see that, when in (5.7) $\left[1-(-)^{p+l+m}\right]=0$, i.e., when $p+l+m$ is even, the corresponding multipole is not present in the exciting field, even when the corresponding amplitude of the incident field is nonvanishing. This vanishing of some amplitudes of the exciting field may thus affect the scattered field up to the suppression of some of the characteristic resonance peaks. Although the considerations above refer to a homogeneous hemisphere, the resonance-suppressing mechanism has no evident dependence on the shape of the particle. Therefore, it would be tempting use this effect to interpret the spectra from particles of arbitrary shape, according to the suggestion of Li and Chýlek [109] and by Johnson [102, 110]. We will see below that this task is rather difficult [78]. Let us, in fact, recall that the scattering amplitude of a hemisphere on a perfectly reflecting surface is

$$f_{\eta\eta}(\hat{\boldsymbol{k}}_{\mathrm{R}}, \hat{\boldsymbol{k}}_{\mathrm{I}}) = \frac{\mathrm{i}}{4\pi k'} \sum_{plm} W^{(p)*}_{\mathrm{R}\eta lm} R^{(p)}_l W^{(p)}_{\mathrm{E}\eta lm}, \qquad (7.1)$$

where the quantities $R^{(p)}_l$, except for the minus sign, are the elements of the transition matrix for the effective (spherical) scatterer, and are defined by (4.6). In turn, the amplitudes $W^{(p)}_{\mathrm{R}\eta lm}$ are defined by (5.3). Note that (7.1) is a particular case of (5.8) for observation in the plane of incidence. The amplitudes $W^{(p)}_{\mathrm{I}\eta lm}$ satisfy the sum rule (3.28) and (3.29) according to whether the field is linearly or circularly polarized [33, 34]. Here we assume linear polarization, so that (3.28) applies. Since the transition matrix of the effective scatterer is independent of m, it is meaningful to define the quantity

$$U^{(p)}_{\eta l} = \sum_m W^{(p)*}_{\mathrm{R}\eta lm} W^{(p)}_{\mathrm{E}\eta lm} = (-)^{\eta-1} 2\pi(2l+1) + \sum_m (-)^{\eta+p+l+m} W^{(p)*}_{\mathrm{I}\eta lm} W^{(p)}_{\mathrm{I}\eta lm}, \qquad (7.2)$$

which has been obtained with the help of (3.28). We can then rewrite (7.1) in the form

$$f_{\eta\eta}(\hat{\boldsymbol{k}}_\mathrm{R},\hat{\boldsymbol{k}}_\mathrm{I}) = \frac{\mathrm{i}}{4\pi k'}\sum_{pl}U^{(p)}_{\eta l}R^{(p)}_l,$$

that shows at once that any resonance associated with a vanishing $U^{(p)}_{\eta l}$ is bound to disappear.

The situation is quite different when the effective scatterer is nonspherical. In this case, indeed, the scattering amplitude is given by (5.8) and the elements of the matrix S depend both on m and on m'. In this case, the resonance suppression it is just due to the vanishing of the amplitude $W^{(p)}_{\mathrm{E}\eta lm}$, which should excite the resonating element of S. In order to show how the resonance suppression mechanism works both for spherical and nonspherical effective scatterers, we will consider a homogeneous hemisphere whose flat face lies on a metal surface, and a binary aggregate of identical hemispheres with their flat faces on the perfectly reflecting surface. In the latter case, the radius of the component hemispheres is chosen to be small while the refractive index is assumed to be rather large. These choices result in a spectrum with a simple structure of the resonances.

7.3.1 Single Hemispheres

We report in Fig. 7.12 the plots of $U^{(p)}_{\eta l}$ for $l \leq 4$, $\eta = 1, 2$ and $p = 1, 2$ as a function of ϑ_I. We notice in Figs. 7.12 (a) and (b) that at $\vartheta_\mathrm{I} = 0°$ all the $U^{(1)}_{1l}$ vanish for even l, whereas all the $U^{(2)}_{1l}$ do vanish for odd l. This result was expected because when $\vartheta_\mathrm{I} = 0°$ the only nonvanishing amplitudes $W^{(p)}_{\mathrm{I}\eta lm}$ are those with $m = \pm 1$, so that

$$W^{(p)}_{\mathrm{E}\eta l,\pm 1} = \left[1 + (-)^{p+l}\right] W^{(p)}_{\mathrm{I}\eta l,\pm 1}.$$

Although the preceding argument is based on the vanishing of the individual amplitudes $W^{(p)}_{\mathrm{E}\eta lm}$ at $\vartheta_\mathrm{I} = 0°$, the usefulness of the quantity $U^{(p)}_{\eta l}$ remains unaffected. In fact, the vanishing of $U^{(2)}_{1l}$ for $l = 2$ at $\vartheta_\mathrm{I} = 45°$ [see Fig. 7.12 (b)] is due to the sum over m in (7.2). The plots in Figs. 7.12 (c) and (d) show that the behavior of $U^{(p)}_{2l}$ is similar to that of $U^{(p)}_{1l}$. We remark that again all the $U^{(1)}_{2l}$ vanish at $\vartheta_\mathrm{I} = 0°$ for even l, whereas all the $U^{(2)}_{2l}$ vanish at the same incidence for odd l and, in particular, $U^{(2)}_{21}$ turns out to be identically zero. These features were expected on the ground of the structure of (7.2).

We report in Fig. 7.13 the quantity

$$\gamma_{\mathrm{R}\eta} = 2k'\mathrm{Im}[(-)^{\eta-1}f_{\mathrm{R}\eta\eta}(\hat{\boldsymbol{k}}_\mathrm{R},\hat{\boldsymbol{k}}_\mathrm{I})],$$

for $\eta = 1$, for a hemisphere of refractive index $n_0 = 3$ on the reflecting surface. The homogeneous medium that fills the accessible half-space was assumed to be the vacuum ($n' = 1$). In fact, according to Sects. 2.7.2 and 5.6, we have

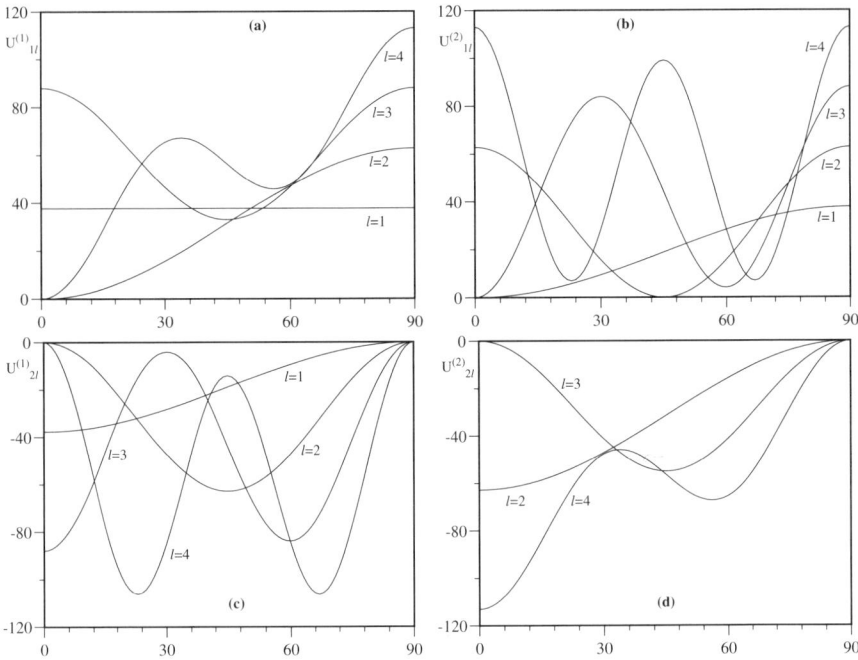

Fig. 7.12a–d. Plot of $U^{(p)}_{\eta l}(\vartheta_\mathrm{I})$ for $l \leq 4$. Note that in (d) the curve for $l=1$ does not appear as it is identically vanishing

$$\gamma_{\mathrm{R}\eta} = \frac{k'^2}{2\pi} \sigma_{\mathrm{T}\eta} \,,$$

a quantity that depends on ρ and k_v only through their product. $\gamma_{\mathrm{R}\eta}$ is plotted as a function of the size parameter $x = k_\mathrm{v}\rho$ in the range from $x=1$ to $x=3$. The angle of incidence is $\vartheta_I = 0°$ in Fig. 7.13 (a), $45°$ in Fig. 7.13 (b), and $70°$ in Fig. 7.13 (c).

Now, it is quite natural to compare the spectrum from the hemisphere on the reflecting surface with the spectrum from the *equivalent sphere*, i.e., from the effective scatterer, illuminated by the actual incident field only. Therefore, in each of Figs. 7.13 (a), (b), and (c) we also report the plot of the quantity

$$\gamma = 2k'\mathrm{Im}[f_{\eta\eta}(\hat{\boldsymbol{k}}_\mathrm{I}, \hat{\boldsymbol{k}}_\mathrm{I})],$$

for the equivalent sphere. In this case, $f_{\eta\eta}$ does not actually depend on the polarization [see also (3.28)] so that the quantity γ need not carry the subscript η.

The strict correspondence between the resonances that disappear and the vanishing of the respective $U^{(p)}_{1l}$ is so evident that, in our opinion, no further comment would be necessary. However, we call the attention of the reader to the simultaneous disappearance, in Fig. 7.13 (a), of the two resonances at

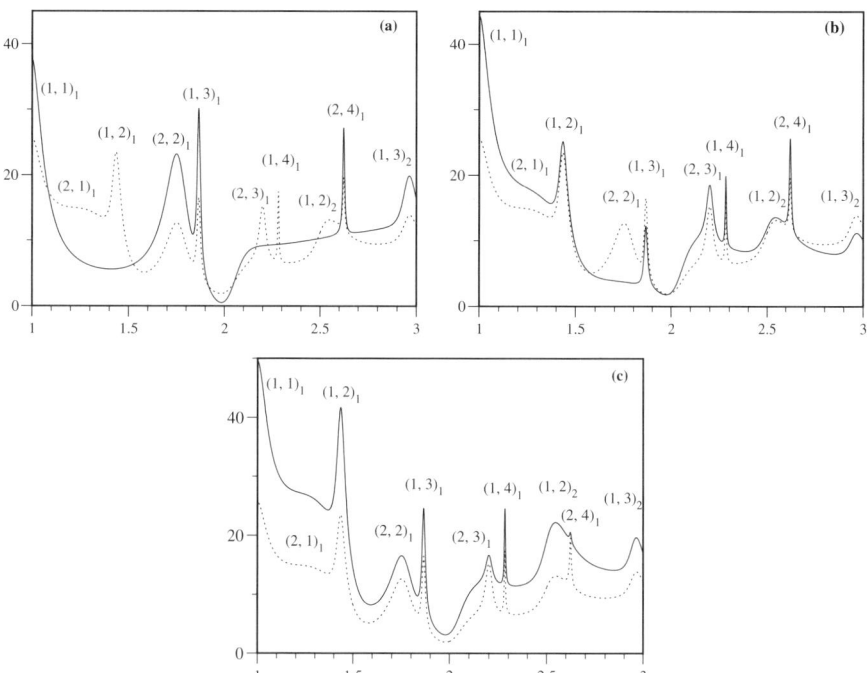

Fig. 7.13a–c. $\gamma_{R\eta}$ (*solid curves*) for a homogeneous hemisphere of radius ρ and refractive index $n_0 = 3$ on the reflecting surface as a function of $x = k_v\rho$ for $\eta = 1$. The angle of incidence is $\vartheta_I = 0°$ in (**a**), $\vartheta_I = 45°$ in (**b**), and $\vartheta_I = 70°$ in (**c**). We also report for the sake of comparison γ for the equivalent sphere (*dotted curves*). The resonances are labelled as $(p, l)_n$ where n distinguishes different resonances with the same value of p and l

$x = 1.3118$, that is associated to $p = 2$ and $l = 1$, and at $x = 1.437$, that is associated to $p = 1$ and $l = 2$. Both peaks belong, in fact, to an odd value of $p + l$. The same mechanism also explains the simultaneous disappearance of the peaks at $x = 2.2$ and at $x = 2.28$. The former peak is associated with $p = 2$ and $l = 3$ whereas for the latter peak $p = 1$ and $l = 4$. Again, both peaks belong to an odd value of $p+l$. Even the resonance spectrum for $\eta = 2$ strictly follows the behavior of the $U^{(2)}_{2l}$ so that we resolved not to report the specific plot that, in spite of its significance, does not add any further information worthy of a separate comment.

7.3.2 Binary Clusters

The lack of a general theory for the resonances of nonspherical particles makes an unambiguous classification of their resonances rather difficult. According to (5.8), the lack of diagonality of the transition matrix prevents a meaningful definition of a function analogous to the quantity $U^{(p)}_{\eta l}$ we defined above. Even in the case of aggregated spheres the transition matrix is not a diagonal matrix so that there is no one-to-one association of the multipole amplitudes of the exciting field to those of the scattered field. Nor there is a simple relation between the resonances of the component spheres and those of the aggregate as a whole. Nevertheless, (5.7) does not depend on the shape of the particles so that the vanishing of any of the amplitudes of the exciting field is expected to affect the resonances even of a nonspherical particle. To show that this is indeed the case, we calculate the resonance spectrum for two contacting spheres of radius ρ and refractive index n_0. We also calculate the spectrum for two contacting hemispheres of radius ρ and refractive index n_0 with their flat face on the reflecting surface. The surrounding medium is the vacuum ($n' = 1$), whereas $n_0 = 10\pi$ has an unusually high value. According to Newton [21] and to our experience [100], this choice makes the resonances of the aggregate as a whole, as well as of its components, occur at so small values of $x = k'\rho$ that fully convergent values of the scattered field are obtained for $l = l' = 1$ only. As a result, the resonances of the aggregate as a whole can surely be associated to the multipole amplitudes with $l = 1$.

In Fig. 7.14 (a) we report, for both values of η, the quantity

$$\gamma_\eta = 2k' \text{Im}[f_{\eta,\eta}(\hat{\boldsymbol{k}}_\text{I}, \hat{\boldsymbol{k}}_\text{I})],$$

for the aggregate of spheres referred to above, whereas in Fig. 7.14 (b) we report $\gamma_{\text{R}\eta}$ for the aggregate of hemispheres on the reflecting surface. All the plots are reported as a function of x and the angle of incidence is $\vartheta_\text{I} = 70°$. Figures 7.15 (a) and (b) are the same as Figs. 7.14 (a) and (b), respectively, except that the angle of incidence is $\vartheta_\text{I} = 0°$.

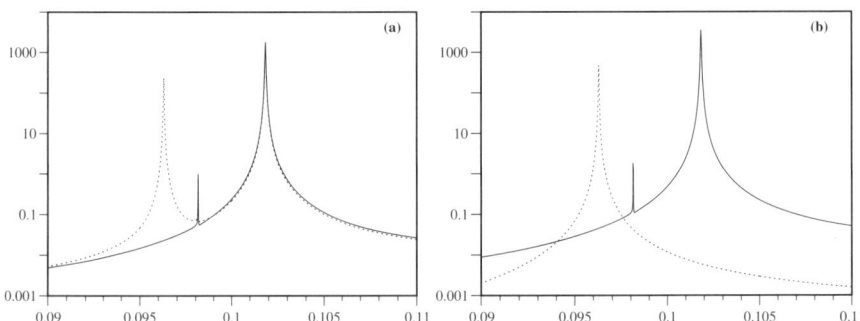

Fig. 7.14a,b. γ_η for the aggregate of two identical mutually contacting spheres of radius ρ and refractive index $n_0 = 31.4$ **(a)**, and $\gamma_{\text{R}\eta}$ for the aggregate of two mutually contacting hemispheres with the same radius and refractive index on the reflecting surface as a function of x **(b)**. The axis of the aggregate lies in the zx plane and is parallel to the x axis. The plane of incidence coincides with the zx plane and the angle of incidence is $\vartheta_\text{I} = 70°$. The *solid* and the *dotted curves* refer to polarization parallel and orthogonal to the plane of incidence, respectively. Note that the spike at $x = 0.09813$ in (a) is double as it appears for both choices of the polarization although this is not visible on the scale of the figure

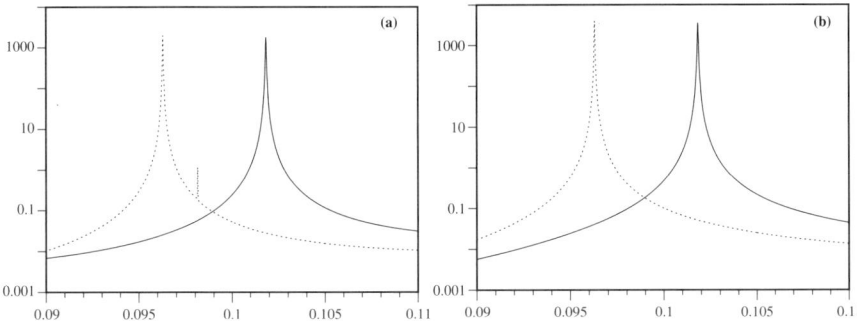

Fig. 7.15a,b. Same as Fig. 7.14 except that the angle of incidence is $\vartheta_I = 0°$

On the whole, Figs. 7.14 and 7.15 present three peaks at $x_1 = 0.09629$, $x_2 = 0.09813$, and $x_3 = 0.10183$. Since for $l = 1$ the component spheres have a single resonance at $x = 0.1$, Figs. 7.14 and 7.15 confirm that the multiple scattering processes within an aggregate produce resonances whose location cannot be related to the locations of the resonances of the component spheres [46, 100]. The dependence of the resonance spectrum of an aggregate on the polarization of the incident light can be understood only through a close examination of the features of the transition matrix and of the amplitudes of the exciting field.

Now, at x_1 the largest elements of S are $\mathcal{S}^{(1\,1)}_{1,\pm1,1,-1} \approx -\mathcal{S}^{(1\,1)}_{1,\pm1,11}$ so that a magnetic resonance ($p = 1$) is expected. At x_2 the leading elements are $\mathcal{S}^{(2\,2)}_{1010}$ and $\mathcal{S}^{(2\,2)}_{1,\pm1,1,-1} \approx \mathcal{S}^{(2\,2)}_{1,\pm1,11}$ so that an electric resonance ($p = 2$) is expected. Finally, at x_3 the leading elements of S are $\mathcal{S}^{(1\,1)}_{1010}$ and $\mathcal{S}^{(1\,1)}_{1,\pm1,1,-1} \approx \mathcal{S}^{(1\,1)}_{1,\pm1,11}$ thus suggesting that this resonance is a magnetic one. Nevertheless, by comparing Figs. 7.14 and 7.15 one sees that not all the possible resonances actually occur. This is due to the dependence of the amplitudes $W^{(p)}_{I\eta lm}$ both on the polarization and on the angle of incidence ϑ_I.

As an example, let us discuss the behavior of the resonance at x_2 that appears in Fig. 7.14 (a) for any choice of the polarization. According to (5.8), the implied amplitudes of the incident field are the $W^{(2)}_{I\eta 1m}$, with $W^{(2)}_{I210} = 0$ and $W^{(2)}_{I\eta 11} = (-)^\eta W^{(2)}_{I\eta 1,-1}$. Therefore, when $\eta = 1$, the peak at x_2 belongs to $m = 0$, whereas for $\eta = 2$ it belongs to $m = \pm 1$. When the reflecting surface is present, the implied amplitudes of the exciting field are the $W^{(2)}_{E\eta 1m}$, but only $W^{(2)}_{E110} \neq 0$. This is enough to explain why the peak at x_2 may appear in Fig. 7.14 (b) for $\eta = 1$ only.

When $\vartheta_I = 0°$ the appropriate resonance spectra are those in Fig. 7.15. The behavior of the peak at x_2 is easily understood when one considers that at $\vartheta_I = 0°$ we have $W^{(2)}_{I110} = W^{(2)}_{E110} = 0$. Therefore this resonance can occur in Fig. 7.15 (a) for $\eta = 2$ only and cannot appear at all in Fig. 7.15 (b). The

192   7. Particles on an Interface

behavior of the resonances at x_1 and at x_3 that appear in Figs. 7.14 and 7.15 can be understood through a quite similar analysis.

In conclusion, the resonance suppressing mechanism should be taken into account when interpreting the spectra from a perfectly reflecting surface sparsely seeded by hemispherical particles such as liquid droplets. As regards the spectra of the single hemispheres, the discussion in Sect. 7.3.1 emphasizes the importance of the function $U_{\eta l}^{(p)}$ to getting insight into the behavior of the resonances as a function of the direction of propagation and of the polarization of the incident wave.

In turn, the discussion above can provide the guidelines for understanding the mechanism through which the combined effect of the boundary conditions on the reflecting surface and of the polarization of the incident field affect the behavior of the resonance spectrum of aggregated hemispheres. In fact, to understand the behavior of the spectrum from aggregated hemispheres, we were forced to examine the behavior of the single elements of the transition matrix as well as of the the multipole amplitudes of the exciting field as a function of ϑ_I and of the polarization. Of course, such an analysis is practicable only when the parameters of the aggregate are so chosen that the order of the transition matrix is small.

7.4 Particles on a Dielectric Substrate

In Chap. 5 we stressed that the techniques that can be used to deal with scattering from particles on a metallic surface are not applicable when the surface has a finite conductivity. For the case of particles on a dielectric substrate, the appropriate approach is described in Sects. 5.3–5.5. This approach, hereafter referred to as multipole reflection theory (MRT), will now be applied to the calculation of the pattern of the scattered intensity from model particles on a dielectric substrate.

7.4.1 Single Spheres

We start remarking that the approximations to which Bobbert and Vlieger [75, 111] and Johnson [112] resorted, as well as any analogous approximation, can be entirely described as specific approximations of MRT [72]. According to the approximation devised by Bobbert and Vlieger [75] (hereafter referred to as B0), the far zone expression of $\boldsymbol{E}_{\mathrm{SR}\eta}$ equals that of the scattered field, calculated for the direction $\pi - \vartheta_S$, where ϑ_S denotes the direction of observation, with its amplitude scaled by the value of the Fresnel coefficient $F_\eta(\pi - \vartheta_S)$. Our variant to B0 (approximation B1) assumes that $\boldsymbol{E}_{\mathrm{SR}\eta}$ equals the field that is obtained by scaling by the factor $(-)^{\eta-1} F_\eta(\pi - \vartheta_S)$, the field that would come from the image sphere with center at O'' (see Fig. 5.4) if the surface were perfectly reflecting. In practice, this is achieved by calculating, for the direction ϑ_S, $\boldsymbol{E}_{\text{SR}\eta}$ as given by (5.33) and (5.34), but scaling by the factor $(-)^{\eta-1}F_\eta(\pi-\vartheta_S)$ the amplitudes $a^{pp'}_{ll';m}$, defined by (5.22) and (5.26), evaluated for a perfectly reflecting surface.

Johnson [112], besides using B0 to get the far zone expression of $\boldsymbol{E}_{\text{SR}}$, introduces a further approximation that affects the imposition of the boundary conditions that yield the equations for the amplitudes $\mathcal{A}^{(p)}_{\eta lm}$, defined by (5.30). Johnson's procedure (hereafter referred to as J0) stems from the equation

$$\sum_{p''l''}\sum_{p'l'm'} J^{(p')}_{l'm'}(\boldsymbol{r}',k')\mathcal{H}^{(p'p'')}_{l'l'';m'}(-2a\hat{\boldsymbol{e}}_z,k')\mathcal{A}^{(p'')}_{\text{R}\eta l''m'}$$
$$= \sum_{plm}\sum_{p'l'} J^{(p')}_{l'm}(\boldsymbol{r}',k')\mathcal{F}^{(p'p)}_{l'l;m}\mathcal{A}^{(p)}_{\eta lm}, \qquad (7.3)$$

that is a consequence of (5.20) and (5.29), and is valid on the surface of the scattering sphere. Equation (7.3) shows that the calculation of the elements of F can be overcome provided that the amplitudes $\mathcal{A}^{(p)}_{\eta lm}$ and $\mathcal{A}^{(p'')}_{\text{R}\eta l''m'}$ can be related through some F-independent equation. Johnson, by invoking the so-called Yousif–Videen approximation [113, 104], states such a relation in the form

$$\mathcal{A}^{(p)}_{\text{R}\eta lm} = (-)^{\eta+p+l+m}\mathcal{A}^{(p)}_{\eta lm}F_\eta(\vartheta = 0°),$$

which becomes exact in the limit of a perfectly reflecting interface. Approximation J0 has been tested together with a variant (approximation J1) that keeps in (5.15) the Fresnel coefficients to the value $F_\eta(\vartheta_k = 0°)$ to get approximate values for the elements of F.

All the approximations that were described above, as well as MRT, where applied to the calculation of the scattered intensity from the systems that were considered by Johnson. Doing this has the further advantage that the results of MRT can also be compared both with experimental data [114] and with results of calculations [115]. Accordingly, the scattered intensity $\mathcal{I}_{\eta\eta}$ (see Sect. 7.1) from a sphere of radius $\rho = 0.27\,\mu\text{m}$ lying on the dielectric surface, illuminated by a radiation of wavelength $\lambda = 632.8\,\text{nm}$, is reported as a function of the angle of observation for parallel polarization in Fig. 7.16 (a) and for perpendicular polarization in Fig. 7.16 (b). Analogous results for a sphere of radius $\rho = 0.38\,\mu\text{m}$ are reported in Figs. 7.17 (a) and 7.17 (b). In all the figures $\vartheta_I = 0°$, $n_0 = 1.59$, $n' = 1$ and $n'' = 3.8$. The angle of observation is measured from the negative z axis. The result that we report in Figs. 7.16 and 7.17 required to extend the multipole expansions up to $l_M = 10$ to achieve an accuracy to three significant digits.

Figs. 7.16 (a) and 7.16 (b) seem to report three curves only. In fact, the pair B0-B1 and J0-J1 yield identical results to a high degree of precision.

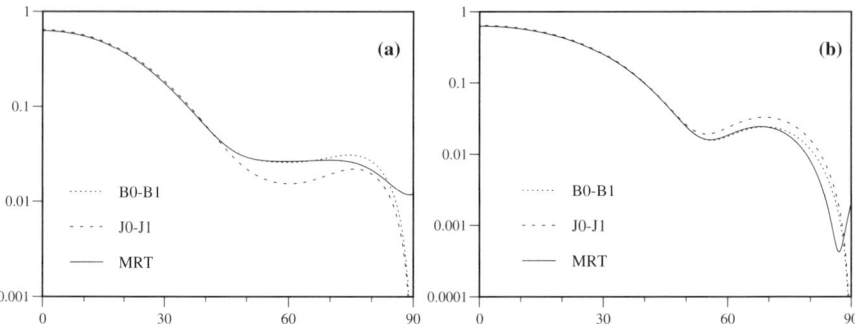

Fig. 7.16a,b. Comparison of the results of the theory of Sect. 5.4 with those yielded by approximations B0, B1, J0 and J1 for a sphere of radius $\rho = 0.27\,\mu\mathrm{m}$ and refractive index $n_0 = 1.59$ in contact with the surface and illuminated by light of wavelength $\lambda = 632.8\,\mathrm{nm}$. The refractive index of the medium beyond the surface is $n'' = 3.8$. We report in $\mu\mathrm{m}^2$ and for normal direction of incidence the quantities $\mathcal{I}_{11}$ in **(a)** and $\mathcal{I}_{22}$ in **(b)** as a function of the angle of observation

This is not surprising when one recalls how these approximations are defined. Anyway, the approximate results compare well with the results of our theory provided that the angle of observation is not near to 90° because MRT predicts the occurrence of a nonvanishing field that propagates along the interface. The situation is analogous in Figs. 7.17 (a) and 7.17 (b). Even in this case, MRT is the only one that gives a nonvanishing field that propagates along the interface. Moreover, the minima of the intensity occur at different angles of observation when different approximations are used. The best coincidence with the results of MRT is attained by B0-B1 for both choices of the polarization. The different result yielded by J0-J1 was to be expected because the latter, besides including approximation B0 to get the far zone expression of $\boldsymbol{E}_{\mathrm{SR}}$, also resort to a further approximation to simplify the imposition of the boundary conditions.

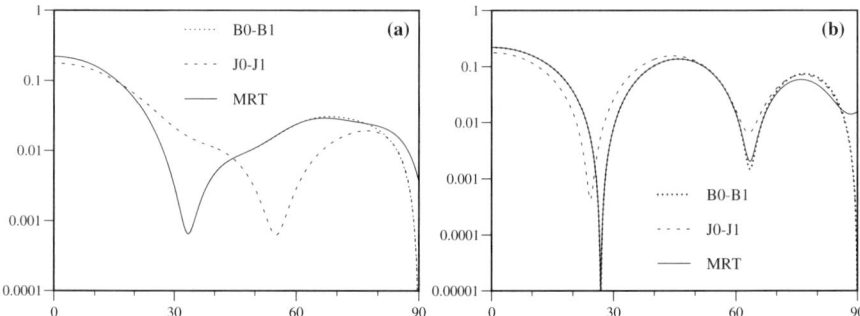

Fig. 7.17a,b. Same as Fig. 7.16 but for a sphere of radius $\rho = 0.38\,\mu\mathrm{m}$

The results in Figs. 7.16 and 7.17 can be compared with the experimental data of Lee et al. [114] and the simulation of Wojcik, Vaughan, and Galbraith [115] that Johnson reports in Figs. 3 and 4 of [112]. A fair agreement is attained not only in the position of the minima but also in our prediction that the field along the surface does not vanish. Of course the calculations needed to complete simulations such as the one in [115] require a computational effort very large with respect to MRT.

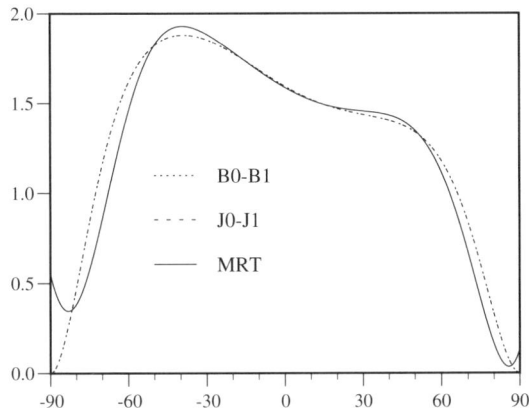

Fig. 7.18. Comparison of the results of the theory of Sect. 5.4 with those of approximations B0, B1, J0, and J1 for a small sphere with polarizability α, whose distance a from the surface is such that $k_\mathrm{v} a = \pi/2$. The quantity that is actually reported is $i = (I_1 + I_2)/(2I_0 \alpha^2 k_\mathrm{v}^4)$ as a function of the angle of observation. The angle of incidence is held fixed at $\vartheta_\mathrm{I} = -45°$

In Fig. 7.18 we report the results of the application of MRT as well as of B0, B1, J0, and J1 to the case of a small particle in the vicinity of a plane interface that Muinonen et al. [116] deal with by means of a variant of image theory designed to take account of the actual refractive index of the substrate, that they call the exact-image theory. In practice, these authors consider a sphere so small that it can be dealt with through the RSA, with polarizability α, for several values of the distance from a plane surface that separates the vacuum from a homogeneous dielectric medium with $\varepsilon = 2.4$. The results in [116] are normalized so as to ensure their independence of the choice of α, the angle of incidence is held fixed at $\vartheta_\mathrm{I} = -45°$ and the incident light is assumed to be unpolarized. The results that we report in Fig. 7.18 refer to a distance such that $k'a = \pi/2$ and their convergence to four significant digits is ensured by including terms up to $l_\mathrm{M} = 6$ both in our exact theory and in the approximations B0, B1, J0, and J1. The need to use so high a value for l_M in spite of the smallness of the scattering particle stems from the fact that the particle itself is not in contact with the surface, as, according to (5.15) and (5.24), $\mathcal{F}^{(pp')}_{ll';m}$ depend on $k'a$ and are independent of the scattering particle.

Actually, Fig. 7.18 seems to report two curves only but this is due to the almost perfect coincidence of all the approximate results that, in turn, agree fairly well with the results that Muinonen et al. report in Fig. 2 of [116]. In particular this coincidence suggests that the further approximation that is included in J0 and J1 beyond the approximation implied in B0 and B1 has little effect on the results on account of the distance of the particle from the surface [112]. The results of our theory have the general shape of the results cited above but for the fact that the observed field does not vanish for grazing angle of observation.

Finally, we come to discuss MRT results for the full scattering pattern from a sphere of radius $\rho = 126$ nm and refractive index $n_0 = 3$ in contact with the surface. We already discussed in Sect. 7.1 the scattering pattern from such a sphere on a perfectly reflecting surface [71] so that, for the sake of comparison, MRT results are plotted in the same frame of reference that is depicted in Fig. 7.1. Even in the present case, we chose $n' = 1$, $\lambda = 628.3$ nm

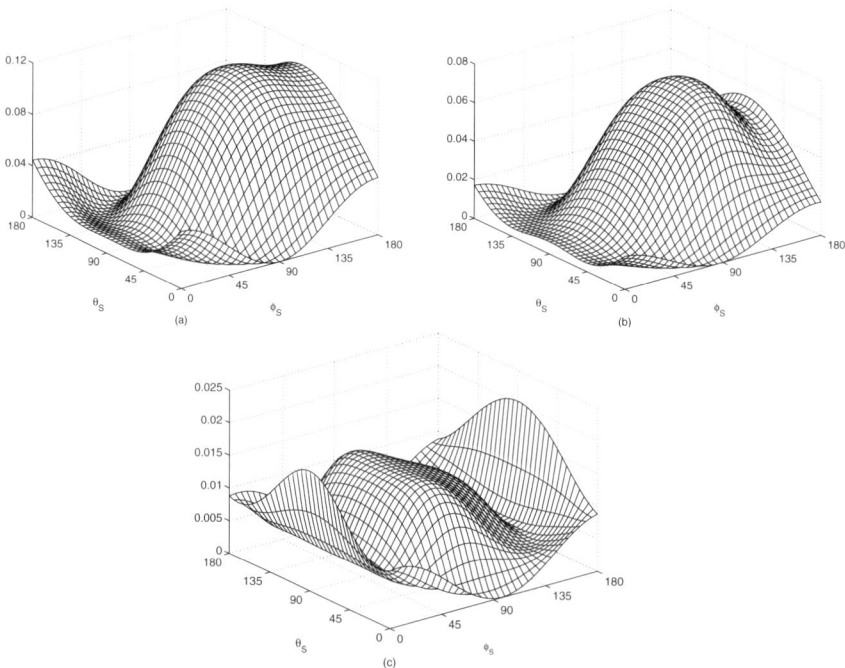

Fig. 7.19a–c. Full scattering pattern for the sphere of radius $\rho = 126.0$ nm and refractive index $n_0 = 3$ in contact with the surface. The quantity that is actually reported is $\mathcal{I}_{\varphi\varphi}$ in µm^2. In **(a)** the surface is perfectly reflecting ($n'' = \infty$); the refractive index of the medium beyond the surface is $n'' = 9.0$ in **(b)** and $n'' = 1.3$ in **(c)**

and $\varphi_I = 225°$. In Fig. 7.19 (a) $n'' = \infty$, in Fig. 7.19 (b) $n'' = 9$, and in Fig. 7.19 (c) $n'' = 1.3$. In all the figures we report $\mathcal{I}_{\varphi\varphi}$, i.e., both the incident and the observed field are polarized along $\hat{\varphi}$. We first notice that the plot in Fig. 7.19 (a) is identical to the one in Fig. 7.2 (a). We stress, however, that in the present case we did not use image theory but calculated the scattered field by putting in (5.30) the elements of the multipole reflection matrix to the value that is obtained from (5.15) when the Fresnel coefficients take on the values $F_\eta = (-)^{\eta-1}$ that are appropriate to a perfectly reflecting surface. The plots in Figs. 7.19 (b) and 7.19 (c) show the evolution of the scattering pattern when the refractive index n'' of the medium that fills the half-space $z > 0$ becomes smaller and smaller.

The case of a nonperfectly reflecting surface shares in common with the case of a perfectly reflecting surface all the features that do not depend on the refractive index but are due to the geometry only. These feature were fully described in Sect. 7.1. We do not report the patterns for the other possible choices of the incident and observed polarization because, at least for a single sphere, they do not show any significant new feature worthy of a separate comment.

MRT proved to yield reliable predictions for the scattering pattern from a sphere on a dielectric surface. On the contrary, on account of the comparisons that we presented above, it can be stated that the approximations that were assumed by other authors may affect the predicted spectrum not only quantitatively but also qualitatively. In this respect, a careful examination of Figs. 7.19 (a), (b), and (c) shows that, when the refractive index of the medium beyond the surface is comparable with the refractive index of the sphere, the intensity of the field that propagates along the surface becomes so relevant that the pattern predicted by approximate approaches may be unacceptable.

MRT may appear to be complicated by the introduction of the multipole reflection matrix F and of H^{-1}. The calculation of these matrices proved, however, to add little to the computational effort. In this respect, calculation of H^{-1} may require some caution (see Appendix A.5). On the contrary, the integrals that are involved in the calculation of F can be computed with high precision through the Gauss–Laguerre method [7, 11]. On the other hand, according to our tests, B0 and J0 are computationally somewhat faster than MRT, but the reliability of the results that they yield need to be carefully checked. We are thus led to conclude that the approximations above do not present substantial computational advantages over MRT. On the other hand, the results of the simulation [115] agree fairly well with MRT predictions but require a heavy computational effort. In fact, what the authors of [115] call *simulation* is actually an ab initio calculation performed through the FDTD method, which we summarized in Sect. 4.6.2. Therefore, both on the side of the computations and on the side of the reliability of the results MRT appears to be more expedient than the other methods.

7.4.2 Aggregates of Spheres in Fixed Orientation

In this section, we apply MRT to the case of aggregates of spherical scatterers deposited on a dielectric plane substrate [72]. Such a move is easily understood when one thinks that aggregation phenomena often occur among the particles on a substrate. In particular, aggregated spheres give origin to particles whose scattering patterns are expected to be rather different from the patterns from single spheres and to depend on the orientation of the aggregate with respect to the incident field. In this respect, we recall that the pattern of the scattered intensity from a tenuous monolayer (in the sense of Sect. 2.6) of identical particles all oriented alike but randomly distributed on the surface is merely proportional to the pattern from a single particle [36]. The case of a monolayer of particles whose orientations are randomly distributed around an axis orthogonal to the surface instead requires considering the average over the orientational distribution. However, in this section we want to focus on the effects of the anisotropy so that we defer to the next section the discussion of the patterns from assemblies of aggregates with random distribution of their orientations.

We deal with binary aggregates of identical mutually contacting spheres embedded in vacuo ($n' = 1$) and deposited on a substrate of Si ($n'' = 3.85 + $ i0.018). We consider spheres of Si_3N_4 with refractive index $n_1 = 2.0$ and radius $\rho_1 = 225$ nm, illuminated by radiation of wavelength $\lambda = 632.8$ nm [73]. The scattered intensity from single spheres of such a morphology deposited on the same substrate has already been calculated by Taubenblatt and Tran through the DDA method [117], which we summarized in Sect. 4.6.1, and by Johnson [112] with the help of the Yousif–Videen approximation [113, 104]. Thus MRT is used to investigate the effect of the aggregation on scatterers whose properties are already well known. We also dealt with the pattern from binary aggregates of contacting spheres of Si_3N_4 with radius $\rho_2 = 178.5$ nm deposited on the surface because the total volume of such an aggregate equals that of a single sphere of radius ρ_1. This choice allows us to study also the effect of the subdivision of the material of a sphere. Hereafter, we will refer to the aggregate of spheres of radius ρ_1 as the case of aggregation and to the aggregate of spheres of radius ρ_2 as the case of subdivision.

The geometry that we adopted to display MRT results is the one depicted in Fig. 7.1. Thus, the surface of the substrate coincides with the zx plane and the normal to the surface coincides with the y axis. The plane of incidence coincides with the xy plane and the direction of incidence forms an angle of 45° with the normal to the surface. More precisely, the polar angles of the direction of incidence are $\vartheta_I = 90°$ and $\varphi_I = 225°$. The incident light is polarized either along the meridians (ϑ-polarization) or along the parallels (φ-polarization) and both the φ-polarized and the ϑ-polarized component of the scattered wave are considered. In practice, the quantity that we calculate is $\mathcal{I}_{\eta\eta'}$ and both co-polarized ($\mathcal{I}_{\varphi\varphi}$ and $\mathcal{I}_{\vartheta\vartheta}$) and cross-polarized ($\mathcal{I}_{\vartheta\varphi}$ and $\mathcal{I}_{\varphi\vartheta}$) patterns are reported. Binary aggregates have their axis perpendicular to the plane of incidence. The numerical results from which all the patterns were drawn achieve a convergence to three significant digits that required us to use $l_M = 12$. Of course, the patterns have all the features that do not depend on the system under study but rather stem from the boundary conditions and from the geometry of the problem [71, 72] and were fully described in Sect. 7.1.

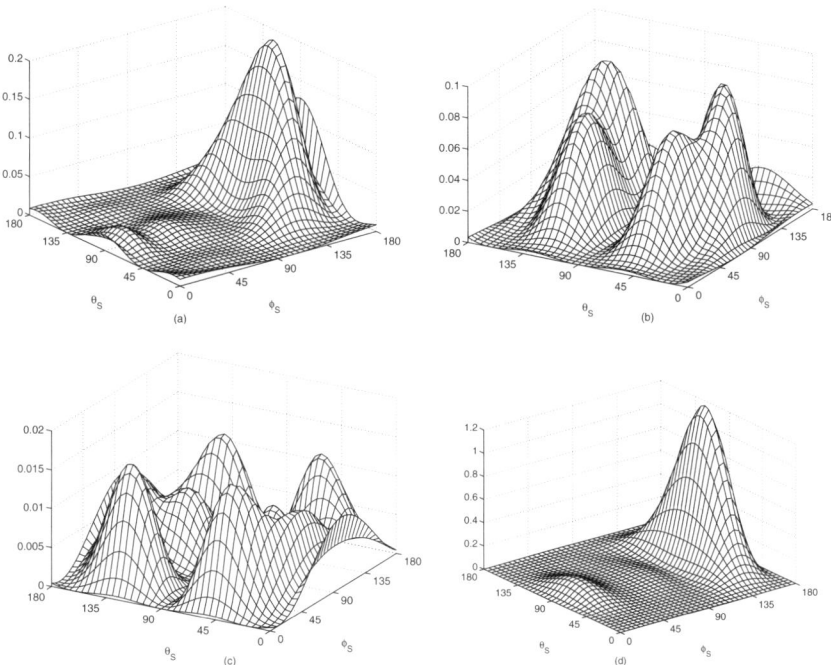

Fig. 7.20a–d. Pattern of the scattered intensity from the single sphere of radius ρ_1. The quantity that is actually reported (in μm^2) is $\mathcal{I}_{\varphi\varphi}$ in (**a**), $\mathcal{I}_{\vartheta\varphi}$ in (**b**), $\mathcal{I}_{\varphi\vartheta}$ in (**c**), and $\mathcal{I}_{\vartheta\vartheta}$ in (**d**)

In Fig. 7.20 we report the pattern from a single sphere of radius ρ_1 in contact with the substrate. The section of the $\mathcal{I}_{\varphi\varphi}$ and of the $\mathcal{I}_{\vartheta\vartheta}$ patterns along the plane $\vartheta_I = 90°$ can be compared with the results of Taubenblatt and Tran [117] and of Johnson [112]. The differences that emerge from this comparison are easily explained along the lines of Sect. 7.4.1, where we stressed how the approximations used by these authors, possibly in conjunction with the approximation used by Bobbert and Vlieger [111] to get the far zone field, may substantially affect even the qualitative features of the scattering pattern [72]. The results in Fig. 7.21 refer to a binary aggregate of spheres of radius ρ_1. The patterns from the aggregate of two spheres of radius ρ_2 are presented in Fig. 7.22.

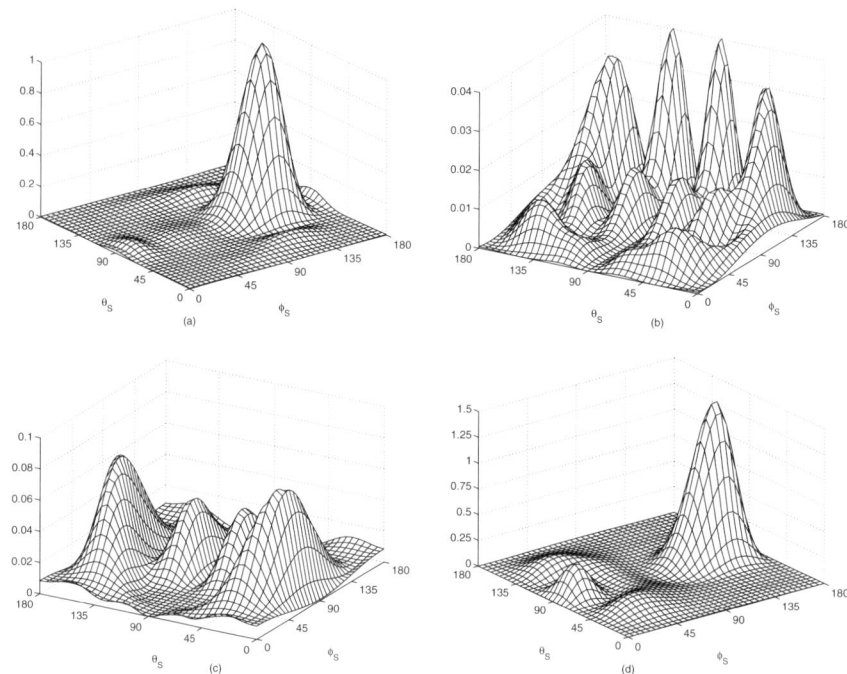

Fig. 7.21a–d. Pattern of the scattered intensity from the binary aggregate of the spheres of radius ρ_1. The quantity that is actually reported (in µm^2) is $\mathcal{I}_{\varphi\varphi}$ in (**a**), $\mathcal{I}_{\vartheta\varphi}$ in (**b**), $\mathcal{I}_{\varphi\vartheta}$ in (**c**), and $\mathcal{I}_{\vartheta\vartheta}$ in (**d**)

We first notice that in agreement with MRT findings, the $\mathcal{I}_{\varphi\varphi}$ patterns show a nonvanishing field that propagates along the surface. Its intensity is smaller than the intensity reported in [72] but this is due to the fact that the refractive index of the substrate has a nonvanishing imaginary part.

To confirm this interpretation, we compared the pattern of $\mathcal{I}_{\varphi\varphi}$ from a single sphere in Fig. 7.20 (a) with the pattern from the same sphere on a perfectly reflecting surface. The latter pattern (not reported here) has a rounded shape in contrast to the rather peaked shape of the pattern in Fig. 7.20 (a). The patterns in Fig. 7.19 for spheres of different size and refractive index and for several choices of the (real) refractive index of the substrate also present a rounded shape. We are thus led to conclude that a nonvanishing imaginary part of n'' produces a strong damping of the scattering patterns. As a result, we are left with a noticeable peak around the direction of reflection. This effect is not restricted to the case of a single sphere because the same peaked shape of the $\mathcal{I}_{\varphi\varphi}$ and of $\mathcal{I}_{\vartheta\vartheta}$ patterns can be observed in the case of the aggregated spheres. In other words, this feature does not depend on the shape of the scatterers and should thus be due to the nonvanishing imaginary part of n''. The height of the main peaks, for φ-polarized incident field, depends

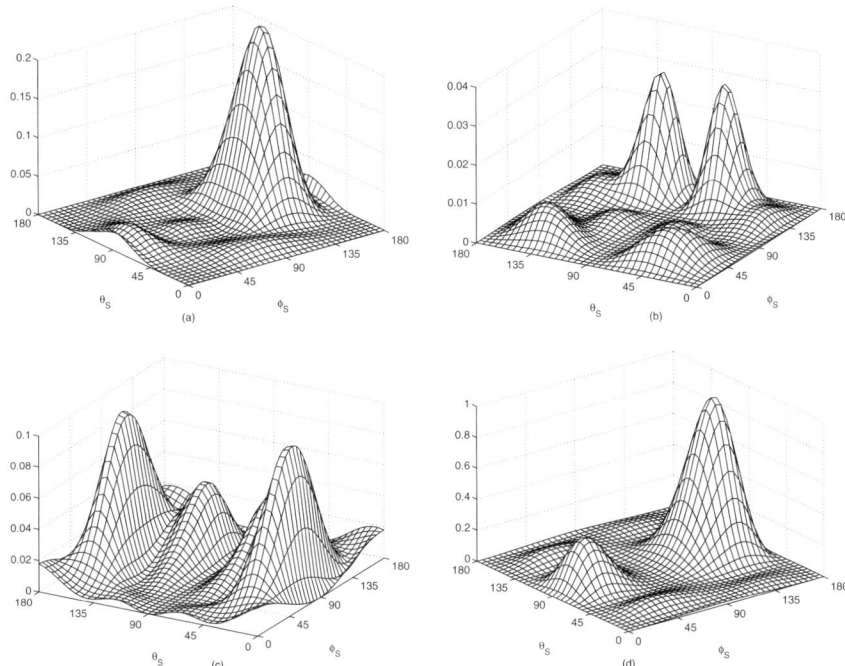

Fig. 7.22a–d. Pattern of the scattered intensity from the binary aggregate of the spheres of radius ρ_2. The quantity that is actually reported (in μm^2) is $\mathcal{I}_{\varphi\varphi}$ in (**a**), $\mathcal{I}_{\vartheta\varphi}$ in (**b**), $\mathcal{I}_{\varphi\vartheta}$ in (**c**), and $\mathcal{I}_{\vartheta\vartheta}$ in (**d**)

more or less on the quantity of refractive material. For instance, the height of the $\mathcal{I}_{\varphi\varphi}$ peak is almost equal for the single sphere and for the case of subdivision, whereas the height for the case of aggregation is about four times larger. Furthermore, the $\mathcal{I}_{\varphi\varphi}$ pattern for the case of aggregation show a noticeable increase, with respect to the case of subdivision, in the direction $\vartheta_S = 90°$, $\varphi_S = 180°$.

Anyway, the largest value of the scattered intensity does not occur exactly in the direction of reflection ($\vartheta_S = 90°$, $\varphi_S = 135°$). This may be surprising because, according to our calculations for the same particles in free space, the largest scattered intensity occurs in the forward direction that, for particles on a surface, coincides with the direction of reflection [78]. Actually, for the single sphere the maximum both of $\mathcal{I}_{\varphi\varphi}$ and $\mathcal{I}_{\vartheta\vartheta}$ occurs at $\varphi_S = 155°$, so that the coupling with the surface is so large as to produce a significant shift of the main peak with respect to the direction of reflection. For the aggregates of spheres of radius ρ_1 and ρ_2 the maximum of $\mathcal{I}_{\varphi\varphi}$ occurs at $\varphi_S = 135°$ but it occurs at $\varphi_S = 150°$ for $\mathcal{I}_{\vartheta\vartheta}$. Even if specific patterns are not reported we can state that a similar behavior occurs even when the axis of the aggregates is parallel to the x axis. This seemingly erratic location of the maximum arises from the coupling of the scattering particles with the substrate and, for the case of aggregated spheres, also on the competing coupling of the component spheres among themselves. The strength of both these couplings is likely to differ for different polarization, whereas the coupling among the component spheres is expected to depend on their size. Hence, the dependence of the location of the maximum on the polarization as well as on the radius of the aggregated spheres can be explained in terms of the ratio between these couplings.

The $\mathcal{I}_{\varphi\varphi}$ and $\mathcal{I}_{\vartheta\vartheta}$ patterns from single spheres as well as from aggregated spheres both of radius ρ_1 and ρ_2 are of rather similar shape, even though the specific values of the scattered intensity may be noticeably different as we noted above with regard to the maximum. This means that an analysis of the scattered intensity restricted into the plane of incidence cannot yield enough information to discriminate the shape and size of the particles. In turn, even a cursory examination of the cross-polarized patterns $\mathcal{I}_{\vartheta\varphi}$ and $\mathcal{I}_{\varphi\vartheta}$ show that they depend both on the shape and on the size of the scattering particles. Thus, any attempt to gain information on the morphology of particles deposited on a dielectric substrate requires considering both the co-polarized and the cross-polarized aspects of their scattering pattern.

The patterns that we discussed above are only a sample of the changes undergone by the scattered intensity when spheres deposited on a dielectric substrate aggregate. However, these results prove that the feature of a nonvanishing field propagating along the interface is distinctive of MRT, which does not imply any approximation. In particular, no approximation is required to get the far zone expression of the scattered field. Strictly speaking, MRT requires the truncation of the multipole expansions on which the formalism is based. Nevertheless, this unavoidable truncation is of numerical and not of physical nature. Therefore, once the convergence has been checked, MRT results remain physically reliable.

7.4.3 Randomly Oriented Aggregates

In this section we report the patterns of the scattered intensity from a dispersion with random orientation of the same aggregates considered in Sect. 7.4.2 [74]. We first stress that the numerical results from which all the patterns were drawn achieved a convergence to three significant digits by using $l_\mathrm{M} = 12$, i.e., the same value of l_M that we had to use in [73] to deal with the same kind of aggregates in fixed orientation. In fact, according to Sect. 3.4, the averaging procedure does not introduce any quantity whose convergence needs to be investigated, its only effect being the disappearance from the final formulas of the rotation matrices. [5]

MRT results for radius ρ_1 are reported in Fig. 7.23, whereas those for radius ρ_2 are reported in Fig. 7.24. We note once again that all the patterns do have all the features that do not depend on the morphology of the particles on the surface but are due to the geometry of the problem. The patterns also show that a nonvanishing field propagates along the dielectric surface. Thus, this distinctive feature of MRT [72, 73] persists even after averaging, as expected, because it occurs for any chosen orientation of a single particle on the surface.

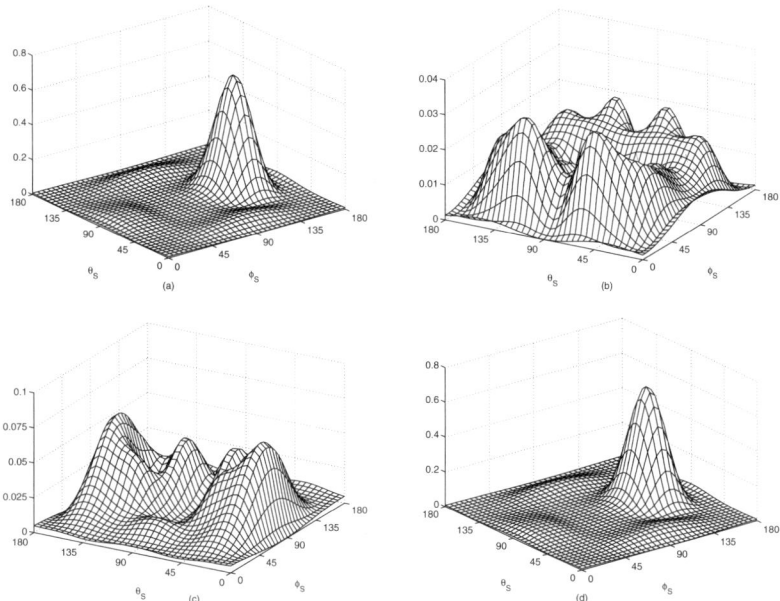

Fig. 7.23a–d. Pattern of the scattered intensity from the assembly of randomly oriented binary aggregates of the spheres of radius ρ_1. The quantity that is actually reported (in μm^2) is $\langle \mathcal{I}_{\varphi\varphi} \rangle$ in (**a**), $\langle \mathcal{I}_{\vartheta\varphi} \rangle$ in (**b**), $\langle \mathcal{I}_{\varphi\vartheta} \rangle$ in (**c**), and $\langle \mathcal{I}_{\vartheta\vartheta} \rangle$ in (**d**)

The effect of the orientational average can be guessed by comparing Figs. 7.23 and 7.24 with Figs. 7.21 and 7.22 of Sect. 7.4.2, respectively, because the latter figures report the patterns from the same binary aggregates on the assumption that they are all oriented alike with their axes perpendicular to the plane of incidence. It is apparent that the averaging procedure has small qualitative effects on the patterns of the observed intensity. In particular, the orientational average does not shift appreciably the location of the main maxima of the co-polarized intensity $\langle \mathcal{I}_{\varphi\varphi} \rangle$ and $\langle \mathcal{I}_{\vartheta\vartheta} \rangle$ in Figs. 7.23 and 7.24 with respect to the location of the main maxima of $\mathcal{I}_{\varphi\varphi}$ and $\mathcal{I}_{\vartheta\vartheta}$ in Figs. 7.21 and 7.22. Thus, the constraint in the sums [see (3.32c)] has only a small influence on the general shape of the patterns. In other words, it does not alter appreciably the ratio between the interaction among the spheres of the aggregates and the interaction of each sphere with the surface. In [72]

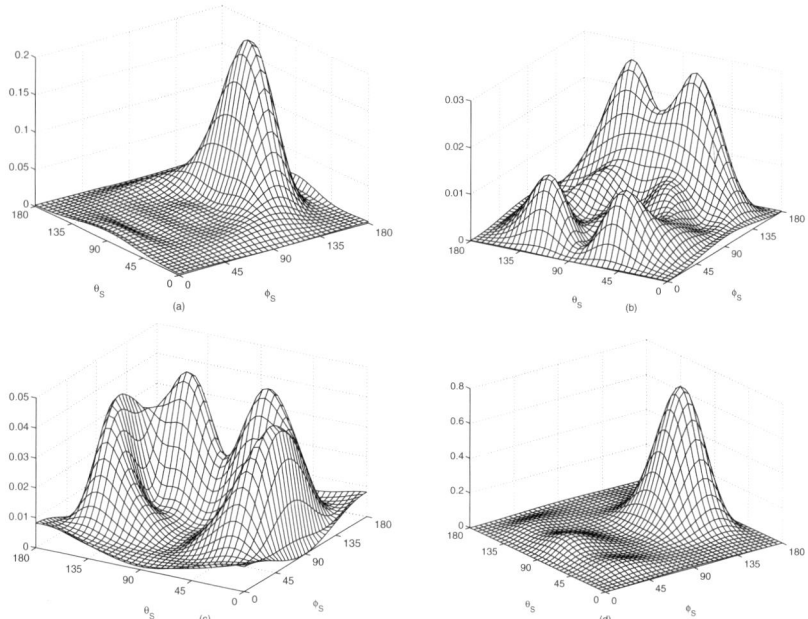

Fig. 7.24a–d. Pattern of the scattered intensity from the assembly of randomly oriented binary aggregates of the spheres of radius ρ_2. The quantity that is actually reported (in μm^2) is $\langle \mathcal{I}_{\varphi\varphi} \rangle$ in (**a**), $\langle \mathcal{I}_{\vartheta\varphi} \rangle$ in (**b**), $\langle \mathcal{I}_{\varphi\vartheta} \rangle$ in (**c**), and $\langle \mathcal{I}_{\vartheta\vartheta} \rangle$ in (**d**)

this ratio was held responsible for the location of the main maxima of the pattern from an aggregate.

In fact, the maxima of $\langle \mathcal{I}_{\varphi\varphi} \rangle$ and of $\langle \mathcal{I}_{\vartheta\vartheta} \rangle$ occur at $\varphi_S = 135°$ and at $\varphi_S = 150°$, respectively, for radius ρ_1 and at $\varphi_S = 135°$ and $\varphi_S = 145°$, respectively, for radius ρ_2. These values turn out to be more or less the same as those for the patterns in Figs. 7.21 and 7.22. Anyway, the location of the main maximum of the scattered intensity from aggregates is nearer to the direction of reflection than for the case of a single sphere. This point has been further investigated by calculating the pattern for single spheres of radius ρ_1 as well as for aggregated spheres of the same radius and with random orientation, on a perfectly reflecting surface. Of course, this can be done either according to Sect. 7.2, or by MRT with $n'' = \infty$, indifferently. The distinctive feature of the resulting patterns (not reported here) is that the maximum of $\mathcal{I}_{\varphi\varphi}$ occurs at $\varphi_S = 180°$ and that of $\mathcal{I}_{\vartheta\vartheta}$ occurs at $\varphi_S = 155°$ for the single sphere, whereas for the aggregates the main maximum occurs at $\varphi_S \approx 140°$ for both choices of the polarization. These results are analogous to those that, in [71], refer to a different choice both of the radius and of the refractive index of the spheres, however. Thus, even in the limit of a perfectly reflecting surface, i.e., when the strongest interaction between the sphere and the substrate seems to occur, the interaction among the spheres is the prevalent one.

The cross-polarized patterns in Figs. 7.23 and 7.24 show another feature that may be useful to gain information on the possible nonsphericity of the particles on the surface. Indeed, we note that, in Fig. 7.20 the cross-polarized intensity from a single sphere do vanish in the plane of incidence ($\vartheta_\mathrm{S} = 90°$). Thus, as expected, a spherical scatterer even in the presence of a substrate does not give rise to depolarization. Of course, we cannot exclude that nonspherical particles may produce no depolarization for particular choices of their orientation. This is suggested by the vanishing for $\vartheta_\mathrm{S} = 90°$ of the cross-polarized intensities reported in Figs. 7.21 and 7.22 that refer to binary aggregates with their axis orthogonal to the plane of incidence. On the contrary, $\langle \mathcal{I}_{\vartheta\varphi} \rangle$ and $\langle \mathcal{I}_{\varphi\vartheta} \rangle$ for the aggregates of spheres both of radius ρ_1 (see Fig. 7.23) and of radius ρ_2 (see Fig. 7.24) do not vanish in the plane of incidence. We are thus led to conclude that a dispersion of randomly oriented nonspherical particles gives rise to depolarization.

8. Applications: Atmospheric Ice Crystals

The aggregates (both external and internal) that were considered in Chap. 6 were not meant to simulate specific kinds of particles, but rather to illustrate the capabilities of the cluster model in association with the transition matrix approach in describing the optical properties of some system of interest. In the present chapter, on the contrary, we describe a specific application to the atmospheric ice crystals. The study of their scattering properties is an important topic in view of the role that they play in the meteorological phenomena and in the thermal budget of Earth. In particular, two problems are currently among the most studied ones: the contribution of the atmospheric ice crystals to the greenhouse effect, and the possibility of discriminating their shape by means of polarimetric radar devices.

In the following sections we use the cluster model to approximate the shape of the ice crystals, although this choice may appear restrictive. Nevertheless, the discussion of our results will make clear that the cluster model may be adequate to give a qualitative description of the scattering features. In fact, it will be inadequate as regards those effects that depend on the precise shape of the scatterers, such as the halos produced by some big atmospheric ice crystals.

8.1 Properties of Atmospheric Ice Crystals in the Infrared

According to recent studies, the atmospheric ice crystals may contribute to the greenhouse effect by scattering back a nonnegligible part of the infrared radiation that is emitted by Earth [118]. In this section a quantitative content is given to this statement by calculating the infrared optical properties of model particles built so as to fit the shape of the ice crystals that are most commonly met in atmospheric studies. In fact, according to Pruppacher and Klett [119], the atmospheric ice crystals occur in several different shapes among which the most common are sticks, crosses, and hexagonal platelets. In this Section we focus on the hexagonal platelets.

Each platelet is modeled as an aggregate of six identical spheres of radius $\rho = 2.5\,\mu m$, whose centers lie at the vertices of a regular hexagon. The neighboring spheres are in mutual contact, so that the resulting aggregate has the overall diameter $d = 15\,\mu\text{m}$ [25]. This choice locates these model particles within the range of size of the hexagonal platelets that are most often met in the atmosphere. They have, in fact, an overall diameter of 10–50 µm and a thickness of 5–15 µm. The refractive index of the component spheres is taken from the paper of Warren [120] whose tabulations cover the wavelength range of interest, i.e., from 8 to 100 µm. In all our calculations we put $n = 1$, but this choice can be changed to simulate, e.g., the water content of the environment.

The transition matrix of the model platelets has been calculated through the procedure of Sect. 4.4 and particular attention has been payed both to the rate of convergence and to the possible occurrence of resonances. Moreover, a few orientational distributions that are likely to occur are considered. According to the discussion at the end of Sect. 4.4, the rate of convergence of our procedure depends both on the size parameter $x_\rho = 2\pi\rho/\lambda$ of the component spheres and on the size parameter $x_\text{d} = \pi d/\lambda$ of the aggregate as a whole. In the present case $1.96 \geq x_\rho \geq 0.157$ and $5.89 \geq x_\text{d} \geq 0.471$, so that, in order to get fair convergence for the transition matrix elements, one has to use $l_\text{M} = 15$ at the shortest wavelengths. We remark that the convergence of the calculations is not affected by the orientational averaging procedure that is necessary to deal with the dispersion of ice crystals. In order to perform the required averages according to Sect. 3.3 we want to assign to each platelet a local reference frame $\bar{\Sigma}$ chosen with regard to their symmetry. Actually, we chose the $\bar{x}\bar{y}$ plane to coincide with the plane of the hexagon on whose center we put the origin of the coordinates; a pair of opposite vertices were chosen to lie on the $\bar{y}$ axis. In turn, the laboratory frame Σ, to which the whole dispersion is referred, is chosen to have the z axis vertical. The orientational distribution assumes that the $\bar{z}$ axes of the platelets form an angle β with the z axis with probability $v(\beta) \propto \cos^4\beta$, whereas the α and γ angles are randomly distributed. This choice of $v(\beta)$ ensures that 50 % of the particles have their $\bar{z}$ axes forming an angle $\beta \leq 30°$ with the vertical. Hereafter, this kind of dispersion will be referred to as constrained dispersion. Two other dispersions are also considered: the dispersion of platelets with random distribution of the orientations and the dispersion of platelets all oriented alike with the axes of their local frame $\bar{\Sigma}$ parallel to the respective axes of Σ. Hereafter, these cases will be referred to as random dispersion and equioriented dispersion, respectively. Finally, in order to assess to what extent the peculiarities of the cluster model affect the results we also calculated the quantities of interest for the equivalent sphere, i.e., the homogenous sphere whose volume equals that of the six spheres that compose the aggregate. The radius of the equivalent sphere is thus $r_\text{e} = 4.543\,\mu\text{m}$ and the size parameter range is $3.57 \geq x_\text{e} \geq 0.285$.

For forward and backward scattering the definition of the plane of scattering is not unique as any plane through the common direction of $\hat{\boldsymbol{k}}_\text{I}$ and $\hat{\boldsymbol{k}}_\text{S}$

8.1 Properties of Atmospheric Ice Crystals in the Infrared   209

is suitable as a plane of reference. Therefore, since we are interested just in forward and backscattered intensities, the plane of scattering is at our choice and it is taken to coincide with the zx plane. Accordingly, both $\hat{\boldsymbol{u}}_{\mathrm{I}2}$ and $\hat{\boldsymbol{u}}_{\mathrm{S}2}$ turn out to be oriented along the y axis.

In Fig. 8.1 we report the quantity $\mathcal{I}_{\mathrm{f}} = (r^2/|E_0|^2)\langle I_{11}(\hat{\boldsymbol{k}}_{\mathrm{I}}, \hat{\boldsymbol{k}}_{\mathrm{I}})\rangle$ as a function of λ for $\hat{\boldsymbol{k}}_{\mathrm{I}}$ parallel to the z axis. This quantity, that is proportional to the transmitted intensity through a dispersion of particles for polarization in the plane of scattering, is reported for the equioriented, for the randomly oriented and for the constrained dispersions. The curve for the equivalent sphere is also included. The calculations were actually performed for $8 \leq \lambda \leq 100\,\mu\mathrm{m}$ but Fig. 8.1, as well as all the other figures in this section, report the resulting curves only in the most significant part of the range of λ. We first notice that the choice of the orientational distribution function does not have much effect on the qualitative features of our results. Moreover, the curve for the equivalent sphere does not differ substantially from the other curves. In fact, the main peak that occurs at $\approx 14\,\mu\mathrm{m}$ for the dispersions of platelets appears to be slightly shifted towards longer wavelengths for the equivalent sphere and to be slightly lower. This suggests that fitting the optical behavior of highly symmetric particles by means of a sphere of suitable radius may lead to conclusions that are not too incorrect, in particular, when the particles are randomly oriented. As expected, the approximate description yielded by the equivalent sphere slightly worsens when the orientational distribution of the actual anisotropic particles becomes more and more different from the

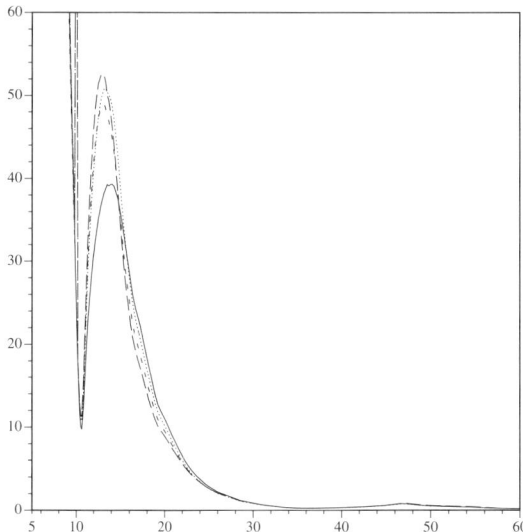

Fig. 8.1. $\mathcal{I}_{\mathrm{f}}$ in $\mu\mathrm{m}^2$ as a function of λ in $\mu\mathrm{m}$ for the equivalent sphere (*solid line*), for the equioriented dispersion (*long dashed line*), for the constrained dispersion (*dashed line*) and for the random dispersion (*dotted line*)

random distribution. Nevertheless, a weak dependence of the transmitted intensity on the choice of the orientational distribution, as well as on the direction of incidence and on the polarization, appears to be an expected result. Therefore we performed calculations also for $\hat{\boldsymbol{k}}_\mathrm{I}$ forming an angle of 30°, 60°, and 90° with the z-axis and for polarization both parallel and perpendicular to the plane of reference. The results for the equivalent sphere and for the randomly oriented dispersion are, of course, rigorously independent both of the direction of incidence and of the polarization. On the contrary, this is not the case either for the equioriented or for the constrained dispersions. A dependence both on the direction of $\hat{\boldsymbol{k}}_\mathrm{I}$ and on the polarization was actually found, but the differences with respect to the case of 0°-incidence were rather small. For this reason the resulting curves do not deserve a separate comment and are consequently not reported. It may be useful to notice, however, that, due to the high symmetry of the platelets around their $\bar{z}$ axis, this weak dependence on the polarization can be easily understood for the equioriented dispersion. The similarity of behavior of the constrained dispersion may, in turn, be attributed to our choice of the orientational distribution function. In fact, the weighting function is strongly peaked for $\beta = 0°$ and β itself is likely to be smaller than 30°. As a result, this kind of orientational distribution may differ little from an equioriented distribution. As a further result, the transmitted intensity should show almost no depolarization. This expectation is confirmed as the ratio $\langle I_{12}(\hat{\boldsymbol{k}}_\mathrm{I}, \hat{\boldsymbol{k}}_\mathrm{I})\rangle / \langle I_{11}(\hat{\boldsymbol{k}}_\mathrm{I}, \hat{\boldsymbol{k}}_\mathrm{I})\rangle$ for any choice of $\hat{\boldsymbol{k}}_\mathrm{I}$ and of the wavelength never exceeds $7 \cdot 10^{-3}$.

All the curves, including the curve from the equivalent sphere, present for $\lambda \approx 10\,\mu\mathrm{m}$ a narrow sharp peak which, on an expanded scale, appears to be superposed on a quite regular curve. Therefore, this is a striking example of a wavelength at which the value of the refractive index plays a dominant role for the forward scattering, whereas the actual structure and orientational distribution of the ice crystals have on the whole little importance.

In Fig. 8.2 we report the quantity $\mathcal{I}_{\mathrm{b}\,1} = (r^2/|E_0|^2)\langle I_{11}(-\hat{\boldsymbol{k}}_\mathrm{I}, \hat{\boldsymbol{k}}_\mathrm{I})\rangle$, that is proportional to the backscattered intensity, for the equioriented, for the randomly oriented, and for the constrained dispersions. Even in this case the curve for the equivalent sphere is included. The direction of incidence is parallel to the z axis. Even a cursory examination shows that the backscattered intensity is rather sensitive to the choice of the orientational distribution function. In this case the maximum occurs at $15\,\mu\mathrm{m}$ both for the equioriented and for the constrained dispersions, but the value at the maximum is four times larger for the equioriented dispersion than it is for the constrained dispersion. The curve for the random dispersion, in turn, reaches its maximum at $\approx 16\,\mu\mathrm{m}$, whereas the equivalent sphere has the least backscattering power with its maximum at $14\,\mu\mathrm{m}$.

The sensitivity of the backscattering to the choice of the orientational distribution suggests also the occurrence of a noticeable dependence on the polarization and on the direction of incidence. To assess this point, we report

8.1 Properties of Atmospheric Ice Crystals in the Infrared   211

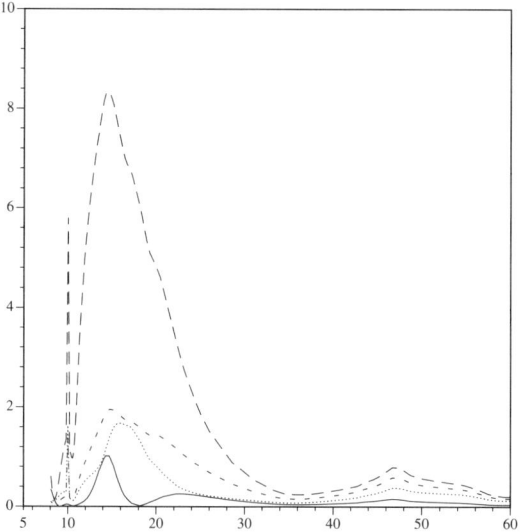

Fig. 8.2. $\mathcal{I}_{b\,1}$ in μm^2 as a function of λ in μm for the equivalent sphere (*solid line*), for the equioriented dispersion (*long dashed line*), for the constrained dispersion (*dashed line*) and for the random dispersion (*dotted line*)

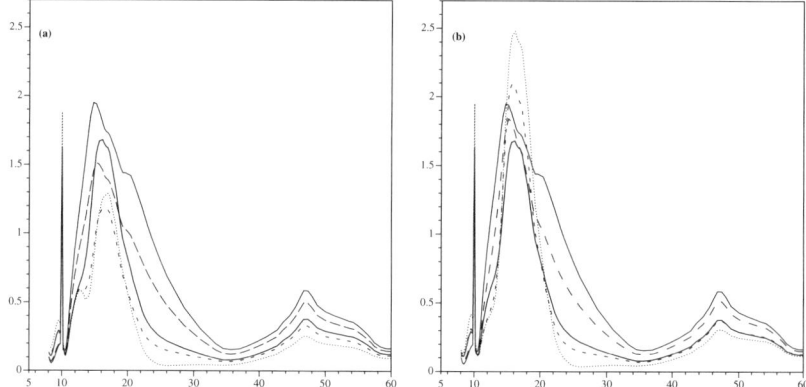

Fig. 8.3a,b. $\mathcal{I}_{b\,1}$ (a) and $\mathcal{I}_{b\,2}$ (b) in μm^2 for the constrained dispersion as a function of λ in μm, for several choices of the direction of incidence. $\vartheta_I = 0°$ (*solid line*), $30°$ (*long dashed line*), $60°$ (*dashed line*), and $90°$ (*dotted line*). For comparison the results for the random dispersion is also reported (*heavy solid line*)

in Fig. 8.3 (a) the quantity $\mathcal{I}_{b\,1}$ and in Fig. 8.3 (b) the quantity $\mathcal{I}_{b\,2}$ for the constrained dispersion and for $\hat{\boldsymbol{k}}_I$ forming an angle of $0°$, $30°$, $60°$, and $90°$ with the z axis. The dependence of the backscattered intensity on the choice of $\hat{\boldsymbol{k}}_I$ and on the polarization is evident and deserves no further comment.

Another quantity of interest that shows a clear dependence on the direction of incidence, is the depolarization ratio of the constrained dispersion,

212   8. Atmospheric Ice Crystals

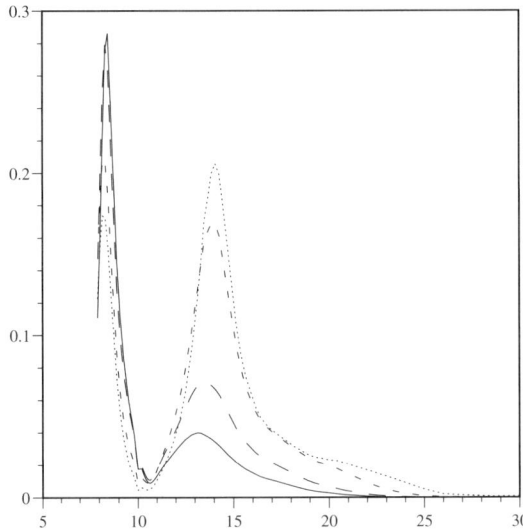

Fig. 8.4. Depolarization ratio D_b for the constrained dispersion as a function of λ in μm, for several choices of the direction of incidence. $\vartheta_I = 0°$ (*solid line*), $30°$ (*long dashed line*), $60°$ (*dashed line*), and $90°$ (*dotted line*)

$D_b = \langle I_{21}(-\hat{\boldsymbol{k}}_I, \hat{\boldsymbol{k}}_I)\rangle/\langle I_{11}(-\hat{\boldsymbol{k}}_I, \hat{\boldsymbol{k}}_I)\rangle$, that we report in Fig. 8.4 for $\hat{\boldsymbol{k}}_I$ forming an angle of $0°$, $30°$, $60°$, and $90°$ with the z axis. For any choice of the direction of incidence the depolarization of the backscattered radiation can be found to be noticeably larger than that of the transmitted radiation D_f that is not reported, however.

The choice of the orientational distribution, has little influence also on the albedo, as is apparent from the curves in Fig 8.5 where we report $W = \langle \sigma_S \rangle/\langle \sigma_T \rangle$. In fact, the similarity of the curves should be noticed. Taking into account that $\sigma_T = \sigma_S + \sigma_A$, one can guess that, with changing distribution, $\langle \sigma_A \rangle$ behaves just as $\langle \sigma_S \rangle$ does. As regards $\langle \sigma_S \rangle$, it may well turn out to be almost insensitive to the choice of any of the distributions considered. In fact, the definition of σ_S implies integration of the flux across a sphere that surrounds the particle of interest. Now, the results that we discussed above show that the forward-scattered intensity, that gives the largest contribution to the integration, is almost independent of the choice of the distribution, whereas the distribution-sensitive backscattered intensity gives a comparatively smaller contribution.

At this stage a few comments are in order. In fact, a reader acquainted with the aerodynamics of atmospheric ice crystals may argue that some of the distribution functions are unrealistic for 15 μm crystals. Actually, we are aware that ice crystals of this size are likely to be randomly distributed in orientation [121]. Nevertheless, once the transition matrix of a single crystal has been calculated, considering a further distribution function does not add

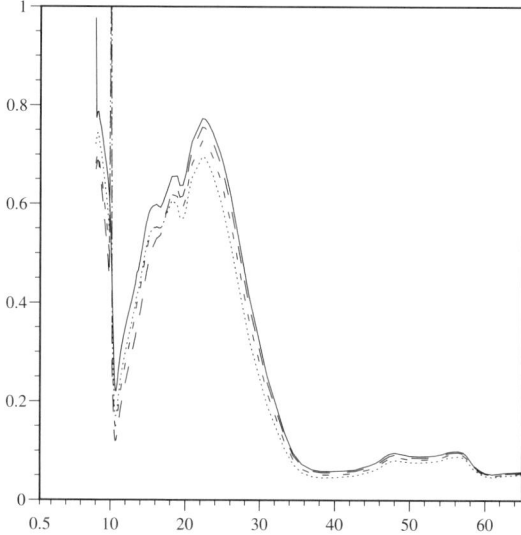

Fig. 8.5. Albedo W as a function of λ in μm for the equivalent sphere (*solid line*), the equioriented dispersion (*long dashed line*), the constrained dispersion (*dashed line*), and the random dispersion (*dotted line*)

much to the computational effort, whereas considering several instances of orientational distributions may help to interpret the experimental data [122]. Moreover, the results that we discussed above can, to a certain extent, be scaled [43] to get information on the behavior of particles of larger size to which equioriented or constrained distribution is more likely to apply. Of course, the principle of the electromagnetic scaling applies when the refractive index is independent of the wavelength. Now, in the infrared, the refractive index of ice depends weakly on the wavelength, at least on the long wavelength side [120]. Therefore, the results that we discussed above may be considered to apply, e.g., to clusters of six spheres of radius 25 μm and thus with an overall diameter of 150 μm in the wavelength range $100 \leq \lambda \leq 1000$ μm. A few sample calculations for the latter cluster confirm this expectation.

8.2 Ice Crystals in the mm Wave Range

In spite of the results mentioned at the end of Sect. 8.1, a complete and reliable extension of our investigation to the millimeter wave range cannot be made by resorting to the principle of optical scalability because of the noticeable difference of the refractive index of ice in the infrared and in the millimeter wave range. Therefore, in this section the investigation of the effectiveness of the cluster model to simulate the scattering properties of atmospheric ice crystals is extended to the millimeter wave range [123]. The results will be compared with those of the analogous calculations performed by Lemke et al. [124]. These authors used the DDA method (see Sect. 4.6.1) that, unlike the cluster model, allows for a better fitting of the actual shape of the crystals. In fact, the question arises as to whether a less coarse model than the one we considered in Sect. 8.1 may give more reliable results. However, experience suggests that the scattering properties of model particles depend more on the quantity of refractive material and on the general symmetry properties than on the detailed fitting of the shape of the actual particles [103].

We calculate the backscattered intensity from aggregates of a few different shapes. The incident radiation is chosen to be a 94 GHz plane polarized wave ($\lambda = 3.2 \cdot 10^{-3}$ m, $k_v = \omega/c = 1.96 \cdot 10^3$ m^{-1}), that is a working frequency of polarimetric radar devices [124]. The refractive index of ice has been taken from the tabulation of Warren [120], i.e., $n = 1.782 + i0.002701$. In all our calculations we assume that the medium surrounding the ice crystals is the vacuum.

The study of scattering properties proceeds in two steps. First, we study the backscattered intensity from single ice crystals of a few selected shapes when they change their orientation; both polarizations are considered. All the crystals are assumed to contain the same quantity of refractive material, namely, the quantity contained into a sphere of radius $\rho_e = 1.0$ mm. The size parameter of this equivalent sphere is $x_e = k_v \rho_e = 1.96$. Second, taking into account that radar returns are determined by simultaneous scattering by particles that may be of different size, we calculate the backscattered intensity and the depolarization ratio from dispersions of model ice crystals, for the same distributions which we considered in Sect. 8.1, as a function of the size parameter x_e of the equivalent sphere.

The shapes of the crystals that we consider are those that are usual in atmospheric studies, namely needles, columns, hexagonal platelets, and compact hexagonal columns [119]. These structures are customarily characterized by their aspect ratio $a = l/d$, where l and d denote the length and the diameter of the crystal, respectively. Each of these crystals is modeled as an aggregate of spheres of appropriate radius and the axes of their local frame $\bar{\Sigma}$ have origin in the center of mass of the aggregate. The orientation of the axes is sketched in Fig. 8.6. Accordingly, the needles are approximated as linear chains of six spheres of radius $\rho = 0.550 \cdot 10^{-3}$ m and thus with an aspect ratio $a = l/d = 6$. Even the columns were approximated as linear chains of three identical spheres with radius $\rho = 0.693 \cdot 10^{-3}$ m and aspect ratio $a = 3$. Of course, using this approximation results in the loss of the details that are due to the hexagonal cross section of the actual column crystals. It will become apparent, however, that these details are rather small and would anyway be smoothed out by the averaging procedure. For both the needles and the columns the centers of the component spheres lie on the $\bar{x}$ axis [see Figs. 8.6 (a) and (b)]. This choice is the most convenient for averaging purposes, although for calculating the transition matrix the choice of the

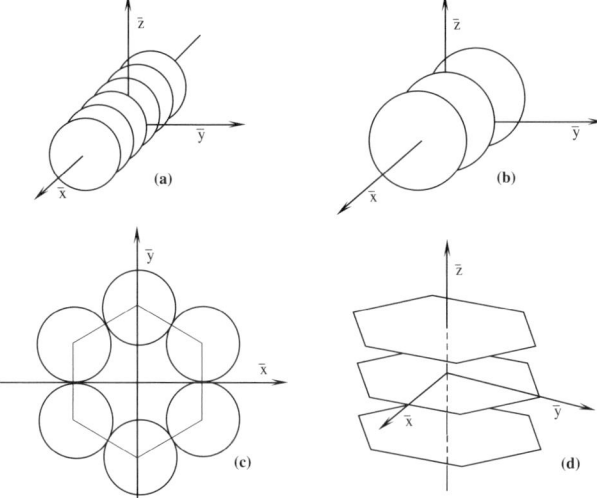

Fig. 8.6a–d. Geometry of the clusters that model the needles (**a**), the columns (**b**), the hexagonal platelets (**c**), and the compact hexagonal columns (**d**). The local axes are also shown and, in the case of the compact hexagonal columns only the supporting hexagons are shown for the sake of clarity

axis of the chains as $\bar{z}$ axis would be highly convenient in order to exploit the cylindrical symmetry of the cluster. As a matter of fact, one calculates the transition matrix putting the centers of the spheres just on the local $\bar{z}$ axis, then transforms the matrix elements to the reference frame $\bar{\Sigma}$ in which the centers lie on the $\bar{x}$ axis [71]. The hexagonal platelets are modeled as clusters of six spheres of radius $\rho = 0.550 \cdot 10^{-3}$ m and, as their centers are at the vertices of a regular hexagon, the resulting aspect ratio is $a = 0.3$. The hexagon lies on the $\bar{x}\bar{y}$ plane with two opposite vertices on the $\bar{y}$ axis [see Fig. 8.6 (c)]. Finally the compact hexagonal columns are approximated as clusters of 18 spheres of radius $\rho = 0.382 \cdot 10^{-3}$ m whose centers lie at the vertices of a stack of three regular hexagons. These hexagons, in turn, have their centers on the $\bar{z}$ axis, lie on planes orthogonal to the $\bar{z}$ axis and have a pair of opposite vertices on the $\bar{y}\bar{z}$ plane. The aspect ratio of this structure is $a = 1$ [see Fig. 8.6 (d)].

In the first step of this study, namely, in assessing the dependence of the backscattered intensity on the orientation of single ice crystals, we consider the approximating clusters with their $\bar{z}$ axis vertical. This choice is not an undue restriction because aerodynamical considerations led to the conclusion that, in calm air, the ice crystals tend to present their largest area perpendicular to the direction of fall [121]. The incident wavevector is assumed to form an angle of $\vartheta_\mathrm{I} = 30°$ with the z axis of Σ. As expected, the backscattered intensity shows all the angular periodicities that are due to the symmetry of the model crystals. In fact, due to the transformation properties of the elements of the transition matrix, any periodicity that is due to the symmetry of the scattering particle is automatically included in the scattered intensity. For this reason we do not report the intensities from all the model particles in Fig. 8.6. We prefer instead to compare the backscattered intensity from the hexagonal platelets and from the compact hexagonal columns, because these structures, in spite of their common hexagonal symmetry have a quite different aspect ratio. As we stated above, the hexagonal platelets were oriented with their local $\bar{z}$ axis vertical and the dependence of the backscattering on the Eulerian angle α was considered. Since the compact hexagonal columns have $a = 1$, we considered these model particles with their $\bar{z}$ axis both parallel to the z axis and lying in the xy plane of Σ, i.e., the backscattered intensity as a function of α was considered not only for $\beta = 0°$ but also for $\beta = 90°$ and $\gamma = 0°$. In Fig. 8.7 we report the adimensional backscattered intensity $i_{\eta\eta} = k_v^2 |f_{\eta\eta}(-\hat{\mathbf{k}}_I, \hat{\mathbf{k}}_I)|^2$. Therefore, the quantities i_{11} and i_{22} coincide with van de Hulst's $|S_2|^2$ and $|S_1|^2$, respectively [3]. Both the platelets and the hexagonal columns with the $\bar{z}$ axis vertical show little dependence on the angle α. This confirms our previous statement that the details due to the hexagonal cross section are negligibly small. On the contrary, when we consider the compact hexagonal columns with their $\bar{z}$ axis horizontal, the dependence on α of the backscattered intensity becomes quite evident. In spite of the fact that this structure has $a = 1$, the α-dependence looks, indeed, rather similar to that of the needles and of the columns. These results suggest that the aspect ratio is not always a good parameter for the classification of the different shapes of the ice crystals, at least in the millimeter wave range.

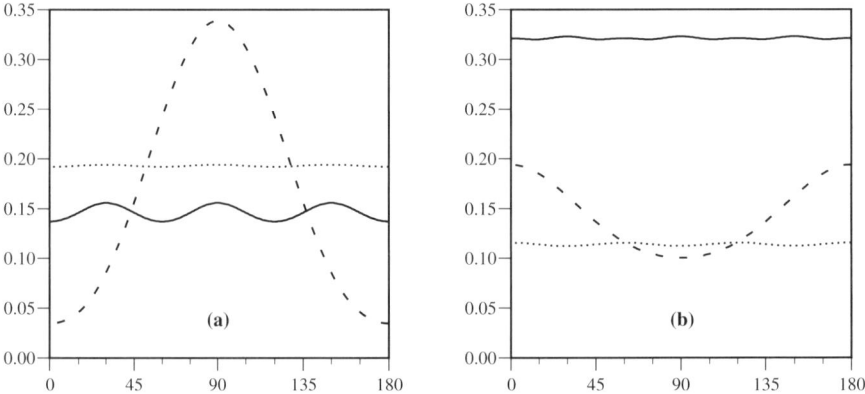

Fig. 8.7a,b. Backscattered intensity $i_{\eta\eta} = k_v^2 |f_{\eta\eta}(-\hat{\mathbf{k}}_I, \hat{\mathbf{k}}_I)|^2$, for $\eta = 1$ in **(a)** and for $\eta = 2$ in **(b)**, as a function of the Eulerian angle α from the hexagonal platelets (*solid line*) and from the compact hexagonal columns with their $\bar{z}$ axis vertical (*dotted line*) and horizontal (*dashed line*)

In the second step of this study, we deal with tenuous dispersions of identical particles whose orientational distribution is assumed to be known. We calculate the backscattered intensity $i_{\eta\eta}$ as a function of the size parameter of the equivalent sphere in the range $0.25 \leq x_e \leq 3.0$ by choosing the direction of incidence parallel to the z axis. In order to grant direct comparison with the DDA results of Lemke et al. [124], in the following figures we report i_{11} again for $\lambda_0 = 3.2$ mm. The results thus apply for the radius of the equivalent sphere that goes from 0.127 mm to 1.53 mm. Nevertheless, since the refractive index of ice does not significantly depend on the wavelength within the millimeter wave range, the principle of electromagnetic scaling applies [43]. As a consequence the same results, when multiplied by the factor $(\lambda_0/\lambda)^2$, can be interpreted as the backscattering from particles of fixed size for varying wavelength. In this case, by assuming an equivalent sphere of 1 mm radius, the wavelength goes from $\lambda = 25$ mm for $x_e = 0.25$ to $\lambda = 2.1$ mm for $x_e = 3$. The usefulness of this twofold possibility of interpretation stems from the fact that the orientational distribution of the atmospheric ice crystal may depend on their size [121]. We report in Fig. 8.8 the results for the case of random distributions, in Fig. 8.9 those for equioriented distributions, and in Fig. 8.10 the results for constrained distributions. The latter distribution has been included to consider the case in which canting of the ice crystals occurs, e.g. as an effect of air turbulence.

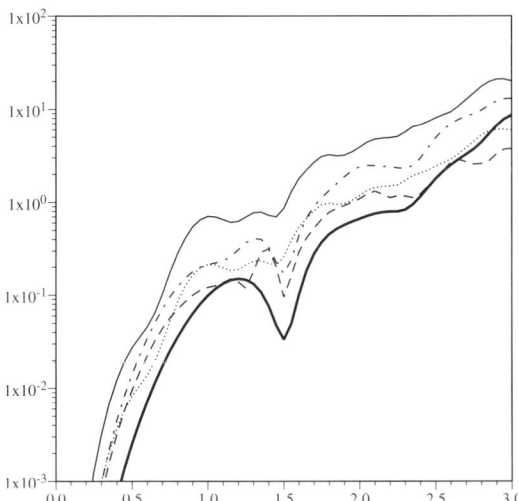

Fig. 8.8. Backscattered intensity i_{11} as a function of the size parameter of the equivalent sphere x_e from randomly oriented dispersions of needles (*dashed line*), columns (*dot-dashed line*), hexagonal platelets (*dotted line*), and compact hexagonal columns (*thin solid line*). The intensity from the equivalent sphere (*heavy solid line*) is also reported

218   8. Atmospheric Ice Crystals

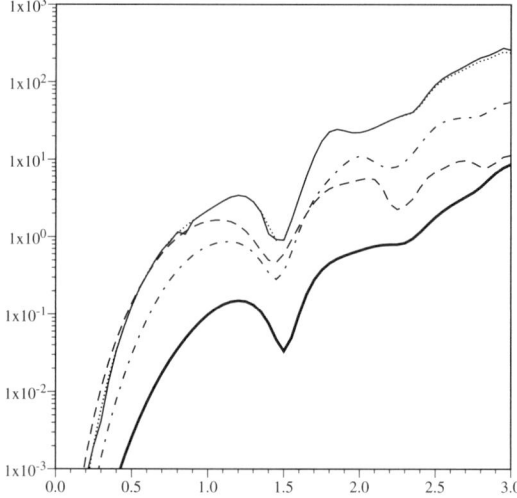

Fig. 8.9. Same as Fig.8.8 but for the equioriented dispersions

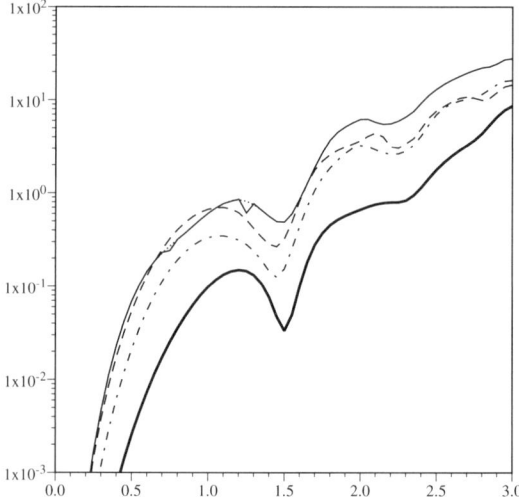

Fig. 8.10. Same as Fig. 8.8 but for the constrained dispersions

Even a cursory comparison of the results in Figs. 8.8–8.10 shows that the backscattering intensity is about one order of magnitude larger for the equioriented dispersions than it is for the random and for the constrained distributions. This result stems from the fact that, in the equioriented distribution, the ice crystals present their largest geometrical cross section orthogonal to the vertical axis that we chose to coincide with the direction of incidence. For all the distributions considered, all the crystal structures display a relatively sharp feature more or less at $x_e \approx 1.5$. A further feature of the random distribution is the separation of all the curves. This result can be understood by considering that the backscattered intensity comes from the collection of contributions, all with the same weight, from particles that are seen under all possible orientations. Thus, in the case of a random distribution, the aspect ratio may play a role for discriminating among the various structures. This conclusion may not hold in the case of other distributions. In fact, the curves for the compact hexagonal columns and for the hexagonal platelets are practically indistinguishable from each other both in the equioriented distribution and in the constrained distribution. For the equioriented distribution, this behavior agrees with the results for i_{11} in Fig. 8.7. On the other hand, the constrained distribution differs little from the equioriented one as we already stressed in Sect. 8.1.

Further insight into the behavior of the ice crystals as a function of their orientational distribution comes from the backscattering depolarization ratio

$$D_\mathrm{b} = I_{21}(-\hat{\boldsymbol{k}}_\mathrm{I}, \hat{\boldsymbol{k}}_\mathrm{I})/I_{11}(-\hat{\boldsymbol{k}}_\mathrm{I}, \hat{\boldsymbol{k}}_\mathrm{I}) \ .$$

This quantity is reported in Fig. 8.11 for the needles, the columns, the hexagonal platelets, and the compact hexagonal columns. In all these figures, the most striking features are the two large peaks that occur at $x_\mathrm{e} \approx 1.5$ and $x_\mathrm{e} \approx 2.3$ for the case of random distribution. The first peak corresponds to the minimum that all the clusters present in Figs. 8.8–8.10, whereas the second peak corresponds to a value of x_e at which dips or shoulders occur. The depolarization for random distribution presents a satellite peak that occurs at $x_\mathrm{e} \approx 1.3$ where the backscattering displays clear features in the form of dips or of shoulders. Note that this satellite peak of the depolarization does not appear for the constrained and for the equioriented distribution, with the exception of a small peak in the constrained distribution of the compact hexagonal columns.

To all these features one could add the vanishing of the depolarization ratio from the hexagonal platelets and from the compact hexagonal columns in the equioriented distribution [see Figs. 8.11 (c) and (d)]. This result was expected on account of the high (hexagonal) symmetry of the crystals around their $\bar{z}$ axis, which yields a weak dependence on α of the co-polarized backscattered intensity and rather small values of the cross-polarization elements of the scattering amplitude. The average over α smooths out any angular dependence to a virtual cylindrical symmetry and yields, as a result, the vanishing of the depolarization referred to above.

Figure 8.7 suggests that, as far as single particles are concerned, the aspect ratio is not always a good parameter for discriminating the scattering properties of ice crystals of different shape. In fact, we have already remarked that, when the compact hexagonal columns are considered with their $\bar{z}$ axis horizontal, the angular dependence of the backscattered intensity looks rather similar to that from more elongated particles such as the needles and the columns. On the contrary, when the $\bar{z}$ axis is vertical, the backscattered

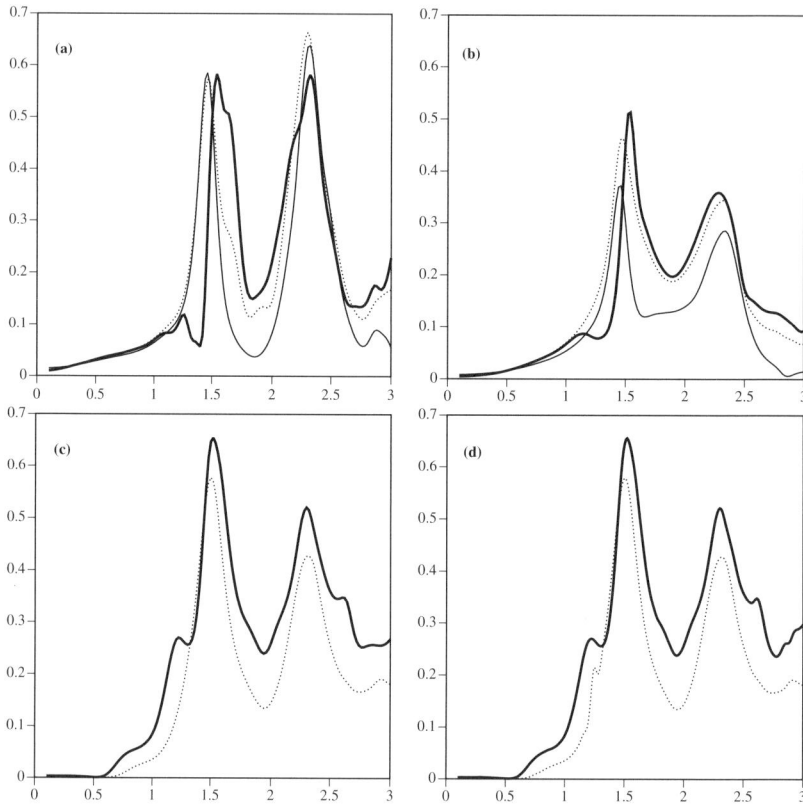

Fig. 8.11a–d. Backscattering depolarization ratio D_b for dispersions of needles (**a**), columns (**b**), hexagonal platelets (**c**), and compact hexagonal columns (**d**) as a function of the size parameter of the equivalent sphere x_e, in cases of random (*heavy solid line*), equioriented (*thin solid line*), and constrained (*dotted line*) orientational distributions

intensity is more akin to that from the hexagonal platelets. Figures 8.8–8.10 show that, when dispersions of compact hexagonal columns and of hexagonal platelets are considered, the curves of the backscattered intensity as a function of x_e coincide, except when the orientational distribution is random, at least as far as i_{11} only is considered. In practice, Figs. 8.8–8.10 show that particles with a different aspect ratio yield well-distinguishable curves only for random distribution. In any case, the curves for a given shape have a spread, with changing orientational distribution, no greater than one order of magnitude, just as large as the spread, for a chosen distribution, of the curves from various shapes. The general trend of all these curves follows the curve for the equal volume sphere that lies generally lower, however. This suggests that a better fitting of the results from model ice crystals could be attained by considering, instead of the equal-volume sphere, the equal-projected-area sphere[125]. In this respect, the conclusions of West on the optical behavior of aggregations of small spheres [126] and the considerations of Zakharova and Mishchenko on the optical properties of ice ellipsoids [127] may be a valuable guide. The analysis of the polarization may be important. For instance, the depolarization ratio in Figs. 8.11 seems to give more information for discriminating the shape of the ice crystals, e.g. because of its vanishing for the equioriented distribution both of the compact hexagonal columns and of the hexagonal platelets. Unfortunately, such a distribution is rather unlikely to occur because of air turbulence. Even when a modest amount of canting occurs the depolarization ratio becomes akin to that from the constrained and from the random distribution. The latter distribution, however, could yet be discriminated provided that the satellite peak at $x_e = 1.3$ could actually be observed.

Even bearing in mind that the aspect ratio is not always a good discriminating parameter, the results presented so far suggest that modeling an actual ice crystal as a cluster of spheres of appropriate geometry yields the correct features of the scattering process that stem from the symmetry of the particles. This, as we already remarked, is a consequence of the transformation properties of the transition matrix under rotation. The regularity of the curves in Fig. 8.7 shows that using the cluster model does not introduce any spurious feature of importance, especially because at 94 GHz the wavelength is noticeably larger that the size of the component spheres so that, for instance, no resonance of the single spheres occurs. Nevertheless, the small dips that occur in the curves for the compact hexagonal columns are certainly specific features of the 18 sphere cluster. Of course, an approach such as the DDA attains a better fitting of the actual shape of the particles of interest. The price to be paid, however, is a noticeable increase of the computational effort because, unlike the transition matrix method, the DDA requires a separate calculation for each direction of incidence and for each direction of observation. Anyway, the results of the transition matrix approach are in substantial agreement with the DDA results of Lemke et al. [124]. Therefore, when a distribution of ice crystals needs to be considered, the conditions outlined in Sect. 1.1 may occur and modeling the single crystals as clusters of spheres is certainly convenient. Ultimately, what computational method is preferable could be established only by comparison of the theoretical predictions with the results of laboratory experiments under controlled conditions.

9. Applications: Cosmic Dust Grains

9.1 Introduction

Although the importance of cosmic dust has long been recognized in astronomy, only a partial understanding of the nature of dust grains has been achieved. In fact, most of the information on the interstellar matter that forms the so-called interstellar medium (IST) comes from spectroscopical analysis and from the study of the extinction of the starlight that propagates through the IST. We can summarize the available information as follows:

- In the range from the visible to the near infrared, the average galactic extinction curve depends more or less linearly on λ^{-1}, with the noticeable exception of the 2175 Å bump. This kind of dependence, which is easily discerned in the curve reported, e.g., by Aiello et al. [128], is consistent with an interstellar medium that includes solid grains whose overall size is of the order of the wavelength [129].
- Since the galactic extinction depends on the polarization, the dust grains should be intrinsically nonspherical. Moreover, the detail of the polarization dependence sets to 0.1–0.2 μm the upper limit for the size of the grains [129].
- The spectroscopic analysis is consistent with a chemical composition of the grains that includes carbon, both in its amorphous and graphitic form, silicates and (possibly organic) ices [130].

Now, the wealth of new data yielded by the observations of the past few years made clear that practically all dust models proposed before 1995 face a serious crisis in explaining interstellar dust phenomenology. All the old models for the cosmic dust were, indeed, based on the assumption that the elemental abundances, i.e., the number of the available elemental atoms per 10^6 H atoms, were more or less constant throughout the galaxy. Therefore, the relative abundances of the elements that form the IST were assumed to be more or less the same that were observed within the solar system. On the contrary, a number of authors in discussing the elemental abundances [131, 132, 133, 134] had to conclude that the solar system is not representative of the cosmic abundances, but is instead enriched in those elements which are supposed prominent in the composition of interstellar dust. As a result, the available quantity of carbon, oxygen, and silicates that is tied up into the average cosmic dust grains must be noticeably smaller than assumed before. This conclusion is commonly referred to as *elemental shortage* or *elemental depletion*.

A few authors addressed the problem of the construction of models able to reproduce the observed extinction while complying with the new observed abundances. For instance, Mathis [135] and Li [109] proposed models for the interstellar dust in which the bulk of the infrared, visual, and ultraviolet extinction arises from composite, core-mantle and multilayered grains with varying degree of porosity. These fluffy, or porous, aggregates can produce more extinction per unit mass than their combined individual constituent components. Such models could be consistent with tighter abundance constraints. However, all previously proposed models did not rigorously treat the scattering and absorption of light by such complex particles, a noticeable exception being the analysis of the interstellar extinction produced by porous grains performed by Vaidya and Gupta [136] in the framework of DDA [59, 61, 62].

We have good indications of the chemical nature of interstellar dust [137]. What is still unclear is how the dust material is assembled to form an interstellar grain. Therefore, in this chapter we exploit the potentialities of the cluster model in describing the optical properties of the cosmic dust grains [138]. To this end, we investigate how and to what extent the extinction of a given mass of astronomical silicates [61] changes when it is subdivided into several clustering spheres. The change of the extinction due to modifications of the chemical composition of the clusters is also investigated. Finally, since the shape of the cosmic grains may play a nonnegligible role, we investigate whether the change of the geometry of the clusters may produce detectable effects on their extinction spectrum.

We also test the transition matrix approach for the description of the extinction from fluffy or porous grains that we propose to model as (possibly coated) spheres containing one or more empty inclusions. Even in this case, the effect of the subdivision of the included voids, for several choices of their geometrical arrangement is considered.

We do not propose a specific model for the dust grains, but we are confident that the techniques that we presented throughout this book will be of help in the definition of new strategies in cosmic dust modeling.

9.2 Modeling Cosmic Dust Grains as Aggregates

In this section, we present a few samples of the relevant information that can be obtained through the theory of Sect. 4.4. Since the problem of the extinction by the available mass in the cosmic dust is a critical one, we consider the extinction produced by the mass of astronomical silicates [61, 139] that would be contained in a homogeneous sphere of radius $\rho = 100\,\mathrm{nm}$ [138]. Hereafter, this sphere will be referred to as the equivalent sphere. We consider also the subdivision of the mass of the equivalent sphere into an increasing number of spherical subunits clustering in a roundish geometry. All the spherical subunits have the same radius that, when N subunits are considered, is given by $\rho_s = \rho/N^{1/3}$, where ρ is the radius of the equivalent sphere. Accordingly, we study the behavior of the extinction cross sections in the following cases:

1. Subdivision of a given mass of astronomical silicates into clustering subunits.
2. Substitution of the material of some subunits in a cluster with amorphous carbon. Of course, in this case the mass is not conserved.
3. Change of the geometry of the clusters.

We use the amorphous carbon dielectric functions derived by Rouleau and Martin (sample Be1) [140]. Clusters of up to 12 spherical subunits are considered and the orientational average is performed, through the formulas of Sect. 3.4, on the assumption of random orientational distribution. In other words, at this stage the possibility of a preferential orientation induced by some external agent is not considered. All the calculations required using $l_\mathrm{M} = 14$ to get an accuracy of four significant digits.

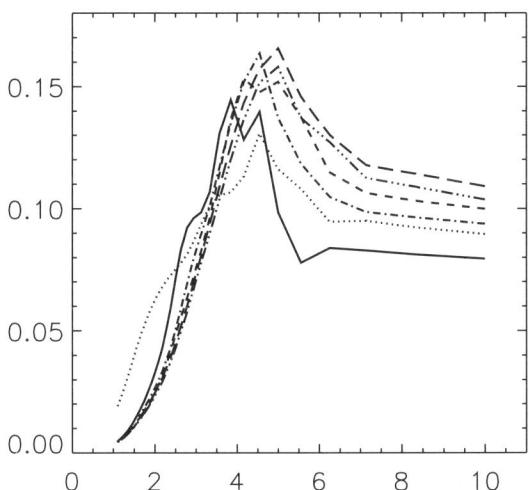

Fig. 9.1. Effects of subdivision on the extinction cross sections. The quantity that is actually reported is σ_T in $\mu\mathrm{m}^2$ as a function of λ^{-1} in $\mu\mathrm{m}^{-1}$ for the equivalent sphere (*solid line*), for a random dispersion of 2-sphere clusters (*dotted line*), of 4-sphere clusters (*short dashed line*), 6-sphere clusters (*dot-dashed line*), 10-sphere clusters (*triple dot-dashed line*), and 12-sphere clusters (*long dashed line*)

In Fig. 9.1 is shown the extinction cross section of the equivalent sphere and that of the clusters containing the same mass of astronomical silicates as a function of λ^{-1} in $\mu\mathrm{m}^{-1}$ in the range $1 \leq \lambda^{-1} \leq 10$. Roundish geometries are assumed [see Fig. 9.3 (a)]. The curves show that the effect of the subdivision and clustering of astronomical silicates may be quite different in the different regions of the spectrum. In fact, the binary clusters produce a larger extinction than that of the equivalent sphere in the visible range, whereas clustering up to 12 subunits produces less extinction in the same spectral region. The effect is reversed in the ultraviolet, where subdivision and clustering yield more extinction than the equivalent sphere, the extent of the effect increasing with increasing degree of subdivision. The origin of this behavior can be tracked back to the multiple scattering processes that occur among the components of a cluster. These processes appear to give a more effective contribution when the wavelength is of the same order of magnitude of the mutual distance of the centers of two subunits. Accordingly, when this distance is comparatively large, as in the case of binary clusters, the largest effect is bound to occur in the visible. The reverse situation occurs, as expected, at shorter wavelengths, i.e., in the ultraviolet. Of course, no such effect can be present in the case of a single homogeneous sphere.

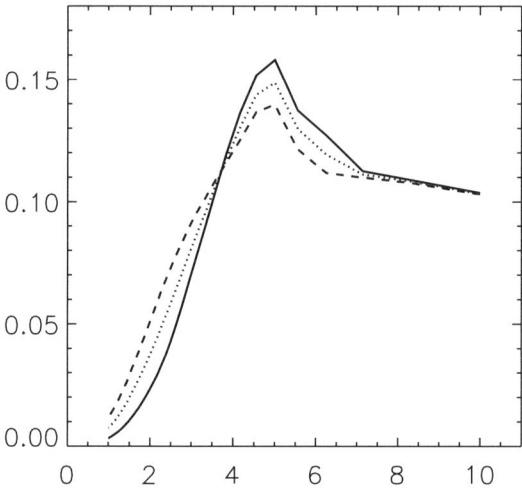

Fig. 9.2. Effects of chemical composition on the extinction cross sections. The quantity that is actually reported is σ_T in μm^2 as a function of λ^{-1} in μm^{-1} for a random dispersion of 10 silicate sphere clusters (*solid line*), of 8 silicate sphere and 2 amorphous carbon spheres clusters (*dotted line*), and 6 silicate spheres and 4 amorphous carbon spheres clusters (*dashed line*)

In Fig. 9.2 we show the effect of changes in chemical composition of dust grains through the partial substitution of silicate subunits by amorphous carbon spheres of the same radii. In this case, the total mass is not conserved. Starting from the same 10-sphere cluster whose extinction is already reported in Fig. 9.1, with the help of a random number generator, first 2 and then 4 spheres are selected and their material is substituted with amorphous carbon, leaving unchanged everything else. Again, the curves in Fig. 9.2

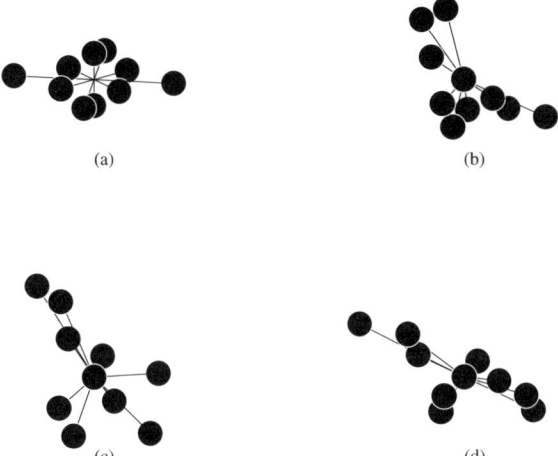

Fig. 9.3a–d. Adopted cluster shapes. **(a)** roundish morphology, **(b)** compact random structure, **(c)** and **(d)** elongated random structures

show that the effect of the substitution is different in the different regions of the spectrum. Substitution of carbon to silicon increases the extinction in the visible. Finally, let us consider the effect of the geometry on the extinction from a dispersion of clusters. In fact, so far we have considered the roundish geometry of Fig. 9.3 (a) for the 10-sphere clusters. By means of a random number generator, we built three further 10-spheres clusters whose geometry is shown in Figs. 9.3 (b)–(d). Note that, in Fig. 9.3 the dots that represent the subunits are well separated to emphasize geometry, but in actual calculations the neighboring subunits were kept in mutual contact. The results are summarized in Fig. 9.4 where the orientational averages of the extinction cross section for the clusters of Figs. 9.3 (b), (c), and (d) are compared with that for the original 10-sphere cluster in Fig. 9.3 (a). The curves in Fig. 9.4 seemingly show that, as a general rule, the randomization of the geometry produces less extinction in the visible but a noticeably larger extinction in the ultraviolet. Close examination of the curves and comparison with the morphologies of Fig. 9.3 suggests that this behavior is mainly due to the fact that the randomization procedure produces somewhat elongated structures. Actually, the results show that the more elongated structures in Figs. 9.3 (c) and (d) produce less extinction in the ultraviolet than the less elongated structure in Fig. 9.3 (b), whereas the latter structure produces more extinction than the roundish structure in Fig. 9.3 (a). Of course, this behavior depends on the strength of the multiple scattering processes that occur among the subunits and contribute to the strength of the scattered field. It is, in fact, expected that these processes are less intense in the more elongated structures in Figs. 9.3 (c) and (d). In general, multiple scattering processes increase the extinction. Nevertheless, the details depend on the

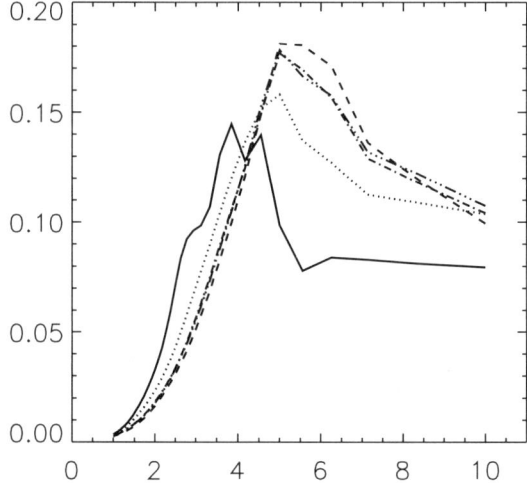

Fig. 9.4. Effects of geometry changes on the extinction cross sections. The quantity that is actually reported is σ_T in μm^2 as a function of λ^{-1} in μm^{-1} for the equivalent sphere (*solid line*) and for random dispersions of the 10-sphere clusters: namely, for the roundish morphology (a) (*dotted line*) and for the random structures (b) (*dashed line*), (c) (*dot-dashed line*), and (d) (*triple dot-dashed line*) in Fig. 9.3

relative position of the subunits, so that a roundish structure, such as the one in Fig. 9.3 (a), might originate processes that, on the whole, are weaker than those occurring within the structure of Fig. 9.3 (b). In any case, these results give a clear indication that, in spite of the random orientational average, the details of the morphology play a relevant role in determining the extinction.

Voshchinnikov and Mathis [141] determined the optical cross sections of composite grains [135] by assuming an approximate description, in which a composite grain consists of many concentric spherical layers of various materials, each with a specified volume fraction. The model calculations show that the ordering of materials in the layering makes some difference to the derived cross sections, but the effect converges rapidly with an increasing number of shells. The net result is a *stiffness* in the extinction cross sections, i.e., a very weak dependence on the assumed geometry. However, real interstellar grains are rough surfaced, irregular, and have their materials arranged in pieces that may occupy appreciable fractions of their size. As a consequence, it is interesting to compare the extinction cross sections for composite grains with relatively large component spheres to those for composite grains for which, according to the method of Voshchinnikov and Mathis, the components can be much smaller than the wavelength. As these authors did, we consider three materials, silicates, amorphous carbon, and vacuum, occupying an equal volume in the grain. Vacuum is not included in the component spheres but it is in the dispersed geometry of the cluster, deriving by the process of aggregation in which subparticles of different types stuck together. As a consequence, the equivalent spheres are two thirds in volume with respect to the ones considered in [141] and contain an equal fraction of silicates and amorphous carbon. We consider clusters of two, six, and ten spheres assembled with the geometries shown in Fig. 9.5 and calculate their extinction cross section at $\lambda = 0.55\,\mu$m. In Fig. 9.6, for the sake of comparison, we display the results of various cross sections following the nonstandard way of [141], so that the quantity that we actually report is $k_v^2 \langle \sigma_{\text{ext}} \rangle / \pi x^{2.5}$ as a function of the size parameter $x = 2\pi R/\lambda$, where R is the radius of the spheres considered in [141]. This choice for R accounts, in an average sense, for the presence of the vacuum, and one has $\rho = (3/2N)^{1/3} R$, where ρ is the radius of each component of the N-sphere clusters in Fig. 9.5. There are several differences in the behavior of the cluster cross sections with respect to the approximate results of multilayered spheres. First of all, the cross sections loose the stiffness shown in the results in [141], and the position and the intensity of the peak of the extinction efficiency depend significantly on the number of components spheres. For instance, the intensity grows steadily with the number of subunits, while the dependence on the size parameter of the position of the peak does not. Secondly, relevant size parameters are shifted to values much larger ($\approx 5-6$) than those derived in [141] ($\approx 2-3$).

Fig. 9.5. Adopted cluster shapes. Silicates are represented in grey, while amorphous carbon materials in black

The results in Fig. 9.1 show that the extinction from a dispersion of clusters may be strikingly different from that yielded by a dispersion of equal-volume spheres even when the clusters are assumed to be randomly oriented. In any case, the optical behavior of a dispersion of clusters turns out to be quite different in the different regions of the spectrum, depending on the number of individual component spheres. This behavior is due to the multiple scattering processes occurring among the subunits. As a consequence, modeling a complex particle as a homogeneous sphere whose refractive index is determined by an effective medium approximation, can lead to somewhat misleading results. Furthermore, the results in this section suggest that it is unlikely that the whole extent of the galactic extinction curve can be described by using clusters of a single morphology. Doubtless, cosmic dust consists of several components. They are typical of the special environments in which they are formed and/or decisively modified. Models of interstellar dust must take into account a variety of observational data that provide constraints on the composition. The results of Fig. 9.2 show that the change in chemical composition may significantly affect the grain cross sections. The results in Fig. 9.4 show, in turn, that the choice of the geometry may also be

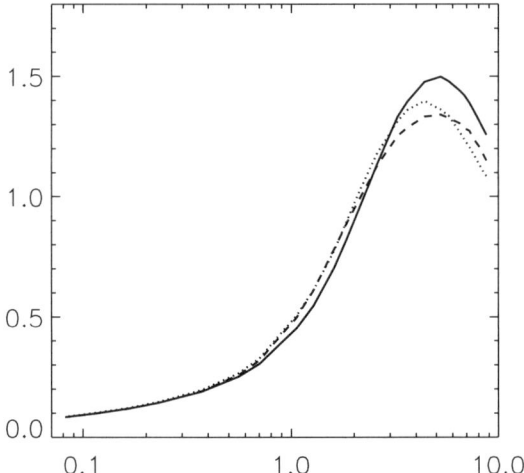

Fig. 9.6. Quantity $k_V^2 \sigma_T / x^{0.5}$ as a function of $x = 2\pi R/\lambda$, where R is the radius of the sphere considered by Voshchinnikov and Mathis, for dispersions of randomly oriented clusters whose geometry is depicted in Fig. 9.5. The *dashed line*, *dotted line*, and *solid line* refer to clusters of 2, 6, and 10 spheres, respectively

important for the extinction in the ultraviolet. This is not surprising because the scale of details of the structure of composite particles is comparable to the wavelength.

9.3 Fluffy Particles

We mentioned in Sect. 9.1 that the existence of solid grains containing empty inclusions has been proposed to overcome the difficulties connected with the elemental abundances constraints of the ISM. Since the existence of porous dust grains is an important and open question, let us study the effect of adding vacuum to spherical and irregularly shaped particles [142]. We first consider the presence of one or more spherical inclusions in a population of spherical grains. Jones [143] has discussed this problem by describing a porous grain as a hollow sphere, whose optical properties were computed by means of the Mie theory for coated spherical particles. We can resort also to the theory in Sects. 4.2, 4.4, 4.5 and 3.4 to calculate the optical properties of dispersions of spheres containing one or more spherical inclusions, of dispersions of multilayered spheres, and of dispersions of clusters.

9.3.1 Optical Properties of Porous Bare Grains

Let us explore the effects of vacuum inclusions in grains on the interstellar extinction curve [128]. The extinction cross sections are computed by modeling the grains as spheres containing a central hole. In the assumed wavelength range (100–1000 nm) the radii of the spheres are of the order of the wavelength and the thickness of the solid mantle around the central void can be a significant fraction of the total particle size. This fact, as we stressed in Sect. 6.3, rules out such popular approximations as the Rayleigh approximation and the Güttler formula for coated particles [3], so that we need to use the theory of Sect. 4.2. We assume a polydispersion of grains of *astronomical* graphite and silicate whose dielectric functions were constructed by Draine and Lee [139], and whose size distribution is governed by the widely accepted MRN model [144]. According to this model the number of grains with radius between a and $a + da$ is

$$dn_\mathrm{d} = K n_\mathrm{H} a^{-q} da, \text{ with } 5\,\mathrm{nm} \leq a \leq 250\,\mathrm{nm}. \tag{9.1}$$

In (9.1) n_H is the number density of hydrogen atoms, and K is a scaling factor related to the number of grains along the line of sight. In the MRN model only two parameters are used to constrain the size distribution: the power-law index q, customarily assumed as $q = 3.5$, and the maximum size allowed a_M. Note that the minimum size a_m is not well constrained because the mass mainly resides in large grains. However, the limiting values of a quoted in (9.1) are the most commonly used. The scaling factors K are related, through the mass of the material, to the number abundance of atoms $\mathcal{A}$ (ppM) of the particular material tied up in grains. The K factors derived by Draine and Lee [139] are $K_\mathrm{C} = 10^{-25.16} \mathrm{cm}^{2.5}/\mathrm{H}$ and $K_\mathrm{Si} = 10^{-25.11} \mathrm{cm}^{2.5}/\mathrm{H}$, which correspond to $\mathcal{A}_\mathrm{C} = 300$ ppM and to $\mathcal{A}_\mathrm{Si} = 32$ ppM. The values used previously would require about twice as much carbon and silicon being available.

At this stage, let us follow the suggestion of Mathis by carrying out a calculation in which the standard MRN model has been modified to allow for the presence of central voids with a volume fraction f_v. The purpose is just to study the effect of such voids on the extinction curve. In Fig. 9.7 is shown the trend of the size average of the extinction cross section

$$\langle \sigma_\mathrm{T} \rangle = \frac{\int_{a_\mathrm{m}}^{a_\mathrm{M}} \sigma_\mathrm{T} dn_\mathrm{d}}{\int_{a_\mathrm{m}}^{a_\mathrm{M}} dn_\mathrm{d}} \tag{9.2}$$

for the graphite grain population. As the degree of vacuum increases, the peaks of the curves move towards longer wavelengths. This behavior is in striking contrast with the well-known stability of the bump at 217.5 nm along all the lines of sight. Therefore, the results in Fig. 9.7 rule out the possibility of constructing models containing porous graphite grains. Accordingly, we explore the effect of introducing porosity in the silicate population only.

In Fig. 9.8 we plot the quantity

$$\eta = \int_{a_\mathrm{m}}^{a_\mathrm{M}} \sigma_\mathrm{T} a^{-q} da. \tag{9.3}$$

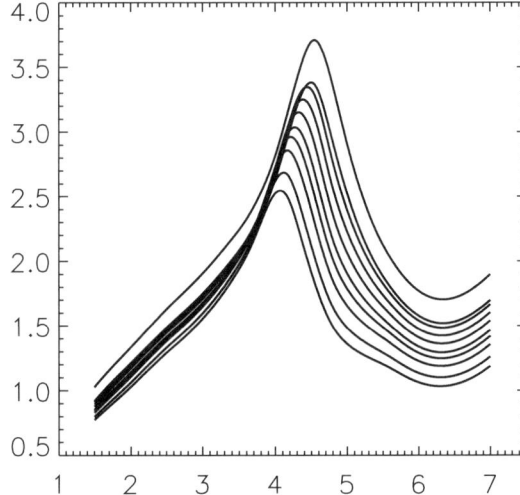

Fig. 9.7. $\langle\sigma_{\mathrm{T}}\rangle$ in $\mu\mathrm{m}^2 \cdot 10^{-4}$ as a function of λ^{-1} in $\mu\mathrm{m}^{-1}$ calculated according to (9.2) for a population of porous graphite grains characterized by a different degree of vacuum. The curves refer, in steps of 0.05, to a degree of vacuum from $f_{\mathrm{v}} = 0.05$ (*topmost curve*) to $f_{\mathrm{v}} = 0.5$ (*bottom curve*). The size distribution is given by (9.1)

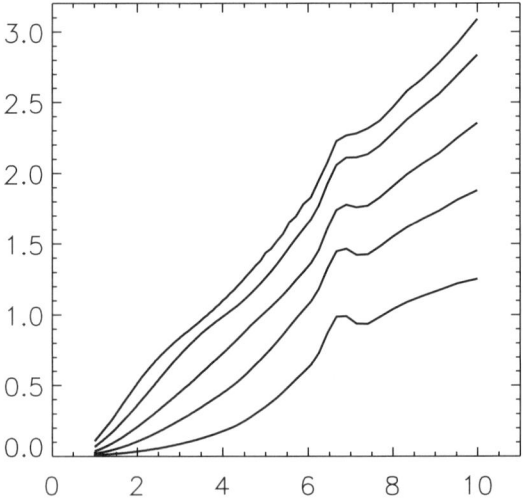

Fig. 9.8. η in $(\mathrm{cm}^{0.5}/\mathrm{H})^{-1} \cdot 10^4$ as a function of λ^{-1} in $\mu\mathrm{m}^{-1}$, calculated according to (9.3) for a population of porous silicate grains characterized by a different degree of vacuum. The curves refer, in steps of 0.2, to a degree of vacuum from $f_{\mathrm{v}} = 0$ (*topmost curve*) to $f_{\mathrm{v}} = 0.8$ (*bottom curve*)

η is a measure of the average cross section per H atom and per unit mass of silicon and depends on the cavity fraction. The presence of a central cavity in the silicate particles does not change the general trend of the extinction curve. The global effect is a lowering of the extinction curve in response to a decrease of the material inside the grains. However, the extinction curve is affected in a selective way. In particular, the curve experiences a larger shift in the visible portion of the spectrum than in the ultraviolet part. In Fig. 9.9 we show the effects of introducing porosity by a centered void with $f_\mathrm{v} = 0.4$ and $f_\mathrm{v} = 0.8$, compared to homogeneous spheres, on the infrared extinction properties of bare silicate grains. The optical properties of grains can change when more than one vacuum inclusion is considered. In Fig. 9.10 (a) a sequence of extinction curves is plotted relative to a silicate grain with a radius $a = 100\,\mathrm{nm}$ and $f_\mathrm{v} = 0.2$, which refer to different numbers of spherical inclusions. The extinction cross sections are obtained from the average over orientations of the inclusions. When the vacuum is more distributed inside the particles the visual extinction is more depressed. In Fig. 9.10 (b) we compare the effects due to different geometrical configurations. The long-dashed curve refers to the case in which the four inclusions are symmetrically located, namely, two inclusions on the z axis and the other two along the x axis. The solid curve refers to the case in which the symmetry in the configuration of the inclusions is broken, as we moved one of the inclusions from the x axis to the y axis. Large differences in the cross sections are evident, since multiple scattering processes among the inclusions have a strong dependence on the mutual geometrical disposition of voids.

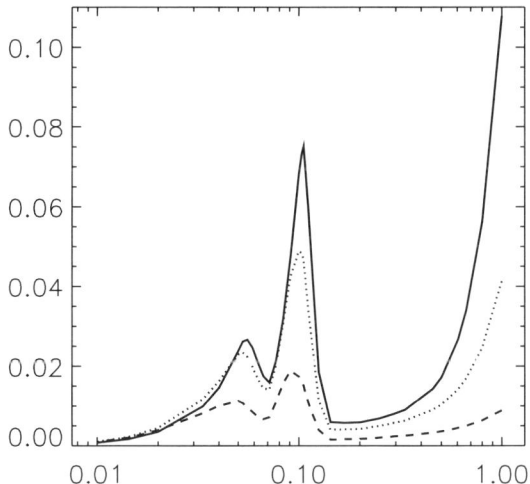

Fig. 9.9. η in $(\mathrm{cm}^{0.5}/\mathrm{H})^{-1} \cdot 10^4$ as a function of λ^{-1} in $\mu\mathrm{m}^{-1}$ in the infrared, calculated according to (9.3), for a population of porous silicate grains characterized by a different degree of vacuum. The curves refer to homogeneous particles (*solid curve*), to a population with a degree of vacuum $f_\mathrm{v} = 0.4$ (*dotted curve*), and to a population with a degree of vacuum $f_\mathrm{v} = 0.8$ (*dashed curve*)

234   9. Cosmic Dust Grains

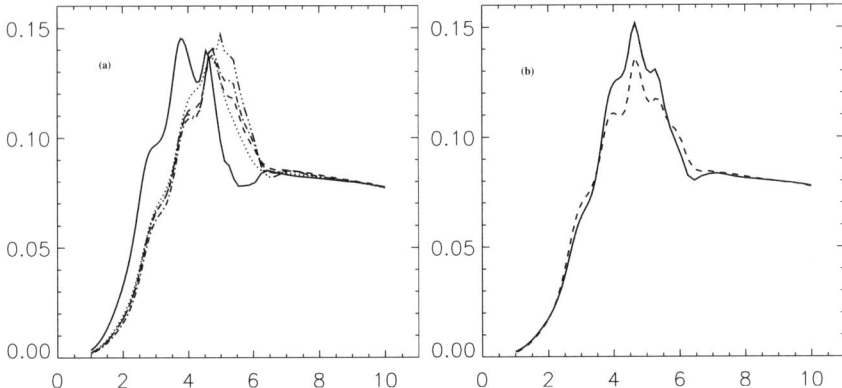

Fig. 9.10a,b. σ_T in μm^2 as a function of λ^{-1} in μm^{-1} for a dispersion of spherical silicate grains with radius $a = 100$ nm and $f_v = 0.2$ with a single central void (*solid line*), two (*dotted line*), three (*dashed line*), four (*dot-dashed line*), and six (*triple dot dashed line*) vacuum inclusions (**a**) and for the same dispersion of grains with two different geometrical configurations of four vacuum inclusions (**b**). The averages are performed on the assumption of random orientational distribution

The results shown in this section suggest that silicate grains with added vacuum yield a somewhat higher extinction than solid particles in the ultraviolet and the infrared. Unfortunately, the visual extinction follows a different trend that prevents matching of the extinction curve with bare porous particles.

9.3.2 Coated Grains and Clustering

Some dust models postulate the presence of coated grains [145, 146], e.g., silicate cores with any icy mantle including hydrocarbons or a covering of hydrogenated amorphous carbon. Early core-mantle models were constructed without any limitation about the elemental depletions. Incorporation of oxygen into models cannot help, since oxygen abundance is facing the same interstellar shortage experienced by carbon [131]. There are indications that core-mantle grains including voids might be able to reproduce the observed extinction within the revised interstellar elemental budget [147]. However, in [147] an approximate description for the dielectric functions is used, i.e., the Bruggeman mixing rule [148]. Furthermore, there is uncertainty in the derived elemental abundances, and so further investigation is warranted.

In fact, when more than two layers are present in a composite particle, the optical properties are often studied by means of the effective medium theory [148]. This kind of mean description could disguise important physical processes, such as resonance effects due to the interaction of the layers. On the contrary, we want to use the technique in Sect. 4.2, which allows

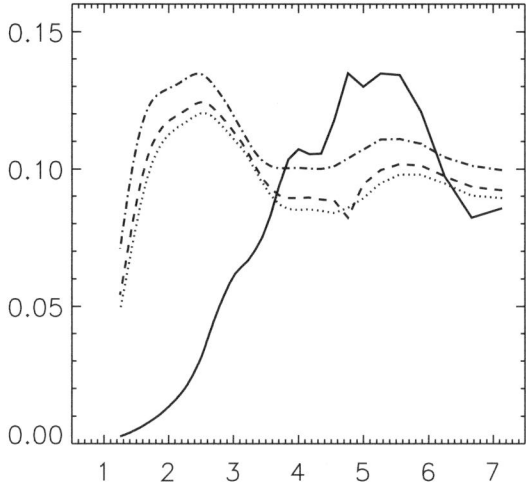

Fig. 9.11. σ_T in μm^2 as a function of λ^{-1} in μm^{-1} for a bare silicate grain of radius $a = 100$ nm, containing a centered vacuum inclusion ($f_v = 0.2$) (*solid line*) and for the same grain coated with a carbonaceous mantle of width $\Delta a = 3$ nm (*dotted line*), $\Delta a = 5$ nm (*dashed line*), and $\Delta a = 10$ nm (*dot-dashed line*)

a realistic description of a multilayered sphere. We apply just that technique to calculate the optical properties of a grain of olivine $(Mg,Fe)_2SiO_4$ with radius $a = 100$ nm with a centered void ($f_v = 0.2$) coated with an external layer of amorphous carbon whose optical constants are taken from Duley [149].

In Fig. 9.11 we compare the cross sections obtained for the bare silicate grain and for the same grain coated with mantles of different widths. The extinction peak shifts towards longer wavelengths and the shape of the curve is noticeably altered. It is very well known that the presence of a mantle produces drastic changes in the optical properties of the particles regarding not only a frequency shift of the extinction peaks but even the possibility of their disappearance [48]. The important point is that the change in the extinction shape is present even for a mantle width of 3 nm, which corresponds roughly to 12 atomic monolayers. Any increase in the mantle thickness produces a smooth enhancement in the extinction but no further change in the overall shape of the curves.

The mantle width is given by

$$\Delta a = \frac{m_C A_C}{\epsilon \rho_C}(1 - f_C) ,$$

where m_C is the mass of carbon atom, $A_C = 2.25 \times 10^{-4}$ the carbon cosmic abundance [132], $\rho_C = 2.25$ g cm^{-3} the carbon density, $\epsilon = 2.1 \times 10^{-21}$ cm^2 (H atom cm$^{-3})^{-1}$ [150], and f_C is the fraction of the total carbon abundance that is in the gas-phase. Thus, the maximum possible mantle width ($f_C = 0$)

is approximately 10 nm. Therefore, one third of the total carbon abundance is enough to cause a sharp increase in the visual extinction, which is just what is required, i.e., larger extinction per unit mass.

The mass contained in the mantle corresponds to an equivalent porous sphere composed of mantle material with an external radius

$$a_{eq} \approx a \left(\frac{3\Delta a/a}{1 - f_v} \right)^{1/3} , \text{ with } \Delta a \ll a ,$$

which for $a = 100$ nm, $\Delta a = 3$ nm, and $f_v = 0.5$ leads to $a_{eq} = 5.6$ nm.

We compute also the extinction cross section for two isolated bare porous particles, and its orientational average for the same two spheres clustered together. The silicate grain has a radius of 100 nm and a vacuum fraction $f_v = 0.2$ (the core of coated grain), while the amorphous carbon grain has a radius of 70 nm and $f_v = 0.5$ (therefore it holds a mass larger than the one inside the mantle). The above choices for particle morphology and composition are made in order to have three extremely different configurations for the available mass, and the size of carbon grain is enlarged to increase the extinction efficiency in the visible. The results of calculations are shown in Fig. 9.12. Clustered spheres produce significantly more extinction in the visible than isolated particles, but the largest contribution (a factor of two) is produced by the layered sphere.

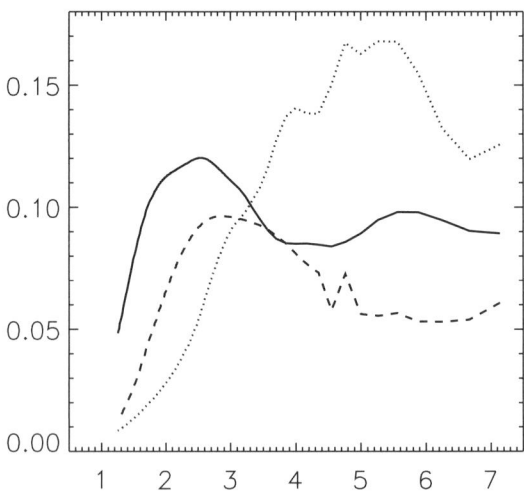

Fig. 9.12. σ_T in µm^2 as a function of λ^{-1} in µm^{-1} for a core-mantle grain with a silicate nucleus of radius $a = 100$ nm and vacuum fraction $f_v = 0.2$ covered by an amorphous carbon mantle of width $\Delta a = 3$ nm (*solid line*); for a pair of independent grains, one of silicate with radius $a = 100$ nm with a fraction of vacuum $f_v = 0.5$, and another of amorphous carbon with radius $a = 70$ nm with the same fraction of vacuum (*dotted line*); σ_T for the preceding two grains clustered together (*dashed line*) averaged on the assumption of random orientational distribution

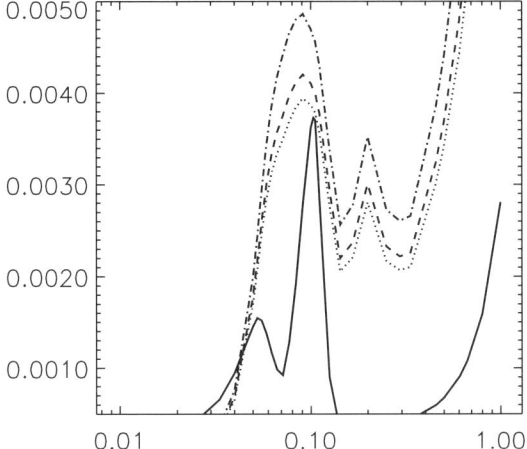

Fig. 9.13. σ_T in μm^2 as a function of λ^{-1} in μm^{-1} in the infrared range for a bare silicate grain of radius $a = 100\,nm$, containing a centered vacuum inclusion ($f_v = 0.2$) (*solid line*) and for the same grain coated with a mantle of width $\Delta a = 3\,nm$ (*dotted line*), $\Delta a = 5\,nm$ (*dashed line*), and $\Delta a = 10\,nm$ (*dot-dashed line*)

Drastic changes similar to those occurring in the extinction in the visible are also found in the infrared range. In Fig. 9.13 we show the extinction cross sections in this range for just the same particles considered in Fig. 9.11. Even in this case, coated grains produce more extinction than bare grains. This result strengthens the need to perform a further analysis to get a wider view of the possible impact on the interstellar extinction curve of porous core-mantle structures. The problem is, however, far from being solved. Even an exact calculation of a particular grain morphology necessarily involves rather arbitrary assumptions about the geometry of each grain constituents, including voids. Furthermore, the use of laboratory optical constants of pure material is not completely appropriate, because the strong environmental processing of grains in the interstellar medium produces large variations in the frequency and the strength of IR absorption and emission bands. Nevertheless, it appears that stratified scatterers show enhanced extinction properties with respect to bare grains or clusters. In this context, dust models based on coated grains could assume a renewed importance, and we hope that significant contributions can be made by increased use of the theoretical machinery that we presented in this book.

A. Appendix

This appendix is devoted to a brief description of the mathematics that is implied by the theory in Chaps. 1–5. In what follows, we will often refer to the quantities l_I and l_E that are defined in Sect. 4.4, and in the present context determine the size of the storage vectors of the codes that were used to perform the calculations presented in Chaps. 6–9.

A.1 Bessel and Hankel Functions

The spherical Bessel and Hankel functions, together with their derivatives, appear in several places of the theory. For instance, they enter the formulas for the Mie coefficients $R_l^{(p)}$, the translation matrix elements for the multipole fields $\mathcal{H}_{\alpha l m \alpha' l' m'}^{(pp')}$ and $\mathcal{J}_{\alpha l m \alpha' l' m'}^{(pp')}$, etc. Let us recall that by definition

$$h_l(z) = j_l(z) + i n_l(z) ,$$

where $n_l(z)$ is a spherical Neumann function. The argument $z = kr$ is, in general, a complex variable. We recall that Bessel and Neuman functions, and thus also the Hankel functions, obey the recursion relation (1.32a) and that their derivatives can be calculated through (1.32b). Forward recursion can be used to calculate the Neumann functions $n_l(z)$ starting from the initial values

$$n_0(z) = -\frac{\cos z}{z} , \quad n_1(z) = -\frac{\cos z}{z^2} - \frac{\sin z}{z} .$$

On the contrary, because of numerical instability, backward recursion must be used to calculate $j_l(z)$. To this end, let us assume that the Bessel functions are needed for at most $l = l_\mathrm{M}$; then we start the backward recursion by choosing $l_\mathrm{s} > l_\mathrm{M}$ and assigning small arbitrary values to j_{l_s} and $j_{l_\mathrm{s}-1}$. The values resulting from the recursion are then so normalized that

$$j_0(z) = \frac{\sin z}{z} , \quad j_1(z) = \frac{\sin z}{z^2} - \frac{\cos z}{z} .$$

The choice of l_s is, in general, effected through some empirical rule. For instance, we used

$$l_s \approx l_M + 32\sqrt{l_M}$$

everywhere, which proved adequate for all the cases we dealt with.

According to the theory of Chaps. 4 and 5, both j_l and n_l may have a complex argument and the maximum values of l are $l_M = l_I$ for calculating the Mie coefficients (see Sect. 4.1), $2l_I$ for calculating matrix H and $2l_{IE}$, for calculating J, where l_{IE} is the larger of l_I and l_E.

A.1.1 Mie Coefficients for Radially Nonhomogeneous Spheres

In Sect. 4.2 we discussed the case of radially nonhomogeneous spheres, i.e., spheres whose refractive index $n_0 = n_0(r)$ is a continuous function, with continuous first derivative, of the distance from the center. The appropriate Mie coefficients are then given by (4.15) and involve the radial functions that are the solution of (4.14). These functions must be calculated by numerical integration. Although more sophisticated methods are available, we found the Runge–Kutta method [77] to be adequate in all the cases we dealt with in this book.

A.2 Spherical Harmonics

According to definitions (1.14) and (5.23), the spherical harmonics are given by

$$Y_{lm} = \frac{1}{\sqrt{4\pi}} \bar{P}_{lm}(z) \exp(im\varphi) ,$$

where $z = \cos\vartheta$ may be complex. The functions $\bar{P}_{lm}(z)$ for $m \geq 0$ are calculated through the recursion relation

$$\bar{P}_{lm} = z \left[\frac{(2l-1)(2l+1)}{(l-m)(l+m)} \right]^{1/2} \bar{P}_{l-1,m}$$
$$- \left[\frac{(2l+1)(l-1-m)(l-1+m)}{(2l-3)(l-m)(l+m)} \right]^{1/2} \bar{P}_{l-2,m} ,$$

which is obtained by the corresponding recursion for the associated Legendre functions. The starting values are

$$\bar{P}_{00} = 1 , \quad \bar{P}_{10} = z\sqrt{3} , \quad \bar{P}_{11} = (1-z^2)^{1/2}\sqrt{3/2} .$$

Of course, in case $z = \cos\vartheta = \pm 1$, it is sufficient to use (1.59).

The spherical harmonics with real argument are used in calculating the amplitudes $W_{lm}^{(p)}$ with $l \leq l_E$, the elements of H with $l \leq 2l_I$ and the elements of J with $l \leq 2l_{IE}$. Spherical harmonics with complex $z = \cos\vartheta$ and $l \leq l_I$ are needed for calculating the elements of the multipole reflection matrix F (see Sects. 5.3–5.5).

A.3 Clebsch–Gordan Coefficients

The Clebsch–Gordan coefficients appear in the matrices H and J that effect the translation of origin of the multipole fields [see (1.58)]. They also appear in the formulas of the orientational averages (Sects. 3.4 and 3.6).

A.3.1 Translation of Origin

The definition of the matrix elements of H and J requires the Clebsch–Gordan coefficients both directly [see (1.58)] and through the Gaunt integrals. The latter, according to definition (1.27), require the Clebsch–Gordan coefficients $C(L, l, l'; 0, 0)$ and $C(L, l, l'; M, m' - M)$, whereas, according to (1.58) $C(1, l-1, l; -\mu, m+\mu)$ and $C(1, l, l; -\mu, m+\mu)$ are necessary. For calculating H, $l, l' \leq l_\mathrm{I}$ and $L \leq 2l_\mathrm{I}$, whereas for calculating J, $l, l' \leq l_\mathrm{IE}$, $L \leq 2l_\mathrm{IE}$. The formulas for $C(1, l-1, l; \mu, m+\mu)$ and $C(1, l, l; -\mu, m+\mu)$ are simple and are reported, e.g., by Rose [12], whereas calculating the Clebsch–Gordan coefficients that enter the Gaunt integrals requires using some caution. Actually, we resort to the relation between the Clebsch–Gordan coefficients and the $3j$-symbols [5]

$$C(J_1, J_2, J_3; M_1, M_2)$$
$$= (-)^{J_1 - J_2 + M_1 + M_2} (2J_3 + 1)^{1/2} \begin{pmatrix} J_1 & J_2 & J_3 \\ M_1 & M_2 & -M_1 - M_2 \end{pmatrix},$$

where the array on the right-hand side denotes the $3j$-symbols, and to the recursion over J_1 for the latter [35]. When the translation occurs along the z axis, we can take advantage of the resulting simplifications by resorting to the recursion over J_1 for the $3j$-symbols

$$\begin{pmatrix} J_1 & J_2 & J_3 \\ 0 & M_2 & -M_2 \end{pmatrix}.$$

Let us remark that, when the array of the $3j$-symbols spans a large interval of allowed values, the recursion may incur in numerical error. Thus one starts the forward recursion from the least admissible value for J_1 and proceeds up to half-way. Then one starts a backward recursion from the largest value of J_1 down to halfway and, after matching the halfway values, at last renormalizes the whole array [151].

A.3.2 Orientational Averages

According to (3.27c) and (3.30) the Clebsch–Gordan coefficients are needed for $l, l', l'' \leq l_\mathrm{E}$, $L, L' \leq 2l_\mathrm{E}$, and $\lambda \leq 4l_\mathrm{E}$. The largest value of λ used in (3.30c) stems from (1.23). The recursion we resorted to calculate these Clebsch–Gordan coefficients is that involving M_2 for the $3j$-symbols [35]. Of course, the actual computational procedure needs to take due account of the recommendations in [151].

A.4 Rotation Matrices $\mathbf{D}^{(l)}$

In order to effect the calculation of the orientational averages, the actual calculation of the rotation matrices is not necessary, provided that the required integral that involves $D^{(l)}$ matrices and the weight function can be performed analytically. Nevertheless, there is a case in which the burden of the calculation can be greatly reduced by calculating the rotation matrices, i.e., the case in which the particle of interest has cylindrical symmetry but the orientational average is required around an axis orthogonal to the cylindrical axis. In this case, it is convenient to calculate the transition matrix $\bar{S}$ by assuming the $\bar{z}$ axis along the cylindrical-symmetry axis. In fact, this choice produces the automatic factorization of the matrix M that greatly reduces the computational effort required by its inversion (see Sects. 4.4 and 4.5). Then, with the help of (3.21), the transition matrix is transformed to a new frame of reference chosen so that the cylindrical axis coincides, e.g., with the x axis. The transformed S can be considered as a new $\bar{S}$ to which all the considerations of Sect. 3.4 apply. In practice, we need

$$\mathcal{D}^{(l)}_{mm'}(\alpha = 0, \beta = \pi/2, \gamma = 0) = \mathrm{d}^{(l)}_{mm'}(\pi/2) ,$$

which can be calculated by using the recursion relations over m' and the specific formulas for $m = l$ and $m = l - 1$. This procedure is particularly convenient because $\beta = \pi/2$. The symmetry properties of $\mathrm{d}^{(l)}$ can be used to further reduce the computational effort, that is rather light, however. Note that $l \leq l_\mathrm{E}$.

A.5 Calculating $\mathbf{M}^{-1}$, $\mathbf{H}^{-1}$, and $\mathbf{H}_\alpha^{-1}$

To perform these calculations, we use the classical LU factorization [77]. Of course, complex algebra is necessary and all the calculations require double precision. In particular, calculating H^{-1} and H_α^{-1}, which is necessary when a particle is in the presence of a dielectric surface (see Sects. 5.3–5.5), may require a specific approach to be used besides the main bulk of the calculations.

A.6 General Approach to the Computational Problem

According to our experience, a code designed to perform the calculations implied by the problem of scattering should be a reasonable compromise to contain both the storage required for the intermediate products of the calculations and the time required to perform the calculations themselves. The preferred procedure to achieve this task consists in calculating once and for all and storing into appropriate vectors all the quantities that are shared with the main stream of the calculations. Among these quantities, which will be picked up according to the need, there are just the Bessel and Hankel functions, the spherical harmonics, and the Clebsch–Gordan coefficients.

References

1. D. W. Mackowski, M. I. Mishchenko: J. Opt. Soc. Am. A **13**, 2266 (1996)
2. J. D. Jackson: *Classical electrodynamics* (Wiley, New York 1975)
3. H. C. van de Hulst: *Light scattering by small particles* (Wiley, New York 1957)
4. J. A. Stratton: *Electromagnetic theory* (McGraw-Hill, New York 1941)
5. E. M. Rose: *Elementary theory of angular momentum* (Wiley, New York 1957)
6. J. L. Powell, B. Crasemann: *Quantum mechanics* (Addison-Wesley, Reading, Mass. 1961)
7. M. Abramowitz, I. A. Stegun: *Handbook of mathematical functions* (Dover, New York 1965)
8. A. R. Edmonds: *Angular momentum in quantum mechanics* (Princeton University Press, Princeton, NJ 1957)
9. P. M. Morse, H. Feshbach: *Methods of theoretical physics* (McGraw-Hill, New York 1953)
10. W. W. Hansen: Phys. Rev. **47**, 139 (1935)
11. E. Fucile, F. Borghese, P. Denti, R. Saija, O. I. Sindoni: IEEE Trans. Antennas Propag. **AP 45**, 868 (1997)
12. E. M. Rose: *Multipole fields* (Wiley, New York 1956)
13. R. Nozawa: J. Math. Phys. **7**, 1841 (1966)
14. F. Borghese, P. Denti, G. Toscano, O. I. Sindoni: J. Math. Phys. **21**, 2754 (1980)
15. A. Ishimaru: *Wave propagation and scattering in random media* (Academic press, New York 1978)
16. M. I. Mishchenko, J. H. Hovenier, L. D. Travis: 'Concepts, terms, notations.' In: *Light scattering by nonspherical particles*, ed. by M. I. Mishchenko, J. W. Hovenier, L. D. Travis (Academic Press, New York 2000), pp. 3–25
17. J. W. Hovenier, C. V. M. van der Mee: 'Basic relationships for matrices describing scattering by small particles.' In: *Light scattering by nonspherical particles*, ed. by M. I. Mishchenko, J. W. Hovenier, L. D. Travis (Academic Press, New York 2000), pp. 61–85
18. J. R. Bottiger: *Notes on 3-D Müller matrices*. Final Technical Report CRDEC-TR-1L162662, August 1989
19. C. F. Bohren, D. P. Gilra: J. Colloid Interf. Sci. **72**, 215 (1979)
20. M. I. Mishchenko, D. W. Mackowski, L. D. Travis: Appl. Opt. **34**, 4589 (1995)
21. R. Newton: *Scattering theory of waves and particles* (McGraw-Hill, New York 1966)
22. A. Ishimaru: *Electromagnetic wave propagation, radiation and scattering* (Prentice-Hall, Englewood Cliffs, NJ 1991)
23. P. C. Waterman: Phys. Rev. D **3**, 825 (1971)
24. F. Borghese, P. Denti, R. Saija, G. Toscano, O. I. Sindoni: Nuovo Cim. **B 81**, 29 (1984)

25. F. Borghese, P. Denti, R. Saija, M. A. Iatì, O. I. Sindoni: J. Quant. Spectr.Radiat. Transfer **70**, 237 (2001)
26. V. K. Varadan: 'Multiple scattering of acoustic, electromagnetic and elestic waves.' In: *Acoustic electromagnetic and elastic wave scattering–Focus on the T-matrix approach*, ed. by V. K. Varadan, V.V. Varadan (Pergamon, New York 1980), pp. 103–134
27. V. V. Varadan, V. K. Varadan: Phys. Rev. D **21**, 388 (1980)
28. N. G. Khlebtzov: Appl. Opt. **38**, 5359 (1992)
29. M. I. Mishchenko: Astrophys. Space Sci. **164**, 1 (1990)
30. M. I. Mishchenko: Kinem. Phys. Celest. Bodies **7**(5), 83 (1991)
31. M. I. Mishchenko: Astrophys. J. **367**, 561 (1991)
32. M. I. Mishchenko: J. Opt. Soc. Am. A **8**, 871 (1991); errata ibid. **9**, 497 (1992)
33. L. Tsang, J. A. Kong, T. T. Shing: *Theory of microwave remote sensing* (Wiley, New York 1985)
34. E. Fucile, F. Borghese, P. Denti, R. Saija: Appl. Opt. **34**, 4552 (1995)
35. D. A. Varshalovich, A. N. Moskalev, V. K. Khersonskii: *Quantum theory of angular momentum* (World Scientific, Singapore 1988)
36. R. Balescu: *Equilibrium and nonequilibrium statistical mechanics* (Wiley, New York 1975)
37. R. Clark Jones: Phys. Rev. B **15**, 93 (1945); errata ibid. 213
38. L. E. Reichl: *A modern course in statistical physics* (University of Texas Press, Austin 1980)
39. L. D. Favro: Phys. Rev. **119**, 53 (1960)
40. E. Fucile, F. Borghese, P. Denti, R. Saija: J. Opt. Soc. Am. A **10**, 2611 (1993)
41. G. Mie: Ann. Phys. **25**, 377 (1908)
42. D. S. Saxon: Phys. Rev. **100**, 1771 (1955)
43. Bo Å. S. Gustafson: 'Microwave analog to light scattering measurements.' In: *Light scattering by nonspherical particles*, ed. by M. I. Mishchenko, J. W. Hovenier, L. D. Travis (Academic Press, New York 2000), pp. 367–390
44. P. J. Wyatt: Phys. Rev. B **127**, 1837 (1962)
45. M. Kerker: *The scattering of light* (Academic Press, New York 1969)
46. F. Borghese, P. Denti, R. Saija, G. Toscano, O. I. Sindoni: J. Opt. Soc. Am. A **4**, 1984 (1987)
47. R. Ruppin: J. Opt. Soc. Am. **66**, 449 (1976)
48. B. R. Johnson: J. Opt. Soc. Am. A **10**, 343 (1993)
49. J. H. Bruning, Y. T. Lo: IEEEE Trans. Antennas Propag. **19**, 378 (1971)
50. B. Peterson, S. Ström: Phys. Rev. D **8**, 3661 (1973)
51. J. M. Gérardy, M. Ausloos: Phys. Rev. B **22**, 4950 (1980)
52. F. Borghese, P. Denti, R. Saija, G. Toscano, O. I. Sindoni: Aerosol Sci. Technol. **3**, 227 (1984)
53. F. Borghese, P. Denti, R. Saija, O. I. Sindoni: J. Opt. Soc. Am. A **9**, 1327 (1992)
54. F. Borghese, P. Denti, R. Saija: Appl. Opt. **33**, 484 (1994); errata ibid. **34**, 5556 (1995)
55. J. G. Fikioris, N. K. Uzunoglu: J. Opt. Soc. Am. **69**, 1359 (1979)
56. N. C. Skaropoulos, M. P. Ioannidou, D. P. Chrissoulidis: J. Opt. Soc. Am. A **11**, 1859 (1994)
57. M. P. Ioannidou, D. P. Chrissoulidis: J. Opt. Soc. Am. A **19**, 505 (2002)
58. M. I. Mishchenko, L. D. Travis: 'Electromagnetic scattering by nonspherical particles.' In: *Exploring the atmosphere by remote sensing*, ed. by R. Guzzi, K. Pfeilsticker (Springer, Heidelberg 2002), in the press
59. E. M. Purcell, C. R. Pennypacker: Astrophys. J. **186**, 705 (1973)
60. B. T. Draine, P. J. Flatau: J. Opt. Soc. Am. A **11**, 1491 (1994)

61. B. T. Draine: Astrophys. J. **333**, 848 (1988)
62. B. T. Draine, J. J. Goodman: Astrophys. J. **405**, 685 (1993)
63. S. K. Yee: IEEE Trans. Antennas Propag. **14**, 302 (1966)
64. T. Levi Civita: *Caratteristiche dei sistemi differenziali e propagazione ondosa* (Zanichelli, Bologna 1992)
65. A. Taflove, M. E. Brodwin: IEEE Trans. Microwave Theor. Technol. **23**, 623 (1975)
66. G. Mur: IEEE Trans. Electromagn. Compat. **23**, 377 (1981)
67. Z. Liao, H. L. Wong, B. Yang, Y. Yuan: Scientia Sinica **27**, 1063 (1984)
68. K. Umashankar, A. Taflove: IEEE Trans. Electromagn. Compat. **24**, 397 (1982)
69. P. Yang, K. N. Liou: J. Opt. Soc. Am. A **13**, 2072 (1996)
70. P. Yang, K. N. Liou: 'Finite difference time domain method for light scattering by nonspherical and inhomogeneous particles.' In: *Light scattering by nonspherical particles*, ed. by M. I. Mishchenko, J. W. Hovenier, L. D. Travis (Academic Press, New York 2000), pp. 174–221
71. F. Borghese, P. Denti, R. Saija, E. Fucile, O. I. Sindoni: J. Opt. Soc. Am. A **12**, 530 (1995)
72. E. Fucile, P. Denti, F. Borghese, R. Saija, O. I. Sindoni: J. Opt. Soc. Am. A **14**, 1505 (1997)
73. P. Denti, F. Borghese, R. Saija, E. Fucile, O. I. Sindoni: J. Opt. Soc. Am. A **16**, 167 (1999)
74. P. Denti, F. Borghese, R. Saija, M. A. Iatì, O. I. Sindoni: Appl. Opt. **38**, 6421 (1999)
75. P. A. Bobbert, J. Vlieger: Physica **137 A**, 209 (1986)
76. H. Weyl: Ann. Phys. **60**, 481 (1919)
77. W. H. Press, S. A. Teukolsky, W. T. Vetterling, B. P. Flannery: *Numerical recipes in fortran* (Cambridge University Press, New York 1992)
78. F. Borghese, P. Denti, R. Saija, E. Fucile, O. I. Sindoni: Appl. Opt. **36**, 4226 (1997)
79. F. Borghese, P. Denti, R. Saija, G. Toscano, O. I. Sindoni: J. Opt. Soc. Am. A **1**, 183 (1984)
80. S. Levine, G. O. Olaofe: J. Colloid Interf. Sci. **27**, 442 (1968)
81. P. W. Barber, D.-S. Wang: Appl. Opt. **17**, 797 (1978)
82. M. I. Mishchenko, D. W. Mackowski: J. Quant. Spectr. Radiat. Transfer **55**, 683 (1996)
83. D. H. Woodward: J. Opt. Soc. Am. **54**, 1325 (1964)
84. D. W. Schuerman, R. T. Wang: *Experimental results of multiple scattering*. Chemical Systems Laboratory Contractor Report ARCSL-CR-81003, November 1981
85. R. T. Wang, J. M. Greenberg, D. W. Schuerman: Opt. Lett. **11**, 543 (1981)
86. F. Borghese, P. Denti, R. Saija, O. I. Sindoni: J. Aerosol Sci. **20**, 1079 (1989)
87. R. Saija, G. Toscano, O. I. Sindoni, F. Borghese, P. Denti: Nuovo Cim. **B 85**, 79 (1985)
88. J. I. Gittelman, B. Abeles: Phys. Rev. B **15**, 3297 (1977)
89. C. Kittel: *Introduction to solid state physics* (Wiley, New York 1976)
90. M. Hammermesh: *Group theory* (Addison-Wesley, Reading, Mass. 1962)
91. F. Wooten: *Optical properties of solids* (Academic Press, New York 1972)
92. F. Abeles: *Optical properties of solids* (North-Holland, Amsterdam 1972)
93. P. Chýlek, G. Videen , D. J. Wally Geldart, J. S. Dobbie, H. C. W. Tso: 'Effective medium approximations for heterogeneous particles.' In: *Light scattering by nonspherical particles*, ed. by M. I. Mishchenko, J. W. Hovenier, L. D. Travis (Academic Press, New York 2000), pp. 274–307

94. K. A. Fuller, G. W. Kattawar: Opt. Lett. **13**, 90 (1988)
95. B. V. Bronk, M. J. Smith, S. Arnold: Opt. Lett. **18**, 93 (1993)
96. R. L. Armstrong, J.-G. Xie, T.E. Ruekgauer, J. Gu, R. G. Pinnick: Opt. Lett. **18**, 119 (1993)
97. S. W. Ng, P. T. Leung, K. M. Lee: J. Opt. Soc. Am. B **19**, 154 (2001)
98. F. Borghese, P. Denti, R. Saija, M. A. Iatì, O. I. Sindoni: J. Opt. (France) **29**, 28 (1998)
99. B. T. Draine: *Planets and protostars* Vol. II, Ed. by D. C. Black, M. Shapley Mattews (University of Arizona Press, Phoenix 1985)
100. F. Borghese, P. Denti, R. Saija, G. Toscano, O. I. Sindoni: Nuovo Cim. **D 6**, 545 (1985)
101. R. Pecora, B. Berne: *Dynamic light scattering* (Wiley, New York 1980)
102. B. R. Johnson: J. Opt. Soc. Am. A **9**, 1341 (1992)
103. F. Borghese, P. Denti, R. Saija, G. Toscano, O. I. Sindoni: Aerosol Sci. Technol. **6**, 173 (1987)
104. G. Videen: J. Opt. Soc. Am. A **8**, 483 (1991); errata ibid. **9**, 844 (1992)
105. I. V. Lindell, A. H. Sihvola, K. O. Muinonen, P. Barber: J. Opt. Soc. Am. A **8**, 472 (1991)
106. M. M. Wind, J. Vlieger, D. Bedeaux: Physica **141A**, 33 (1987)
107. P. A. Bobbert, M. M. Wind, J. Vlieger: Physica **146A**, 69 (1987)
108. P. Denti, F. Borghese, R. Saija, M. A. Iatì, O. I. Sindoni: J. Opt. (France) **29**, 78 (1998)
109. A. Li, J. M. Greenberg: Astron. Astrophys. **323**, 566 (1997)
110. B. R. Johnson: J. Opt. Soc. Am. A **11**, 2055 (1994)
111. P. A. Bobbert, J. Vlieger, R. Greef: Physica **137A**, 243 (1986)
112. B. R. Johnson: J. Opt. Soc. Am. A **13**, 326 (1996)
113. H. Yousif: *Light scattering from parallel tilted fibers.* Ph. D. dissertation, Department of Physics, University of Arizona, Tucson, Ariz. (1987)
114. H. S. Lee, S. Chac, Y. Ye, D. Y. H. Pui, G. L. Wojcik: Aerosol Sci. Technol. **14**, 177 (1991)
115. G. L. Wojcik, D. K. Vaughan, L. K. Galbraith: 'Calculation of light scatter from structures on silicon surfaces.' In: *Lasers in microlithography*, ed. by J. S. Batchelder, D. J. Ehrlich, J. Y. Tsao, Proc. Soc. Photo-Opt. Instrum. Eng. **774**, 21 (1987)
116. K. O. Muinonen, A. H. Sihvola, I. V. Lindell, K. A. Lumme: J. Opt. Soc. Am. A **8**, 477 (1991)
117. M. A. Taubenblatt, T. K. Tran: J. Opt. Soc. Am. A **10**, 912 (1993)
118. B. Saltzmann: 'Climatic system analysis.' In Advances in geophysics, vol. 25, ed. by B. Saltzmann (Academic Press, New York 1983)
119. H. R. Pruppacher, J. D. Klett: *Microphysics of clouds and precipitation* (Kluwer, Dordrecht 1997)
120. S. G. Warren: Appl. Opt. **23**, 1206 (1984)
121. K. Sassen: J. Opt. Soc. Am. A **4**, 570 (1988)
122. K. Sassen: 'Lidar backscatter depolarization technique for cloud and aerosol research.' In *Light scattering by nonspherical particles*, ed. by M. I. Mishchenko, J. W. Hovenier, L. D. Travis (Academic Press, New York 2000), pp. 393–416
123. R. Saija, M. A. Iatì, P. Denti, F. Borghese, O. I. Sindoni: Appl. Opt. **40**, 5337 (2001)
124. H. Lemke, M. Quante, O. Danne, E. Raschke: 'Backscattering of radar waves by non-spherical atmospheric ice crystals: An application of the discrete dipole approximation.' In *Proc. 3rd workshop on electromagnetic and light scattering: Theory and applications* (University of Bremen, Bremen 1998), pp. 183–192

125. V. Vouk: Nature (London) **162**, 330 (1948)
126. R. A. West: Appl. Opt. **30**, 5316 (1991)
127. N. T. Zakharova, M. I. Mishchenko: Appl. Opt. **39**, 5052 (2000)
128. S. Aiello, B. Barsella, C. Chlewicki, J. M. Greenberg, P. Patriarchi, M. Perinotto: Astron. Astrophys. Suppl. **73**, 195 (1988)
129. D. C. B. Whittet: *Dust in the galactic environment* (Institute of Physics Publishing, Bristol, UK 1992)
130. L. Spitzer: *Physical processes in the interstellar medium* (Wiley, New York 1985)
131. T. P. Snow, A. N. Witt: Science **270**, 1455 (1995)
132. T. P. Snow, A. N. Witt: Astrophys. J. **468**, L65 (1996)
133. U. J. Sofia, E. Fitzpatrick, D. M. Meyer: Astrophys. J. **499**, 951 (1998)
134. D. M. Meyer, M. Jura, J. A. Cardelli: Astrophys. J. **493**, 222 (1998)
135. J. S. Mathis: Astrophys. J. **472**, 643 (1996)
136. D. B. Vaidya, R. Gupta: Astron. Astrophys. **348**, 594 (1999)
137. D. A. Williams: Astr. Geophys. **41**, 3.8 (2000)
138. R. Saija, M. A. Iatì, F. Borghese, P. Denti, S. Aiello, C. Cecchi-Pestellini: Astrophys. J. **559**, 993 (2001)
139. B. T. Draine, H. M. Lee: Astrophys. J. **285**, 89 (1984)
140. F. Rouleau, P. G. Martin: Astrophys. J. **377**, 526 (1991)
141. N. V. Voshchinnikov, J. S. Mathis: Astrophys. J. **526**, 257 (1999)
142. M. A. Iatì, C. Cecchi-Pestellini, D. A. Williams, F. Borghese, P. Denti, R. Saija, S. Aiello: Montly Not. R. Astr. Soc. **322**, 749 (2001)
143. A. P. Jones: Mont. Not. R. Astr. Soc. **234**, 209 (1988)
144. J. S. Mathis, W. Rumpl, K.M. Nordsiek: Astrophys. J. **217**, 425 (1977)
145. S. S. Hong, J. M. Greenberg: Astron. Astrophys. **88**, 194 (1980)
146. A. P. Jones, W. W. Duley, D. A. Williams: QJRAS **31**, 567 (1990)
147. C. Cecchi-Pestellini, D. A. Williams: Mont. Not. R. Astr. Soc. **296**, 414 (1998)
148. C. F. Bohren, D. R. Huffman: *Absorption and scattering of light by small particles* (Wiley, New York 1983)
149. W. W. Duley: Astrophys. J. **287**, 694 (1984)
150. W. W. Duley, D. A. Williams: *Interstellar chemistry* (Academic Press, London 1984)
151. K. Schulten, R. G. Gordon: J. Math. Phys. **16**, 1961 (1975)

Index

Angle of scattering 35
Angular momentum 8
– coupling 14
– eigenvalue equations 8
– representations 9
Asymmetry parameter 60
Atmospheric ice crystals
– albedo 212
– aspect ratio 214, 221
– backscattering 210, 214, 217
– depolarization ratio 211, 219
– forward scattering 209
– models 207, 214
– orientational distributions 208

Clebsch–Gordan coefficients 14
– calculation 241
– – $3j$-symbols 241
– inverse coupling 16
– orthogonality relations 15
– series 16
– – Gaunt formula 16
– symmetry properties 15
– triangular relation 15
Correlation spectroscopy 66, 166
Cross section 33
– absorption 33
– additivity 43
– differential 33
– extinction 34
– scattering 33

DDA, discrete dipole approximation 89

Efficiencies 34
– for nonspherical scatterers 121
Eulerian angles 9

FDTD, finite difference time domain method 91
Field equations 2
– boundary conditions 4

– constitutive 3
– for harmonic field 3
– Helmholtz 4
– Maxwell 2
Finite elements methods 88
– discrete dipole approximation 89
– finite difference time domain 91

Güttler formula 142

Hansen's vectors 22, 23
– addition theorem 28
Helmholtz equation 19
– vector solutions 19

Interstellar dust 223
– abundance constraints 231
– change of chemical composition 226
– change of geometry 227
– effect of subdivision of silicates 225
– elemental shortage 224
– galactic extinction curve 223
– model of Voshchinnikov and Mathis 228
– – comparison with cluster model 229
– models for dust grains 224
– porous grains model 231
– – effect of coating 234
– – effect of mass constraints on the coating 236
– – effect of porosity distribution 233
– – size distribution 231

Legendre functions 12

MRT, multipole reflection theory 192
Müller matrix 40
Multipole expansion
– of homogeneous plane wave 49
– of inhomogeneous plane wave 51
– of scattered wave 51
– of scattering amplitude 52

252   Index

Multipole reflection rule 103
– reflection matrix 105

Optical theorem 41, 119
Orientational averages 56
– for particles in motion 66
– for particles on a surface 101
– in an electrostatic field 62

Plane of scattering 35
– for forward and backward scattering 36
Polarizability tensor 62
– and transition matrix 65
– of axially symmetric particles 63
– of particles of arbitrary shape 64
Polarization 36
– linear basis 36
– of the incident wave 36
– – circular 37
– – linear 36
– of the scattered wave 38
Propagation constant 3

Refractive index 3
– of an assembly of particles 44
– – birefringence 47
– – dichroism 47
– – of polydispersions 46, 55
Resonances 77
– from eccentrically coated spheres 161
– of hemispheres on a perfectly reflecting surface 186, 190
– – resonance suppression 186, 190
– of homogeneous spheres 77
– of nonhomogeneous spheres 77
Rotation matrices 10
– orthonormality conditions 10
– – on the unit sphere 11
– unitarity properties 10
Rotation operators 8
– Euler rotations 9
RSA, Rayleigh scattering approximation 64

Scalar Helmholtz harmonics 26
– addition theorem 26
Scattering
– electromagnetic scaling 74, 81, 126
– from aggregates 78
– – comparison with experiments 125
– – effect of an electrostatic field 135
– – effect of structural changes 128

– – shadow effect 123
– from homogenous spheres 71
– from particles on a dielectric substrate 109, 114, 192, 198, 202
– – comparison with approximate methods 192
– – comparison with perfectly reflecting surface 196
– from particles on a perfectly reflecting surface 98, 173
– – effective scatterer 98
– – exciting field 98
– from particles on a surface
– – dielectric half-space 103
– – Fresnel coefficients 97
– – incident and reflected fields 96
– from radially nonhomogenous spheres 74
– – layered spheres 144, 146, 147
– – spheres with soft surface 143
– from spheres containing inclusions 84
– – cavities 154
– – metallic inclusions 151, 155
– Mie theory 71
Scattering amplitude 31
– normalized 32
– transversality 32
SPA, small particle approximation 167
Spherical Bessel and Hankel functions 19
– calculation 239
– limiting expressions 20
– recursion relations 20
– Wronskian 20
Spherical harmonics 11
– addition theorem 12
– as spherical tensors 12
– complex conjugation 12
– eigenvalue equations 11
– orthonormality 12
– parity 12
Spherical multipole fields 25
– and spherical tensors 25
– magnetic and electric 26
Spherical tensors 11
– irreducible 11, 18
– – complex conjugation 18
– – eigenvalue equations 18
– – orthogonality 18
– – parity 18
– – transformation properties 18
Stokes parameters 39

– circular basis 40
– for actual beams 40
– for monochromatic waves 40
– linear basis 39

Transition matrix 53
– and scattering amplitude 53
– convergence 73, 81, 87, 119, 148
– for a sphere on a dielectric substrate 112
– for aggregated spheres on a dielectric substrate 118
– for aggregates of spheres 81
– for homogeneous spheres 73
– for layered spheres 76
– for radially nonhomogeneous spheres 76
– for spheres containing inclusions 86
– transformation properties 53

Vector fields 5
– Cartesian 5
– spherical components 13
– – dot product 13
– spin 7
– transformation 6
Vector Helmholtz harmonics 20
– addition theorem 28
– divergence and curl 21
Vector spherical harmonics 24
– spherical components 24
– transverse 24

You are one **click** away
from a **world of physics** information!

Come and visit Springer's
Physics Online Library

Books
- Search the Springer website catalogue
- Subscribe to our free alerting service for new books
- Look through the book series profiles

You want to order?   Email to: orders@springer.de

Journals
- Get abstracts, ToC´s free of charge to everyone
- Use our powerful search engine LINK Search
- Subscribe to our free alerting service LINK *Alert*
- Read full-text articles (available only to subscribers of the paper version of a journal)

You want to subscribe?   Email to: subscriptions@springer.de

Electronic Media
- Get more information on our software and CD-ROMs

You have a question on
an electronic product?   Email to: helpdesk-em@springer.de

● Bookmark now:

www.springer.de/phys/

 Springer

Springer · Customer Service
Haberstr. 7 · 69126 Heidelberg, Germany
Tel: +49 (0) 6221 - 345 - 217/8
Fax: +49 (0) 6221 - 345 - 229 · e-mail: orders@springer.de

d&p · 6437.MNT/SFb

Printing (Computer to Plate): Saladruck Berlin
Binding: Stürtz AG, Würzburg